Advances in
Sonar Technology

Advances in Sonar Technology
Edited by Sergio Rui Silva

This is a reprint of the book published in 2009 by In-Tech

Republished by InTech
Janeza Trdine 9, 51000 Rijeka, Croatia

Technical Editor Teodora Smiljanic

Cover Designer Dino Smrekar

Additional hard copies can be obtained from orders@intechopen.com

Advances in Sonar Technology, Edited by Sergio Rui Silva
p. cm.
ISBN 978-3-902613-48-6

For EU product safety concerns:
IN TECH d.o.o., Prolaz Marije Krucifikse Kozulić 3, 51000 Rijeka, Croatia,
info@intechopen.com or visit our website at intechopen.com.

Preface

The demand to explore the largest and also one of the richest part of our planet, the advances in signal processing promoted by an exponential growth in computation power and a thorough study of sound propagation in the underwater realm, lead to remarkable advances in sonar technology in the last years.

Since the use of imaging system that rely on electromagnetic waves (optical, laser or radar) is restricted to only very shallow water environments, and given that the good propagation of sound waves in water is known from at least the writings of Leonardo da Vinci, the sonar (sound navigation and raging) systems are the most widespread solution for underwater remote sensing.

Sonar systems can be divided into two major types: passive sonar systems that enable detection of a sound emitting target and active sonar systems that use the properties of a signal reflected on the targets for its detection and image formation.

As system complexity increases, the study of the way sound is used to obtain reflectivity and bathymetry data from targets and submersed areas becomes fundamental in the performance prediction and development of innovative sonar systems.

Because of the many similarities between sonar and radar, algorithms created for the latter found application in sonar systems which made use of the advances in signal processing to overcome the barriers of the problematic underwater propagation medium and to challenge the resolution limits. In particular, synthetic aperture methods, applied with so much success in radar imagery, were adapted to sonar systems. This in turn enabled a considerable increase in sonar image quality and system robustness. Target detection developments lead to the use of multiple transducer sensors and sophisticated beam forming techniques with also excellent results.

High quality sonar imagery with reduced noise and enhanced resolution enables more complex applications. Leaving the traditional real of military applications, sonar civil applications arise for the study of biology ecology and related fields. Moreover integration and data fusion of different sensors is becoming more and more common, being it navigation data integration and enhancement for synthetic aperture, sonar systems with different propagation characteristics or optical image integration for the improvement of object detection.

But, not unlike natural evolution, a technology that matured in the underwater environments is now being used to solve problems for robots that use the echoes from air-acoustic signals to derive their sonar signals.

The work on hand is a sum of knowledge of several authors that contributed in various different aspects of sonar technology. This book intends therefore to give a broad overview of the advances in sonar technology of the last years that resulted from the research effort of the authors in both sonar systems and its applications. It is destined to scientist and engineers from a variety of backgrounds and, hopefully, even those that never had contact with sonar technology before will find an easy entrance in the topics and principles exposed here.

The editor would like to thank all authors for their contribution and all those people who directly or indirectly helped make this work possible, especially Vedran Kordic who was responsible for the coordination of this project.

Editor

Sergio Rui Silva
University of Porto

Contents

Side-looking Sonar

1

Simulation and 3D Reconstruction of Side-looking Sonar Images

E. Coiras and J. Groen
NATO Undersea Research Centre (NURC)
Italy

1. Introduction

Given the limited range and applicability of visual imaging systems in the underwater environment, sonar has been the preferred solution for the observation of the seabed since its inception in the 1950s (Blondel 2002). The images produced by the most commonly used side-looking sonars (side-scan and, more recently, synthetic aperture sonars) contain information on the backscatter strength recorded at every given range. This backscatter strength mainly depends on the composition and the orientation of the observed surfaces with respect to the sensor.

In this chapter, the relations between surface properties (bathymetry, reflectivity) and the images resulting when the surface is observed by side-looking sonar (backscatter strength) are studied. The characterization of this sonar imaging process can be used in two ways: by applying the forward image formation model, sonar images can be synthesized from a given 3D mesh; conversely, by inverting the image formation model, a 3D mesh can be estimated from a given side-looking sonar image. The chapter is thus divided in two main parts, each discussing these forward and inverse processes. The typical imaging sensor considered here is an active side-looking sonar with a frequency of hundreds of kilohertz, which usually allows for sub-decimetre resolution in range and azimuth.

2. Sonar simulation

Simulation is an important tool in the research and development of signal processing, a key part of a sonar system. A simulation model permits to study sonar performance and robustness, giving the analyst the opportunity to investigate variations in the sonar results as a function of one system parameter, whilst keeping other parameters fixed, hereby enabling sensitivity studies. A sonar simulator can be used as well for image data base generation, as an addition to costly measured data of which there is typically a shortage. A data base with sufficient actuality and variability is crucial for testing and developing signal processing algorithms for sonar image analysis, such as object detectors and classifiers. An example is illustrated in Fig. 1, where a measured synthetic aperture sonar (SAS) image of a cylinder sitting on the seafloor and a simulated image of a similar object at the same range are shown.

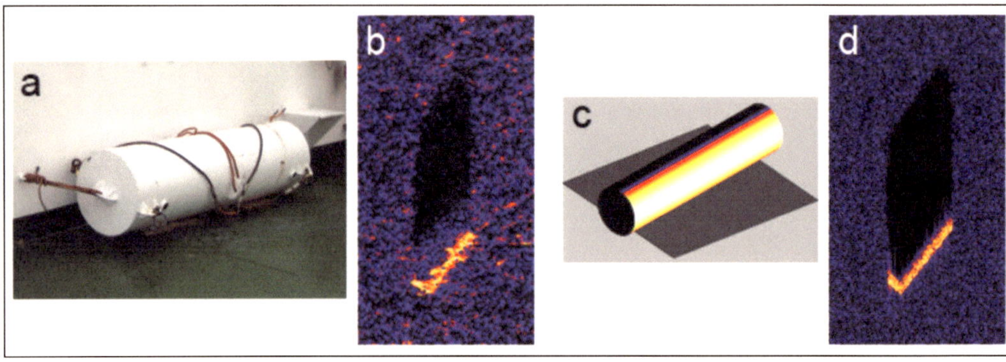

Fig. 1. (a) NURC's test cylinder. (b) Image of the cylinder measured with MUSCLE's synthetic aperture sonar (SAS). (c) 3D computer model of a cylinder and (d) its corresponding sonar image simulated with the SIGMAS model.

2.1 Sonar fundamentals

The basic idea behind any sonar system is as follows: an acoustic signal (or ping) is emitted by the sonar into an area to be observed; the sonar then listens for echoes of the ping that have been produced when bouncing back from the objects that might be present in the area. Typically, sonar images are produced by plotting the intensity measured back by the sonar versus time, and since the speed of sound underwater is known (or can be measured), the time axis effectively corresponds to range from the sonar.

In this way, just as light illuminates a scene so that it can be perceived by an optical sensor, the acoustic ping "ensonifies" the scene so that it can be perceived by an acoustic sensor. Also, as it happens in the optical field, imaging can be approached as a ray-tracing or a wave propagation problem.

2.2 The acoustic wave equation

The propagation of acoustic waves is described by the acoustic version of the wave equation (Drumheller 1998), a second order differential equation for acoustic pressure p, which is a function of time (t) and space (x, y, z). Assuming constant water density and constant sound speed (c) it can be written as:

$$\frac{\partial^2 p}{\partial x^2}+\frac{\partial^2 p}{\partial y^2}+\frac{\partial^2 p}{\partial z^2}-\frac{1}{c^2}\frac{\partial^2 p}{\partial t^2}=-\delta\left(x-x_s,y-y_s,z-z_s\right)s\left(t\right) \tag{1}$$

The physical process starts with a normalized acoustic wave signal $s(t)$ emitted by a source located at (x_s, y_s, z_s). In the equation the source is modelled as a point source, with a Dirac delta (δ) spatial distribution function.

When the propagation of sound is described by Eq. 1, the expression for $p(\mathbf{x}; t) = p(x, y, z; t)$ in the case of an infinite water mass around the source is given by:

$$p\left(\mathbf{x};t\right)=\frac{s\left(t-\frac{r}{c}\right)}{4\pi r} \tag{2}$$

Where r is the range from the sonar's acoustic source:

$$r = \sqrt{\left(x - x_s\right)^2 + \left(y - y_s\right)^2 + \left(z - z_s\right)^2} \tag{3}$$

From Eq. 2 it is clear that the acoustic pressure level is reduced according to the reciprocal of the distance to the source. This loss in acoustic pressure and energy is referred to as spherical spreading loss, and in the case of sonars it is to be applied twice: a signal travels from source(s) to target(s) and then back from target(s) to receiver(s). The signal received back at the sonar is a delayed and attenuated version of the initially transmitted signal.
It should be noticed that Eq. 2 is obtained with the assumption that the acoustic source is a monopole and has no dimensions. If this is not the case, p becomes frequency dependent.

2.3 Practical approaches to sonar simulation

From the implementation point of view, several approaches to sonar simulation are possible and frequently hybrid models are implemented. The most common are as follows:

Frequency domain models

In this approach the Fourier transform of the acoustic pressure that is measured back at the sonar receiver is expressed in terms of the Fourier transform of the acoustic pulse used for ensonifying the scene. This is the approach used in NURC's SIGMAS simulator and is discussed in detail in section 2.4. This implementation has the advantage of simplifying the inclusion of several processes that are easier to represent in Fourier space, such as the matched filtering of the received signal or the inclusion of the point spread function (PSF) of the sonar transducers.

Finite difference models

The wave equation given in Eq. 1 can be solved numerically by applying finite difference modelling, which imposes discretizations in time and space dimensions. Using, for instance, a forward difference scheme permits to approximate the time derivative of the pressure as follows:

$$\frac{\partial p}{\partial t} \approx \frac{p\left(x,y,z;t+\Delta t\right) - p\left(x,y,z;t\right)}{\Delta t} \tag{4}$$

where Δt is the temporal discretization step. For the spatial derivatives (with respect to the x, y, z coordinates) a similar formula is used. Starting the computation with initial conditions, i.e. the acoustic field at $t = 0$, permits to estimate the pressure field as any other point of time and space. The problem with finite difference models when applied to the side-looking sonar case is the dimensions of the computation. In order to obtain an accurate acoustic pressure field the sampling in both space and time is required to be on the order of a fraction of the reciprocal of the frequency and the wavelength, respectively. Even when avoiding parts of the computation—for instance solving only the wave equation around the location of the object of interest—the problem cannot be practically approached for frequencies higher than several kilohertz.

Finite Element (FEM) and Boundary Element models (BEM)

The finite element models and boundary element models are alternatives to finite differences that discretize the problem in a more optimized way. These approaches are complex to implement but typically generate more stable and accurate results with lower computational costs. However, even with these more sophisticated numerical techniques, no reasonable computation times have been achieved for sonar image modelling for frequencies much higher than ten kilohertz.

Ray tracing
Ray tracing (Bell 1997) is a method to calculate the path of acoustic waves through the system of water, sea bottom, sea surface and objects of interest. When the sound speed cannot be assumed constant in the water column refraction of the rays results in bent rays focused to certain places. The paths of the rays are advanced until they hit an object, where the particular contribution of the ray to the returned signal is then computed. Reflection, refraction and scattering events can be accurately modelled by computing a large number of rays, and these can account for complex phenomena observed in sonar imaging, such as multi-path effects or the behaviour of buried targets. Generally speaking, ray tracing is capable of rendering very accurate images but at a high computational cost.

Rasterization
Most current computer graphics are generated using rasterization techniques, which are based on decomposing the scene in simple geometrical primitives (typically triangles) that are rendered independently of each other. This permits fast generation of synthetic images, although effects that require interaction between primitives (such as mirror-like objects) can be complicated to simulate. A big advantage of raster methods is that most current computers include specialized hardware (Graphical Procesing Units, or GPUs) that greatly accelerate raster computations. NURC is currently working on a GPU-based implementation of its SIGMAS sonar simulator, in order to achieve faster simulation performance.

2.4 The SIGMAS sonar simulator

Using the frequency domain approach followed by the SIMONA model (Groen 2006) the SIGMAS simulator calculates the acoustic pressure for every pixel in the sonar image at the same time. In this sense, the signal processing, i.e. the imaging algorithm, is included in the model. In order to develop a realistic but sufficiently fast model some assumptions have been made. The sound speed in the water column is assumed to be constant, which means that acoustic paths follow a straight line. The surfaces of simulated objects are assumed discretized into facets to which the Kirchhoff approximation to the scattered field is applied.

The general expression in frequency domain for the acoustic pressure at the receiver $\mathbf{x_r}$ outside of an object's surface A can be derived using Green's theorem (Clay 1977, Karasalo 2005):

$$P(\mathbf{x}_r;f) = \iint_{\mathbf{x}\in A} \left[G(\mathbf{x}_r,\mathbf{x};f)\nabla P(\mathbf{x};f) - P(\mathbf{x};f)\nabla G(\mathbf{x}_r,\mathbf{x};f) \right] \cdot \mathbf{n}(\mathbf{x})\, dA \tag{5}$$

In the expression, $\boldsymbol{n}$ is the surface normal and G is Green's function, which for a homogeneous medium is given by (Neubauer 1958):

$$G(\mathbf{x}_r,\mathbf{x};f) = \frac{e^{ik|\mathbf{x}-\mathbf{x}_r|}}{4\pi|\mathbf{x}-\mathbf{x}_r|} \tag{6}$$

Where $k = 2\pi f/c$ is the wave number.

On hitting a surface, part of the pressure wave will be scattered back (reflected) and some of it will be refracted into the surface material or absorbed as heat. The fraction of pressure that is returned is measured by the reflectivity (or reflection coefficient) R of the surface material. The surface boundary conditions that relate the incident (P_i) and scattered (P) waves are:

$$P(\mathbf{x};f) = \left(1 + R(\mathbf{x};f)\right) P_i(\mathbf{x};f) \tag{7}$$

$$\frac{\partial P(\mathbf{x};f)}{\partial n} = 1 - R(\mathbf{x};f) \tag{8}$$

Where n indicates variation in the direction normal to the surface.
For our simulation purposes, an object with an acoustically rigid surface can be assumed, which means the reflectivity R is set to unity and therefore elastic effects do not play a role. This simplifies the boundary conditions on the scattering surfaces and is the approach used by SIGMAS. Substituting R for 1 in the boundary conditions (7) and (8), and using Green's function (6) twice to propagate from source $\mathbf{x}_s$ to surface A and then to receiver $\mathbf{x}_r$ yields the expression:

$$P(\mathbf{x}_r;f) = \iint_{\mathbf{x}\in A} -\frac{i\, e^{ik|\mathbf{x}-\mathbf{x}_r|} e^{ik|\mathbf{x}-\mathbf{x}_s|}}{\lambda |\mathbf{x}-\mathbf{x}_r||\mathbf{x}-\mathbf{x}_s|} P(\mathbf{x}_s;f) \frac{\mathbf{x}-\mathbf{x}_r}{|\mathbf{x}-\mathbf{x}_r|} \cdot \mathbf{n}(\mathbf{x})\, dA \tag{9}$$

Knowing that in sonar the source and the receiver are in the same position—which from now on we assume to be at the coordinate origin—and assuming the surface is discretized on small facets of area a^2, the integral can finally be expressed as the following summation:

$$P(f) = \frac{a^2}{\lambda} \sum_k \frac{\hat{\mathbf{n}}_k \cdot \hat{\mathbf{r}}_k}{\|\mathbf{r}_k\|^2} S(f) e^{i\frac{2\pi f}{c}\|\mathbf{r}_k\|} \tag{10}$$

Where $\mathbf{r}_k$ is the vector from the sonar to the kth surface element, $S(f)$ is the Fourier transform of the transmit signal $s(t)$, and where the hats indicate unit vectors.
Application of Eq. 10 produces results as those presented in Fig.2, where a barrel sitting on flat sand and a truck wheel on a bumpy clay floor have been simulated.

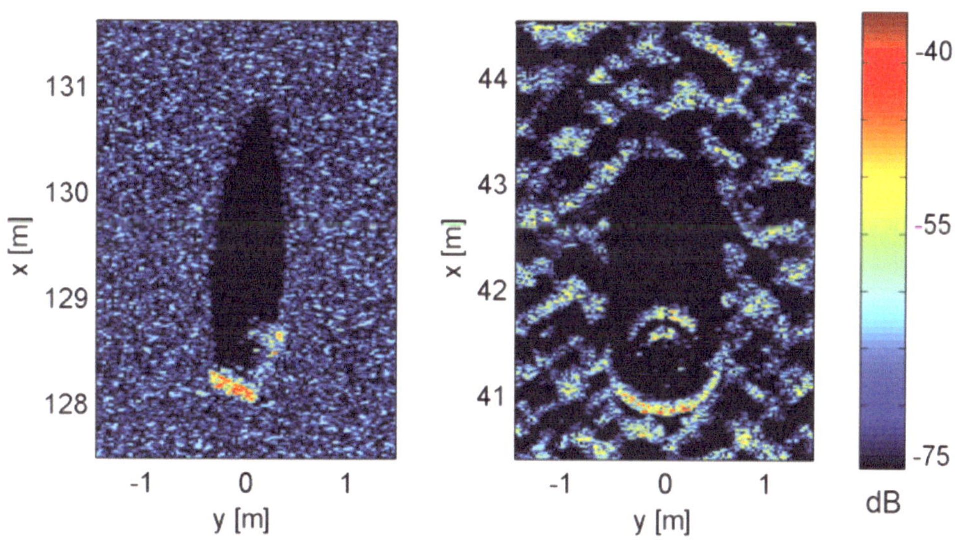

Fig. 2. Two examples of the output produced by the SIGMAS simulator: a barrel sitting on a sandy seafloor at 128 meters distance, and a truck wheel at 41 meters distance on a bumpy clay seabed.

2.5 Model simplification

The approach to simulation used by SIGMAS produces very accurate results that correlate well to experimental data. This accuracy comes with a computational cost, since the coherent model used requires sampling the objects to be simulated at scales smaller than the wavelength of the ensonificating signal. In the examples shown, for instance, the number of discrete point scatterers is around one million per object.

For very smooth objects or if elasticity effects are relevant, coherence has to be considered. On the other hand, if surfaces are rough at wavelength scale, the complex exponentials in Eq. 10 can be dropped and the discretization can use bigger surface elements with an equivalent reflectivity value R. The constructive and destructive interferences described by the complex exponentials can also be replaced by a noise distribution (Bell 1997). Furthermore, since most sonars perform some kind of Time-Varying Gain (TVG) intensity corrections to compensate for the spherical spreading loss, that contribution can also be dropped and replaced by a final image level scaling. Computations can be performed directly in image space, removing also the need for the FFT when working in frequency domain, resulting in the following expression for the observed pixel intensity at surface point $\mathbf{r}$:

$$I(\mathbf{r}) = K(\mathbf{r}) \sum_k (\hat{\mathbf{n}}_k \cdot \hat{\mathbf{r}}) R_k(\mathbf{r}) \chi_k(\mathbf{r}) \tag{11}$$

Where the sonar is assumed at the coordinate origin, K is a normalization constant that groups all scaling factors and unit conversions, R_k is the reflectivity of the kth surface patch and χ_k is the characteristic function of the patch (one if the circle of radius $|\mathbf{r}|$ intersects the patch, zero if not). Note that Eq. 11 basically corresponds to the Lambertian illumination model for diffuse surfaces (Zhang 1999), where the perceived brightness of a surface point depends on the relative direction of illumination and is independent of the direction of observation.

All these simplifications greatly reduce the complexity of the computations and even permit to use standard computer graphics renderers, such as OpenGL (OpenGL ARB 2004), to create the simulated images. Standard 3D models for objects to be simulated, like the VRML barrel shown in Fig. 3, can also be used directly. The final result is much faster scene

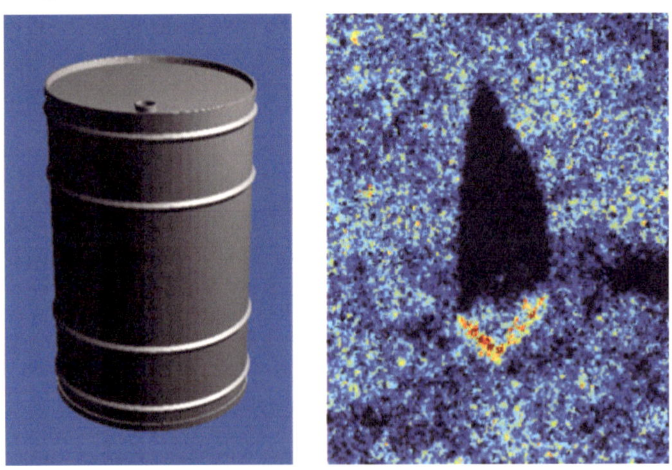

Fig. 3. A three-dimensional VRML model of a barrel and the result of a simplified sonar simulation using standard computer graphics rendering. The barrel is assumed to sit on a rough seafloor at 130 meter distance.

composition and rendering, at the cost of losing the significance of the resulting image values, which are no longer in correspondence with the actual pressure levels.

3. 3D reconstruction of sonar images

This second part of the chapter is dedicated to the sonar inversion process, which allows to infer a computer CAD model from a given sonar image, thus recovering the underlying 3D surface geometry of the observed scene.

3.1 Side-looking sonar image formation model

The geometry of the image formation process for a side-scan sonar is briefly sketched in Fig. 4. The sensor's acoustic source at **o** produces an ensonification pulse that illuminates the seafloor. Some of the acoustic energy reaching any seabed point p is scattered back and can be measured by the sensor. The intensity of the corresponding pixel on the side-scan image will be proportional to the amount of energy scattered back from the surface point. The illuminating pulse is not isotropic, but follows a particular beam-profile Φ that depends on the grazing angle α subtended by the vector **r** from **o** to surface point p.
In the case of synthetic aperture sonar (SAS) the side-looking image is formed differently, by emitting wider acoustic pulses and integrating the returned signals over time (Belletini 2002). For our inversion purposes, however, SAS images can still be regarded as having been produced by a sophisticated side-scan sonar, and the following discussion applies to both types of sensors.

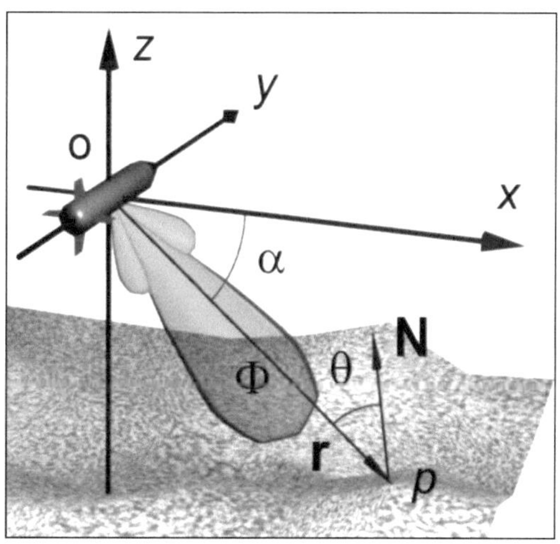

Fig. 4. Side-looking sonar imaging geometry (adapted from (Coiras 2007)).

In order to model the scattering process we use the traditional Lambertian (Zhang 1999) model already described in Eq. 11, which permits one to derive the returned intensity from the parameters defining the observed scene. This simple model for diffuse scattering assumes that the returned intensity depends only on the local angle of incidence θ of the illuminating sound pulse, and not on the direction of observation or on the frequency of the pulse. For the problem to be manageable the surface describing the observed scene has to be univalued, which forces to replace the expression in Eq. 11 for the following simpler one:

$$I(\mathbf{r}) = K\Phi(\mathbf{r})R(\mathbf{r})\frac{\mathbf{n}\cdot\mathbf{r}}{\|\mathbf{n}\|\|\mathbf{r}\|} = K\Phi(\mathbf{r})R(\mathbf{r})\cos(\theta(\mathbf{r})) \tag{12}$$

Where Φ represents the intensity of the illuminating sound wave at point p, R is the reflectivity of the seafloor, θ is the incidence angle of the wave front and K is a normalization constant. Since most logged side-looking images already include some kind of intensity correction, all the intensity variations caused by the sensor's beam-profile, the spherical spreading loss and the TVG and other corrections are supposed to be grouped under the beam-pattern Φ.
Following (Coiras 2007), with the coordinate system centered at the sensor in **o**, the x axis being the across-track ground distance and y pointing along the sensor's trajectory, we have:

$$\begin{aligned}\mathbf{r} &= (x, 0, Z(x,y)) \\ \mathbf{n} &= \left(-\frac{\partial Z}{\partial x}(x,y), -\frac{\partial Z}{\partial y}(x,y), 1\right)\end{aligned} \tag{13}$$

Where the y coordinate in **r** is 0 because the side-scan sonar pulse Φ is shaped so that only the points contained in the x-z plane are illuminated. Note that although this does not directly apply to the sonar pulses used for SAS imaging, the resulting SAS images are to all practical purposes equivalent to a side-scan image with constant resolution in the range direction.
Combination of expressions (12) and (13) yields the forward model for the computation of the intensity I at any point p, given the model parameters R, Z and Φ in ground range coordinates x, y from the sensor:

$$\begin{aligned}I(x,y) &= K\Phi(x,y)R(x,y)\cdot \\ &\frac{Z(x,y) - x\cdot\frac{\partial Z}{\partial x}(x,y)}{\sqrt{x^2 + Z^2(x,y)}\cdot\sqrt{\left(\frac{\partial Z}{\partial x}(x,y)\right)^2 + \left(\frac{\partial Z}{\partial y}(x,y)\right)^2 + 1}}\end{aligned} \tag{14}$$

Where the surface gradients can be approximated by finite differences (as shown for Eq. 4) and where the normalization value K is:

$$K(x,y) = \frac{Z\cdot\sqrt{x^2+Z^2}\cdot\sqrt{1+\left(\frac{\partial Z}{\partial y}\right)^2 + \frac{x^2}{Z^2}\left(1+\left(\frac{\partial Z}{\partial y}\right)^2\right)^2}}{-Z^2 - x^2\cdot\left(1+\left(\frac{\partial Z}{\partial y}\right)^2\right)} \tag{15}$$

Where the explicit dependencies on (x, y) have been dropped for clarity.

3.2 Sonar inversion

Equation 14 provides a direct formula for estimating the returned intensity given the model parameters. But the inverse problem—obtaining the model parameters from the observed

intensities—is clearly under-determined, since we only have one observation (of *I*) at each point to compute the values of the three model parameters.
In order to solve this problem and following (Coiras 2007) we take a least mean squares (LMS) approach to minimize the sum *E* of the squared differences between the points of the observed image, *I*, and those rendered by the model in Eq. 14, $\hat{I}$:

$$E = \sum_{x,y} E(x,y) = \sum_{x,y} \left(I(x,y) - \hat{I}(x,y) \right)^2 \tag{16}$$

And the following optimization problem needs to be solved:

$$(Z, R, \Phi) = \arg\min(E) \tag{17}$$

A solution can be found by using Expectation-Maximization (EM) (Dempster 1977), which will iteratively converge to an optimal set of modeling parameters. Every iteration of the EM method consists of two stages: in the Expectation stage, the current estimates for the model (R, Φ, Z) are substituted in Eq. 14 to obtain an estimation for the intensity $\hat{I}$. In the Maximization stage gradient descent is used to locally minimize *E*, by updating the model parameters as follows:

$$\begin{aligned} R(x,y) &\leftarrow R(x,y) - \lambda \cdot \frac{\partial E}{\partial R}(x,y) \\ \Phi(x,y) &\leftarrow \Phi(x,y) - \lambda \cdot \frac{\partial E}{\partial \Phi}(x,y) \\ Z(x,y) &\leftarrow Z(x,y) - \lambda \cdot \frac{\partial E}{\partial Z}(x,y) \end{aligned} \tag{18}$$

Where λ is a small constant value used to control the rate of change. Direct operation using Eq. 14, 16 and 18 yields:

$$\begin{aligned} R(x,y) &\leftarrow R(x,y) + 2\lambda \cdot \frac{\hat{I}(x,y)}{R(x,y)} \left(I(x,y) - \hat{I}(x,y) \right) \\ \Phi(x,y) &\leftarrow \Phi(x,y) + 2\lambda \cdot \frac{\hat{I}(x,y)}{\Phi(x,y)} \left(I(x,y) - \hat{I}(x,y) \right) \\ Z &\leftarrow Z - 2\lambda \hat{I} \left(I - \hat{I} \right) \cdot \\ &\cdot \left(\frac{-\partial Z/\partial x - \partial Z/\partial y}{1 + (\partial Z/\partial x)^2 + (\partial Z/\partial y)^2} + \frac{1+x}{x(\partial Z/\partial x) - Z} + \frac{Z}{x^2 + Z^2} \right) \end{aligned} \tag{20}$$

Where the explicit dependence of the parameters on (*x*, *y*) has been removed in the last equation for clarity.
The expressions in Eqs. 20 are iterated until the variation in the error *E* is below a given threshold.

Regularization

As the method is pixel-based, a regularization scheme is needed to smooth the overall solution. A very simple regularization is performed at the end of every iteration by filtering the estimated reflectivity and beam-pattern maps. Reflectivity values for the points in shadowed areas are set to that of their nearest illuminated neighbors by hexadecagonal dilation (Coiras 1998) of non-shadowed areas. Whereas values of Φ for all the points subtending the same angle *α* to the sensor are set to their median value, since the beam profile of the sensor is supposed to be constant for each grazing angle *α*:

$$\Phi(x,y) = Median\left\{ \Phi(x_0,y_0) \middle| \alpha(x_0,y_0) = \alpha(x,y) \right\} \tag{21}$$

Initialization
The optimization procedure starts by initialization of the *R*, *Z* and Φ maps. The reflectivity is set to a constant value (typically 0.9), and the elevation of every point (*x*, *y*) is set to that of the first return at (0, *y*), corresponding to the altitude of the sonar over the seafloor at that point of its trajectory. The initial beam-pattern Φ is set to the original image values *I* and then regularized using Eq. 21.
Multi-resolution
A multi-resolution implementation of the method described on the paragraphs above results in better convergence and improved results. Implementation of the multi-resolution version starts by the construction of a multi-resolution pyramid by iterated sub-sampling of the source side-looking image. Processing starts at the smallest level (coarser resolution), using the initialization and regularization procedures described in the previous sections. The resulting *R*, *Z* and Φ maps from one level are used as initial maps for the next resolution level. The process finishes when the final stage—corresponding to the full resolution image—is processed. Typically 3 levels of multi-resolution are used.

3.3 Sonar inversion results
During the MX3 2005 experiment carried out by NURC and partner nations in La Spezia (Italy), extensive surveys of an area of seabed were performed using several different vehicle and sensor combinations. The Sea Otter autonomous underwater vehicle (AUV), which was equipped with a Klein 2000 side-scan sonar, and one of the seabed images it produced is shown in Fig. 5.

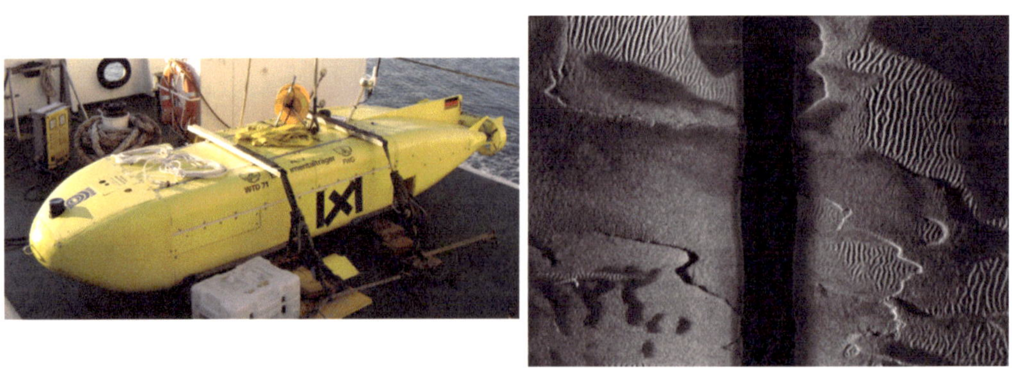

Fig. 5. The Sea Otter AUV and one of the images it collected during the MX3 trials with its Klein 2000 side-scan sonar.

Application of the inversion procedure to the slant-range sonar image of Fig. 5 permits to derive the projection of the sonar's beam-profile, the reflectivity map of the seabed and an estimation of the seabed topography. The maps for these three components are shown in Fig. 6.

The recovered reflectivity, beam-pattern and elevation maps are frequently noisy, which is due to the ill-posed nature of the reconstruction problem; most intensity variations can be caused by changes of any of the three forward model components (Φ, R, Z). Nevertheless, the reflectivity map in Fig. 6 suggests the presence of two different materials in the seafloor: one brighter, responsible for the rippled areas, and a less reflective one on the darker smooth areas. The reconstructed elevation map also looks satisfactory, although its accuracy is difficult to evaluate without actual measures of the area's bathymetry. Additional views of the interesting complex region at the top of Fig. 6(c) are shown in Fig. 7.

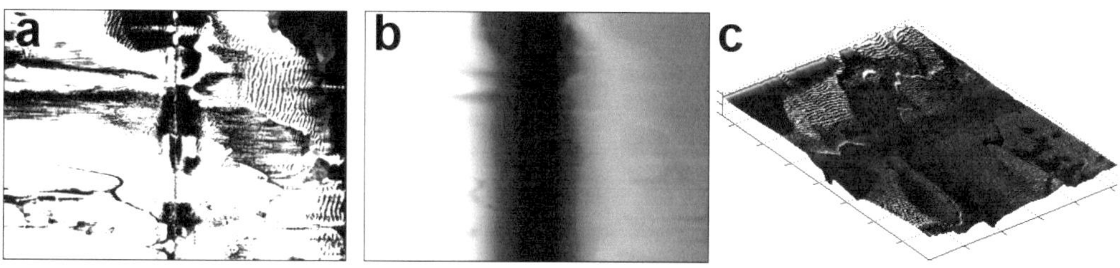

Fig. 6. Results of the 3D reconstruction procedure applied to the sonar image shown in Fig. 5. (a) Reflectivity map, (b) projection of the sonar's beam-profile, (c) textured 3D surface reconstruction using the recovered elevation map.

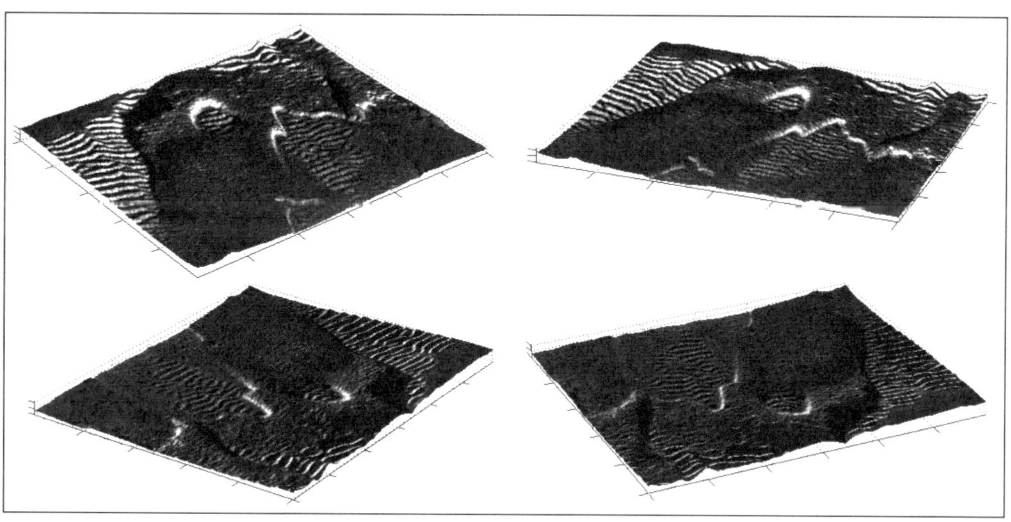

Fig. 7. Zoomed view of a selected area of the reconstructed seafloor, using four different points of view.

Apart from the most visual output of the sonar inversion procedure—the textured elevation map—the different recovered components can also be used for other purposes. As discussed, the reflectivity map hints at the material composition of the seabed. The recovered elevation map can also be used to produce more accurate ground-range projections of side-looking sonar images, one of whose coordinates is originally slant-range; this is relevant when sonar images are to be tiled to produce seamless seafloor mosaics. The recovered beam-profile—which we assume includes as well the intensity corrections (TVG) made during the image formation—has a special interest, in the sense that it is unique for each particular sonar. This means that after processing several sonar files, an estimate of the sensor's beam-profile can be computed and stored for later reference, reducing future inversion problems to the determination of two functions (the reflectivity and the elevation) instead of three.

The projection of the beam-profile is also useful to produce ground-range images with even intensity. As we mentioned already (section 2.2) the spherical spreading loss is responsible for a decrease of signal intensity with range, which is in principle compensated for by a TVG function. Yet there is an additional source of intensity reduction: since points that are far from the sonar subtend lower grazing angles—and therefore lower incidence angles—the average intensity (governed by Eq. 12) further decreases with range. The average incidence angle at a ground range x is:

$$\langle I \rangle(\mathbf{r}) \propto \cos(\langle \theta \rangle(\mathbf{r})) = \langle \hat{\mathbf{n}} \rangle \cdot \frac{\mathbf{r}}{\|\mathbf{r}\|} = (0,0,1) \cdot \frac{(x,0,z)}{\|(x,0,z)\|} = \frac{z}{\sqrt{x^2 + z^2}} \tag{22}$$

Where the angular brackets indicate expected value, and where the expected value of the unitary surface normal is assumed vertical because the seafloor is mostly flat. Setting the beam-profile to the inverse of Eq. 22 produces the result shown in Fig. 8(c), which features even illumination at all average incidence angles.

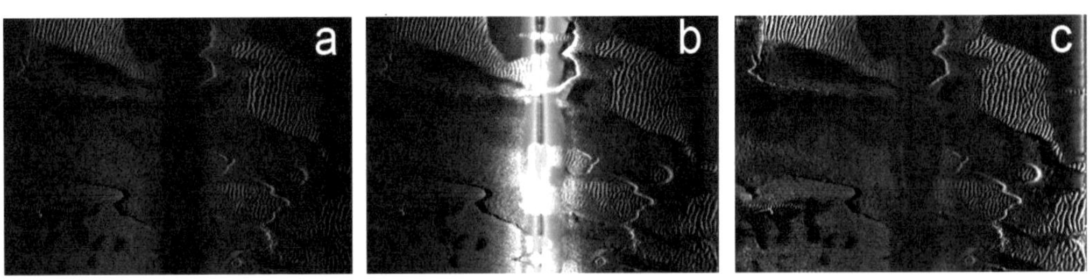

Fig. 8. The recovered beam-profile projection can be used to level the intensities of the ground-range version of the sonar image. (a) The original TVG-corrected sonar image in ground-range coordinates after proper projection using the recovered elevation map. (b) How would the image look if the beam-profile of the sonar was isotropic (had the same intensity on all grazing angles). (c) Ground-range image with a modified beam-profile that ensures an even illumination at all average incidence angles; this is the texture map that has been used on Figures 6(c) and 7.

4. Conclusions

The image formation process for side-looking sonar has been studied in this chapter. Both the forward and inverse realizations have been considered, and their application to simulation and 3D reconstruction of sonar images has been shown.

5. Acknowledgements

The authors would like to acknowledge the work of E. Bovio of NURC as scientist in charge of the MX3 experiments, where the Sea Otter vehicle was operated by Atlas Electronik.

6. References

J.M. Bell, L.M. Linnett (1997). Simulation and analysis of synthetic sidescan sonar images. *IEE Proceedings on Radar, Sonar and Navigation*, Vol. 144, No. 4, pp. 219–226.

A. Bellettini, M.A. Pinto (2002). Theoretical accuracy of synthetic aperture sonar micronavigation using a displaced phase-center antenna. *IEEE Journal of Oceanic Engineering*, Vol. 27, No. 4, pp. 780-789.

P. Blondel and B. J. Murton, (1997). *Handbook of Seafloor Sonar Imagery*. Wiley, New York.

C. Clay, H. Medwin (1977). *Acoustical Oceanography : Principles and Applications*. John Wiley & Sons Inc.

E. Coiras, J. Santamaria, C. Miravet (1998). Hexadecagonal Region Growing. *Pattern Recognition Letters*, Vol. 19, pp. 1111-1117.

E. Coiras, Y. Petillot, D.M. Lane (2007). Multi-Resolution 3D Reconstruction from Side-Scan Sonar Images. *IEEE Transactions on Image Processing*, Vol. 16, No. 2, pp. 382-390.

A.P. Dempster, N.M. Laird, D.B. Rubin (1977). Maximum Likelihood from incomplete data via the EM algorithm. *Journal of the Royal Statistical Society B*, Vol. 39, No. 1, pp. 1-38.

D.S. Drumheller (1998). *Introduction to Wave Propagation in Nonlinear Fluids and Solids*. Cambridge University Press, UK.

J. Groen (2006). *Adaptive Motion Compensation in Sonar Array Processing*. Netherlands Organization for Applied Scientific Research (TNO), The Netherlands.

I. Karasalo, P. Skogqvist (2005). Identification of Object Parameters from Transient Bistatic Scattering Data. *Proceedings of the Boundary Influences in High Frequency Shallow Water Acoustics Conference*, pp. 357-365. Bath, UK.

W.G. Neubauer (1958). *Acoustic Reflection from Surfaces and Shapes*. Naval Research Laboratoty.

OpenGL Architecture Review Board (2004). *OpenGl Reference Manual* (4th Edition), Dave Shreiner. Addison-Wesley Professional.

R. Zhang, P. Tsai, J. Cryer, M. Shah (1999). Shape from Shading: A Survey. *IEEE Transactions on Pattern Analysis and Machine Intelligence*, Vol. 21, No. 8, pp. 690-706.

Synthetic Aperture Sonar

2

Synthetic Aperture Techniques for Sonar Systems

Sérgio Rui Silva, Sérgio Cunha, Aníbal Matos and Nuno Cruz
University of Porto, Faculty of Engineering
Portugal

1. Introduction

Today a good percentage of our planet is known and well mapped. Synthetic aperture techniques used in space and airborne systems has greatly aided to obtain this information. Nevertheless our planet is mostly covered by water and the level of detail of knowledge about this segment is still very far away from that of the land segment.

Synthetic aperture is a technique that enables high resolution through the coherent processing of consecutive displaced echo data. Instead of using one static large array of transducers, it uses the along-track displacement of the sensors to synthesize a large virtual array. The resolution thus obtained is in the order of the transducer size and, most importantly, independent of the range between sensor and target. While a modern high frequency real-aperture sonar system can have a beam width below 1°, this translates into a resolution of half a meter at a range of just 25m. A synthetic aperture system using the same transducer can obtain a resolution of about 5cm across the whole range. Moreover the transducers used for synthetic aperture can be much simpler and so a good deal less expensive. Because there is no need to have a small real aperture, the frequency employed can be considerably lower, which enables longer reach due to the better propagation of lower frequencies in water.

This potential resolution increase comes at the cost of algorithm complexity in the image formation. The sonar must also describe a movement with tight tolerances with respect to deviations from known velocity and path. Also, the platform maximum velocity is a function of the pulse repetition rate and sensor size. This limit relates to the resolution of the image that if not respected will lead to aliasing.

The most used platform for synthetic aperture sonar is the tow-fish. Good designs enable smooth motion, but the inability to use satellite navigation technology leads to expensive solutions that integrate high grade inertial navigation units and data extracted from the sonar array itself. This only works for arrays with a high count of elements that operate at the nominal or above the nominal pulse repetition frequency. The sonar can also be mounted on the hull of a ship, providing access to high precision GPS navigation that can be integrated with data from moderate cost inertial systems to further refine the navigation solution. Nevertheless a ship is seldom easy to manoeuvre and presents considerable operation and maintenance costs of the sonar system itself. An autonomous boat arises as an interesting solution for these problems. It can be used as standalone or with the support of a ship. It enables the use of GPS and inertial navigation units efficiently. Moreover, its path and velocity can be easily controlled for better motion stability. The operation and maintenance costs are low and its availability is very high. It can be used both for sporadic missions and for regular security check of harbours, river navigability assessment, infrastructure inspection, etc.

The position of the sonar must be known to a 1/8 of the wavelength for proper synthetic image formation. Traditional synthetic aperture image formation techniques assumes a straight path for the sonar motion (typically for method that operate in the frequency domain) and treat deviations from this straight path as motion errors. A newer approach uses time-domain methods (such as back-projection) that don't rely on the assumption that the sonar follows any particular path. Instead, the information obtained by the navigation system is used at each sonar sampling position to form the virtual array. Here, only the position uncertainties are considered as errors.

High frequency systems require navigation precision bellow the centimetre level. This level of precision is not feasible to be obtained with the navigation systems of today. Therefore, the image formation starts with the available navigation solution and then a global auto-focus algorithm (that searches for an optimum measurement of image quality) refines the image formation parameters to mitigate the unavailable necessary navigation precision. Instead of using redundancy in the data (that comes at a high cost), global auto-focus algorithms parameterizes the image formation process enclosing navigation errors and medium fluctuations. Because of the large number of parameters that result from this, an efficient process of focusing synthetic aperture images is necessary.

2. Overview of current systems

Active sonar systems enable underwater imaging through angular and range echo discrimination. When a single beam is used to illuminate a swath as the sonar platform moves it is said that the sonar is a side-scan. In these systems a single echo line is obtained at each time with the angular discrimination being given by the beam width. Thus a narrow beam is desirable for high angular discrimination or along-track resolution. Typical beam widths are in the order of 1°. In these systems the along-track resolution is dependent of the range and a large array has to be used to obtain suitably low beam widths at the desired range. This type of sonar enables high area coverage speed. Alternatively, several narrow beams can be used to spatially sample the swath obtaining range and intensity information for each angular position. This is called multi-beam sonar. In this case the footprint of each beam is also dependent of the range. This type of sonar requires expensive and complex hardware to achieve a high number of sampling narrow beams. The area coverage speed is also limited by the area covered by the beams.

Synthetic aperture enables a high resolution/high area coverage binomial not possible with other sonar techniques. Instead of using a long physical array, a large virtual array is synthesised through the coherent combination of the echoes in the along track dimension of a side-scan sonar. Range independent along-track resolution is in this way obtained. Moreover the obtained along-track resolution is not influenced by the frequency of the signals employed and is in the order of the transducer physical dimensions. Lower frequency signals can thus be employed to extend the sonar range. Also because of the processing gain, the necessary transmitting power is lower when compared to its real aperture counterpart.

While it is possible to apply synthetic aperture techniques to multi-beam sonar, these have not been of great dissemination and the focus has been on obtaining high resolution sonar systems with large area coverage speed.
It is possible to obtain height estimation through the use of interferometric techniques on side-scan sonar images or synthetic aperture sonar images ([Saebo, T.O. (2007); Silva, S. et al (2008 c)]). Nevertheless multi-beam sonar obtains height measurements directly.
Synthetic aperture sonar systems can operate in strip-map or spot-light mode. In spot-light mode the beam is steered so a single zone is illuminated as the sonar platform moves, while in strip-map mode an area is sequentially sampled. Since sonar applications normally strive for area-coverage, strip-map is the more used synthetic aperture mode.
Synthetic aperture techniques are of common use in radar systems (typically know as SAR, [Tomiyasu, K. (1978)]). Here the signal propagation velocity through the medium is much higher, the wavelength long in comparison to radar platform motion uncertainties and medium phase fluctuations. Moreover the scene width is short when compared to the centre range which enables the use of simplifying approximation in the image formation algorithms. Also the bandwidth to centre frequency ratio is small. This means that SAR algorithms are not suitable for direct application in sonar data.
Medium stability for synthetic aperture sonar system is a major issue, but has been proved to be not restraining factor for the application of this technique ([Gough, P. T. et al (1989)]).
Most active synthetic aperture sonar systems are still in the research and development stage ([Douglas, B. L.; Lee, H. (1993); Neudorfer, M. et al (1996); Sammelmann, G. S. et al (1997); Nelson, M. A. (1998); Chatillon, J. et al (1999)]) but this technology is starting to emerge as a commercially advantageous technology ([Hansen, R.E. et al (2005); Putney, A. et al (2005)]).
Nevertheless most of these systems rely on complex and costly underwater vehicle configurations with multiple receivers, making them not attractive cost wise. This is because to correctly focus a synthetic array one needs to know the position with a very high degree of precision. Since it is not possible to use a high precision navigation system, such as a GPS, in the underwater environments, these systems have to rely on positioning schemes that use sonar data itself. This usually means having to use a complex array with multiple receivers, limit the range to use higher pulse repetition frequency or greatly limit the area coverage speed. Inertial navigation systems are of hardly any utility in standalone mode and must be corrected by other navigation sources. The underwater vehicles can be simply towed by a ship or be autonomous.
Surface autonomous vehicles are easier to operate and maintain. Plus it is possible to use readily available high precision differential GPS navigation system. Other navigation data sources can also be effectively used to enhance the navigation solution. A simple transducer array can be used since there is no longer mandatory to have a multiple element system for navigation. This makes high resolution synthetic aperture sonar a cost effective possibility.

3. Problem statement

Synthetic aperture techniques can be used to obtain centimetre level resolution in the along-track direction. Nevertheless to obtain this level of resolution in the cross-track direction, one needs to use large bandwidths and thus high frequency signals. This makes the position accuracy issue even more problematic for synthetic aperture sonar system since the necessary accuracy is directly related to the wavelength of the signal used.

Further more a sonar platform is subject to several undesirable motions that negatively affect the synthetic aperture performance. Normally a synthetic aperture sonar platform would move through a straight path with constant velocity. But this is seldom the case. Low frequency oscillations around the desired path and high frequency motion in heave, sway and surge directions adversely affect the sonar platform. This is due to water surface motion, currents and platform instability.

In the case of a tow-fish, there is little or no control on its motion. For an autonomous under water vehicle it is difficult to maintain a path reference since there are no reliable navigation sources.

An autonomous boat offers automatic position and velocity following capabilities through the use of a highly available differential GPS system (DGPS-RTK). Furthermore, the sonar sampling position and attitude is known to a high precision level which can than be integrated in the image formation algorithms.

The synthetic aperture sonar is then thus only limited by the unknown portion of the platform motion and medium phase fluctuations. Typical DGPS-TRK systems can operate with an error in the centimetre level. This mean that a synthetic aperture image with a wavelength of about 8 cm can be directly formed using only this position estimation. Furthermore, in this way, the obtained images can be easily and accurately integrated with other geographic information systems. Nevertheless, shorter wavelengths require auto focus procedures to be applied to the data for successful image formation.

Traditional synthetic aperture image formation algorithms treat deviations from a linear path as motion errors. They do not integrate well large deviation from a theoretical linear constant velocity path, and so new algorithms had to be developed to cope with this information.

Auto-focus algorithms are essential to successfully produce high-resolution synthetic aperture images. Their role is to mitigate phase fluctuations due to medium stability and more importantly reduce the navigation precision requirements to acceptable levels.

Most auto-focus algorithms require either large pulse repetition frequencies (PRF), several receiver transducers or both. In practical terms is not always possible to use high pulse repetition frequencies due to range ambiguity. With several receiver transducers the PRF can be lower, but the system cost is considerably higher not only due to the transducer array itself but also because of the data acquisition and signal processing system increased complexity.

4. System description

The complete system setup ([Silva, S. (2007a)]) is constituted by a control base station, a surface craft and the sonar itself (Fig. 1). Because of its size and modular construction, this setup is easily portable and has low deployment time. The boat it self is modular and easy to assemble in site without the need of any special tools.

The sonar platform is an autonomous boat (Fig. 2). This is a small catamaran like craft, proving high direction stability, smooth operation and several hours of unmanned operation, which can be command manually from the base station or fulfil a pre-defined mission plan. It was built using commonly available components to lower cost and simplify maintenance. It has two independent thrusters for longitudinal and angular motion that provide high manoeuvrability at low speeds and a maximum speed of 2 m/s. Its size is suitable for, together with the navigation system, executing profiles and other manoeuvres with sub-meter accuracy. The boat carries three GPS receivers, for position and attitude calculation, a digital compass and a inertial navigation system. The data from the digital compass and inertial system is integrated with the GPS data to provide a better navigation solution. It also embodies an on-board computer for system control, as well as for acquisition and storage of data.

The communication to the boat is obtained using an ethernet radio link (Wi-Fi). The boat and base station antennas were studied as to minimize the effect of the water reflective surface and maximize radio reach which is in the kilometre range.

The sonar system is carried by the boat as payload and transducers are placed at the front of the vessel, rigidly coupled to the boats structure.

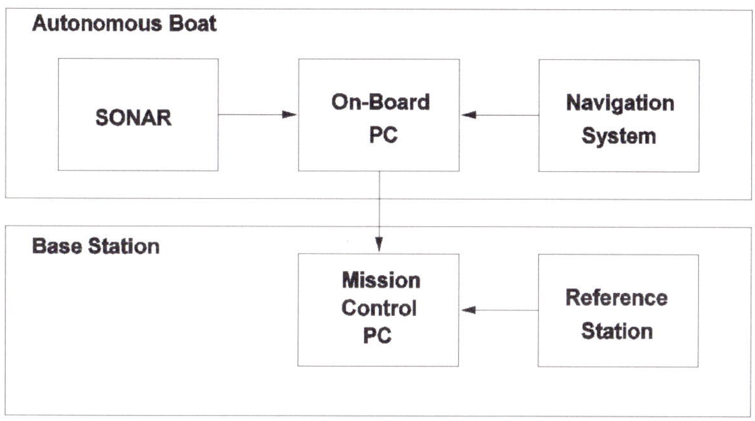

Fig. 1. Sonar system overview.

Fig. 2. Autonomous boat based synthetic aperture sonar.

Fig. 3. Autonomous boat and base station before a mission.

Compared to a solution based on a tow-fish, this sonar platform was mainly chosen because of the possibility to use a GPS positioning system. The velocity and position following is of great importance too. Low frequency errors such as deviations from the desired linear path or inconstant mean velocity are in this way limited. Other surface crafts could be used, but it would still have to provide some degree of motion control.

Having the sonar placed at the surface of the water, limits its use to shallow water realms such as rivers, lakes, dams harbours, etc, because of the maximum possible range. These are nevertheless an important part of the synthetic sonar possible application scenarios. It also makes the sonar more vulnerable to undesirable heave, roll and pitch motion that has to be measured by the boats navigation system and integrated in the sonar image formation algorithms.

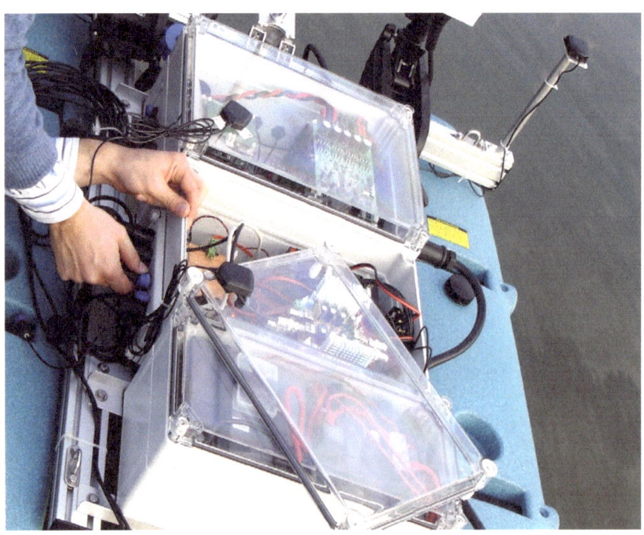

Fig. 4. Autonomous boat in preparation for a mission.

The base station is constituted by a portable PC that runs the boat and sonar control software, a GPS reference station and a high speed digital radio link to the boat (Fig. 3). The purpose of the base station is to control the mission and provide real-time visualization of the sonar data.

The synthetic aperture sonar system is it self constituted by two major components: the transducer array (Fig. 5) and the digital signal processing system for signal generation and acquisition (Fig. 6).

The floating platform transports the acoustic transducer array, placed beneath the waterline. The transducer array is formed by two displaced sets of 4 transducers separated by 0.5 m. Although only one transducer is necessary for transmission and reception of the sonar echoes, the vertical arrangement enables interferometric height mapping. The horizontal set of transducers is for future use and will enable image enhancement through the use of micro-navigation techniques. Each individual set can be regulated for an angle suitable to illumination of the swath at the expected underwater surface depth.

The transducers operate at a centre frequency of 200 kHz, corresponding to a wavelength of 0.75 cm. As appropriate for synthetic aperture operations, their real aperture is large (approximately 18 degrees), but have a strong front-to-back lobe attenuation ratio which is fundamental to minimize the reflections on the near water surface. The effective transducer diameter is 5 cm, which allows for synthetic images with this order of magnitude of resolution in the along-track direction.

The usable bandwidth of the transducers (and of the signals employed) is explored thought the use of amplitude and phase compensation to obtain the highest possible range resolution from the system, thus enabling the use of bandwidths of 20 to 40 kHz with the current configuration.

The sonar signal acquisition and generation system is made up of four principal components: the digital processing and control system, the power amplifier, the low noise amplifier, and a GPS receiver for time reference (Fig. 7).

This system is tailored for a high resolution interferometric synthetic aperture sonar and so special attention was paid to the different analogue components in terms of bandwidth, frequency amplitude response and phase linearity.

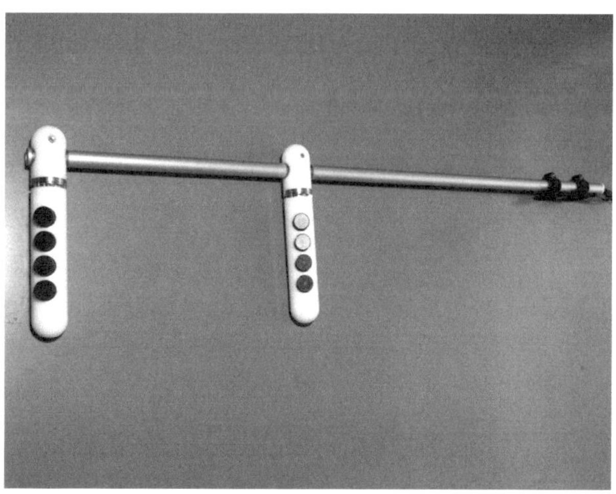

Fig. 5. Transducer array and support system.

The power-amplifier is a linear amplifier which can have a bandwidth as high as 10 MHz, and has a continuous output power rate of 50 W (RMS). A trade-of was made between output power and bandwidth. Because we are interested in using our system in shallow waters, an output power of 50W was found to be adequate for this 200 kHz sonar system. In exchange it was possible to build a power-amplifier with very flat amplitude and linear phase response in addition to low distortion (THD < -60 dB) in the spectral band of interest. The power amplifier only operates through the transmission period: this saves energy and lowers the electric noise levels during echo reception. Traditionally, switch amplifiers are use to drive the transducers. A linear amplifier design of this type enables the use of amplitude modulated signals and typically provides lower noise and spectral interference. The transmitted signal can be of any kind: linear chirp, logarithmic chirp or pseudo-random sequence. It can also have any suitable windowing function applied. At this moment, a linear chirp of 30 kHz bandwidth is being used for system tests. The system is prepared to output a pulse rate between 1 to 15 Hz. Because we are interested in shallow water surveys, short ranges enable higher pulse rates. Nevertheless, because of the broad beam-width of the transducer and its effective diameter, a pulse rate of 15 Hz still imposes very low moving speeds to the autonomous boat (0,16 m/s) that in turn has to couple with higher precision trajectory tracking constrains.

Each receiving channel has a low noise amplifier and a controllable gain amplifier to handle the high dynamic range of echo signals. The pre-amplifier system has a bandwidth of 10 MHz, a noise figure of 2 dB and a 50-115 dB gain range. It can also recover rapidly from overdrive caused by the transmitting pulse. To reduced aliasing, a linear phase analogue filter is used in combination with a high sample rate A/D (up to 40 MSamples/s). This anti-aliasing filter is interchangeable and can have a cut-off frequency of, for example, 650 kHz or 2,3 MHz. Both the D/A and A/D have 12 bit resolution and are capable of 200 MSamples/s and 40 MSamples/s respectively. The 12 bit resolution of the A/D (effective 72 db SNR) was found to be high enough to cope with the front-end performance that is dominated by the transducers noise.

The sonar signal processing system uses a direct to digital system architecture. This means that only the front-end elements of the sonar, like the power amplifier and the low-noise amplifier, use analogue electronics and all other functions such as signal generation, frequency down/up-conversion, filtering and demodulation are performed in the digital domain.

Fig. 6. Sonar signal generation and acquisition system.

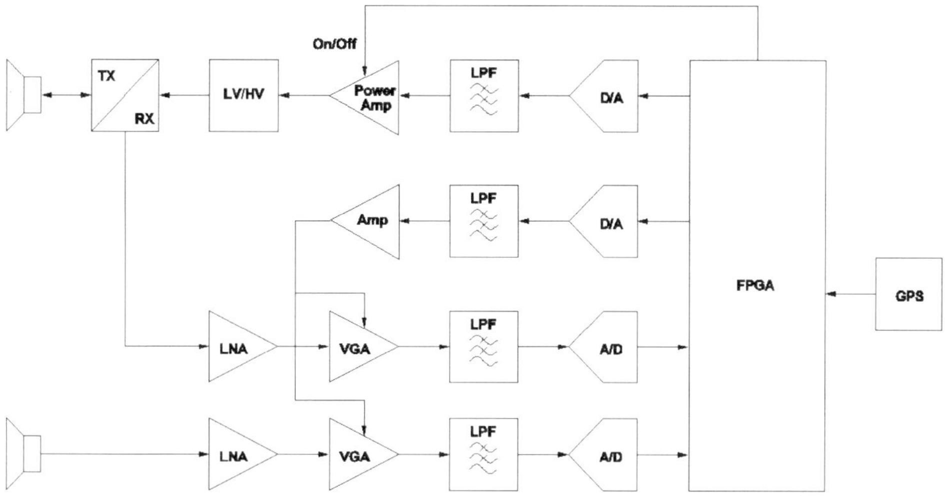

Fig. 7. Sonar schematic system overview.

This greatly simplifies the electronic hardware at the cost of more complex digital signal operations and hardware. But these in turn are powerfully performed in an field programmed array system (FPGA), providing a low cost solution.

A FPGA is a device that enables the implementation of customized digital circuits. This device can perform several tasks in parallel and at high speed. The system complexity stands, therefore, in the digital domain, enabling more flexible and higher quality signal acquisition and processing through this implementation. The use of this technology results in a low power consumption system that fits a small box, compatible with the autonomous boat both in size and energy consumption.

The FPGA system is responsible for the generation of complex acoustic signals, for controlling the transmitting power amplifiers and the adaptive gain low noise receiving amplifiers, for demodulating and for match filtering the received signals with the transmitted waveform (Fig. 8).

The first task of the FPGA is therefore to control the digital-to-analog (D/A) and analog-to-digital interface (A/D). At each transmission pulse trigger, the FPGA reads the signal wave form from memory (that is stored in base-band, decimated form), converts the signal to pass-band (frequency up conversion), interpolates and filters the signal and finally supplies it to the D/A.

During the receiver stage, the samples are read from the A/D, converted to base-band (frequency down-conversion), filtered, decimated and finally placed in a FIFO memory.

Each transmit pulse is time-stamped using a real-time clock implemented in the FPGA system that is corrected using the time information and pulse-per-second trigger from a GPS receiver. This enables precise correlation with the navigation data.

The processor embedded in the FPGA bridges the low level hardware and the control PC, providing an ethernet access to the sonar.

The results are then supplied to an embedded computer for storage and acoustic image computation. The shore base station accesses this data through a high-speed digital radio-link making available the surveyed data.

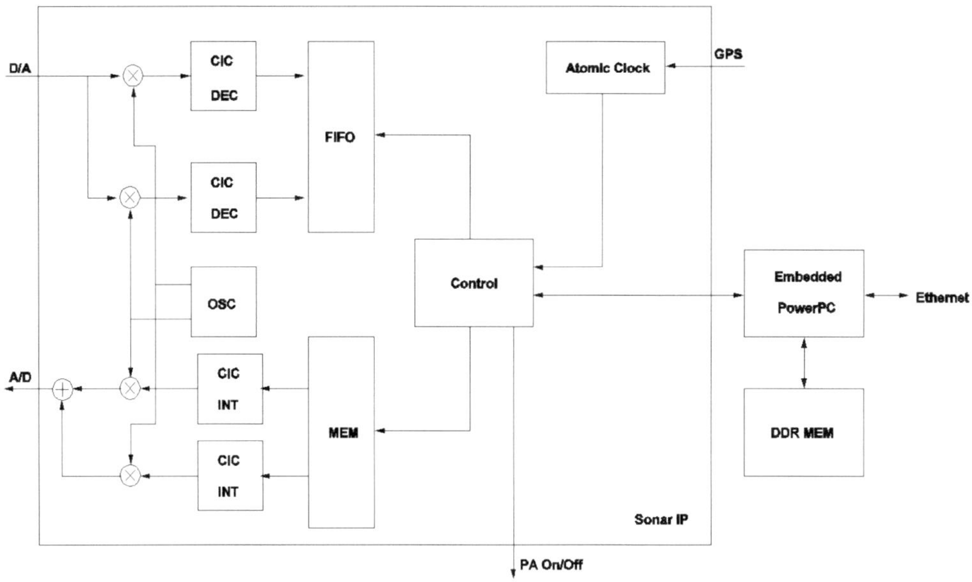

Fig. 8. FGPA system detail.

5. Navigation system

The accurate target area image formation process depends on the precise knowledge of the sensor position as it traverses the aperture. One of the major strengths of using a surface vessel is the use of satellite systems for navigation. Since the used wavelengths are in the order of centimetres, this is the level of relative accuracy that is required for the vessel along the length of the synthetic aperture. This is possible with GPS navigation if operated in differential mode using carrier phase measurements. The long term errors of a CP-DPGS solution for a short baseline to the reference station is in the order of the decimetre; the short term errors, due to noise in the carrier phase tracking loops, is in the order of the centimetre. This error can further be smoothed out through integration of the GPS solution with inertial measurements. For height computation through In-SAS processing of pairs of images, precise attitude data has to be obtained so to precisely transform the lever between two sensors into target position estimate. For this purpose inertial sensor play a crucial role: through blending with the GPS data, it supplies pitch and roll estimates with levels of accuracy in the order of arcminutes; the heading error estimate is slightly poorer, but has

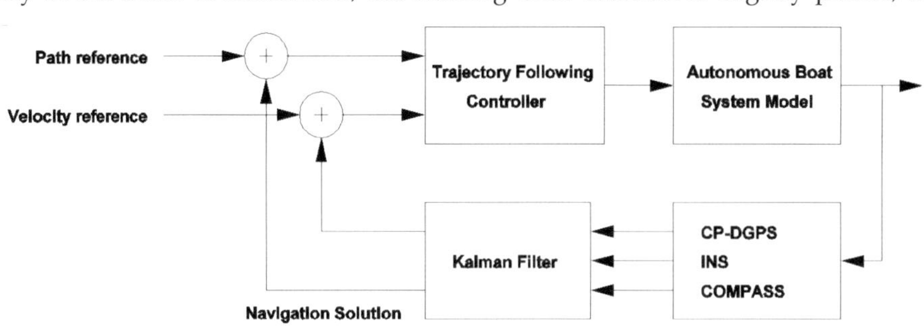

Fig. 9. Navigation system diagram.

less significant impact in the In-SAS processing algorithms. The Navigation subsystem is basically a Kalman filter mixing the dynamic behaviour of the vessel through the inertial sensor with the GPS reading. The major states are positioning, velocity and altitude (alignment) errors, but also includes inertial sensor biases. The output is a complete navigation solution, including position velocity and attitude information at a high data rate.

6. Autonomous boat control

To achieve the best results, the boat should follow the predefined paths at constant speed over ground and with minimal roll and pitch motions. Furthermore, it is desirable that the intended boat trajectories are described with minimal deviation to allow multiple sweeps in close parallel lines which are mandatory for dual passage interferometric height mapping.
To accomplish these goals, a control system that automatically drives the vehicle along user specified trajectories at given speeds was developed. The control system is organized in two independent control loops, one for the velocity and the other for the horizontal position of the vehicle. The velocity loop determines the common mode actuation of the boat while the horizontal plane loop determines the differential mode actuation. These values are then combined to produce the commands for the starboard and port thrusters.
The velocity loop is based on a proportional plus integral controller that assures that the velocity of the boat is, in steady state, the one defined by the user. The controller parameters where tuned to assure a smooth motion by rejecting high frequency noise from the navigation sensors. Different controller parameters can be used to obtain the best behaviour at different velocity ranges.
The horizontal plane loop implements a line tracking algorithm. It is composed by an outer controller that computes a heading reference based on the cross track error (distance of the boat to the desired straight line) and an inner controller that drives the vehicle heading to the given reference ([Cruz, N. et al (2007)]). This two stage control loop assures zero cross track error in steady state regardless of the water current and the tuning of the controller parameters took into account a dynamic model of the boat.
The whole system has already been tested in operational scenarios with great success. In particular, it has been observed that the vehicle describes intended trajectories with average cross tracking errors below 10 cm. For illustration purposes, Fig. 10 presents a histogram of the cross tracking error for a 50 m straight line described by the autonomous boat while collecting SAS data.

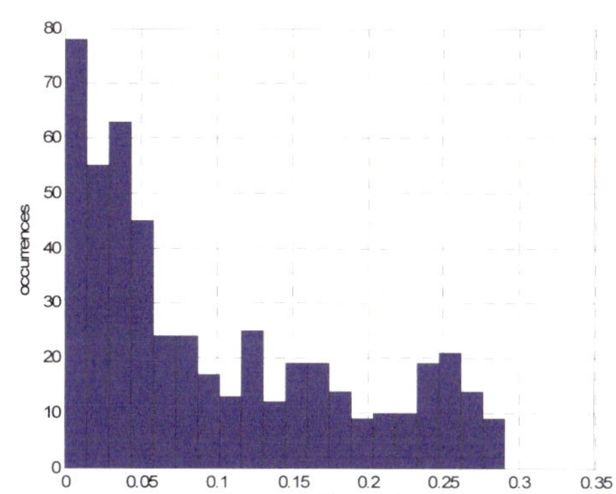

Fig. 10. Line tracking performance.

In Fig. 10 it is illustrated the path following ability of the autonomous boat, where it was programmed to execute two parallel profiles with a distance of 1 m of each other.

Fig. 11. Example path.

Fig. 12 shows the boat velocity graph, measured by the navigation system, during the execution of a profile.

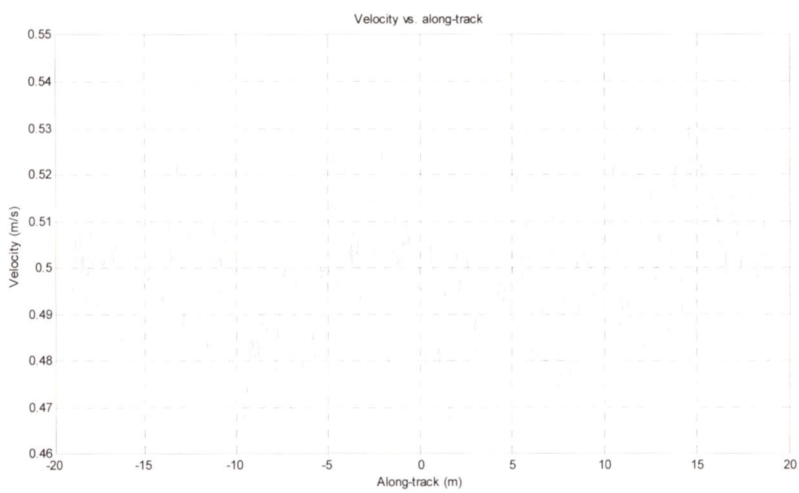

Fig. 12. Autonomous boat velocity during a mission.

7. Synthetic aperture sonar system model

Consider a platform that moves trough a nominal straight path. Since the boat is programmed to follow straight lines at a constant speed, this is a good assumption. The sonar transducers are rigidly coupled to the sonar structure, through which the acoustic signals are sent and their respective echoes are received (Fig. 13).

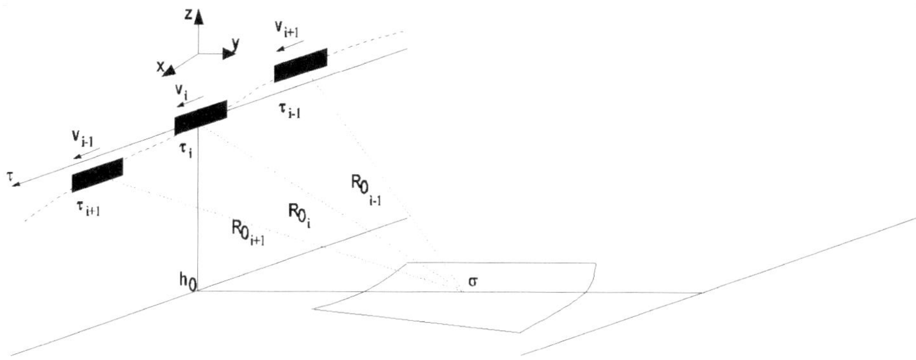

Fig. 13. Autonomous boat based SAS model.

Each echo contains indistinctive information of an area corresponding to the radiation pattern of the transducers. The transducers are adjusted so that a swath of about 50 m is at the boat left broadside, which is the strip-map configuration.
At each along-track sampling position, the echoes caused by the swath reflectivity are received and recorded along the track.
The echo ($ee(\tau, t)$) data is formed trough the reflection of the transmitted pulse ($p_m(t)$) on the swath ($ff(x,y_s)$):

$$ee(\tau,t) = \iint ff(x,y_s)\,p_m\left(t - \frac{2}{c}\sqrt{(v\tau - x)^2 + y_s^{\,2}}\right)dxdy \tag{1.1}$$

Transducer beam pattern effects and signal propagation loss have not been included in the equation for simplicity but should be accounted for in the image reconstruction process.
The echo data is converter to base-band though multiplication with the signal carrier frequency:

$$ee_b(\tau,t) = ee(\tau,t)e^{-j2\pi f_0 t} \tag{1.2}$$

Let $p_b(t)$ be the baseband transmitted signal, the cross-track pulse compressed echo image will be:

$$ss_b(\tau,t) = ee_b(\tau,t) * p_b(t) \tag{1.3}$$

And thus resulting in the following equation:

$$ss_b(t,\tau) = \iint ff(x,y_s)\,p_c\left(t - t_v(\tau)\right)e^{-j2\pi f_0 t_v(\tau)}dxdy \tag{1.4}$$

Where $p_c(t)$ is the autocorrelation of the transmitted baseband signal and $t_v(\tau)$ is the time-of-flight and is given by:

$$t_v(\tau) = \frac{2}{c}\sqrt{\left(x_p(\tau) - x_0\right)^2 + \left(y_p(\tau) - y_0\right)^2 + \left(z_p(\tau) - z_0\right)^2} \tag{1.5}$$

The point-target has coordinates (x_0, y_0, z_0) and $(x_p(\tau), y_p(\tau), z_p(\tau))$ are the coordinates of the sonar platform position at each along-track sampling instant τ. In the absence of motion errors, $y_p(\tau)$ and $z_p(\tau)$ will be zero and $x_p(\tau) = v\tau$, where v is the platform velocity.

The task to retrieve the estimated image ($\hat{ff}(x, y_s)$) is done through the inversion of this model, coherently combining the along-track samples to form an image:

$$\hat{ff}(x, y_s) = \iint ss_b(\tau, t)\left(t - t_v(\tau)\right) e^{j2\pi f_0 t_v(\tau)} d\tau dt \tag{1.6}$$

The sonar platform seldom dislocates through a straight line, and so the sonar model must account for the irregular along-track sampling positions. Therefore the polar coordinate model must be abandoned and a broader expression for the echo travel time must contain the estimated position of the transducers and a rough estimation of the bottom height (a flat bottom is enough for a first approximation). Moreover, different transducer geometries mean that the travel time will not be exactly the simple double of the distance to reach the target, but the distance from the transmitting transducer to the target and back to the receiver which is given by:

$$t_v(\tau) = \frac{1}{c}\left(\left\|T_{TX}^{'}(\tau) - X_\sigma\right\| + \left\|X_\sigma - T_{RX}^{'}(\tau)\right\|\right) \tag{1.7}$$

Attitude variations must also be accounted for in the calculation of the transducer positions. Knowing the arm between the boat navigation centre (reference for the navigation system) and each transducer, and also the roll, pitch and yaw angles, the positions of the transducers can be correctly calculated. The transducers attitude also influences the swath footprint. During image synthesis this information should be used to calculate the correct weights of the combined echoes.

The frequency domain model of the imaging system can be obtained through Fourier transformation of equation (1.4) and the Principle of Stationary Phase ([Gough, P. T. (1998)]). The image reconstruction task can, as a result, be done by multiplication of the frequency domain inverse sonar imaging model.

For calculation of the obtained along-track resolution and sampling frequency, consider that the target will be seen by the sonar during the time it is inside the aperture (3dB lobe) of the transducer, which is approximately given by:

$$\theta_{3dB} \approx \frac{\lambda}{D} \tag{1.8}$$

The array spacing from Nyquist spatial sampling and classical array theory is $\lambda/2$ (two way equivalent 2π phase shift), which means that for angles of arrival of a wave-front the inter-element phase difference must be less than 2π ([McHugh, R. et al (1998)]).

This is also true for motion errors. To correctly form a synthetic aperture the platform position must be known within 1/8 of a wavelength so the echoes can be coherently combined with negligible image deterioration ([Cutrona, L. J. (1975); Tomiyasu, K. (1978); Fornaro, G. (1999)]).

We can also consider the array spacing to be given by a pulse repetition frequency (PRF) that is at least equal the maximum Doppler shift experienced by a target. The Doppler shift f_D is related to the radial velocity v_r by:

$$f_D = \frac{2v_r}{\lambda} = \frac{2v\sin\theta(\tau)}{\lambda} \tag{1.9}$$

The maximum radial velocity is obtained at the beam edge and so the lower bound for the PRF is ([McHugh, R. (1998)]):

$$PRF \geq \frac{2v\sin\theta_{3dB}}{\lambda} \approx \frac{2v\lambda/D}{\lambda} = 2\frac{v}{D} \tag{1.10}$$

The minimum synthetic array spacing is thus:

$$d_{SA} = v\frac{1}{PRF} = \frac{D}{2} \tag{1.11}$$

The along-track resolution is independent of the range and wavelength. This results from the fact that for a transducer with a fixed length D, the synthetic aperture length DSA will be given, approximately, by:

$$D_{SA} \approx 2R_0\theta_{3dB} = 2R_0\frac{\lambda}{D} \tag{1.12}$$

Where R_0 is the distance to the center of the scene.
This than gives the classical synthetic aperture along-track resolution δ_{AT} formula:

$$\delta_{AT} \approx R_0\theta_{SA} = R_0\frac{\lambda}{D_{SA}} = R_0\frac{\lambda}{2R_0\frac{\lambda}{D}} = \frac{D}{2} \tag{1.13}$$

We see here that the phase relations that enable the synthetic array formation are tightly related to the wavelength of the signal and the effective synthetic array length. Normally these two values are interconnected due to the transducers real aperture width, but can be explored to mitigate some of the problems inherent to synthetic aperture.
The image formed in this way has a cross-track resolution of $c/2BW$ and an along-track resolution of $D/2$ (where c is the speed of sound, BW is the transmitted signal bandwidth and D is the effective transducer diameter). More importantly, the along-track resolution is independent of the target range. To correctly synthesize an image without aliasing artefacts in the along-track dimension, it is necessary to sample the swath with an interval of $D/2$ (considering the use of only one transducer for transmission and reception). This constraints, together with the maximum PRF defined by the longest distance of interest and the along-track sampling restrictions, imposes a very speed to a sonar platform ([Cutrona, L. J. (1975); Gough, P. T. (1998)]).

8. Image formation process

The sonar acquires the data in pass-band format which is then converted to base-band and recorded. Starting with this uncompressed base-band recorded data, the first step in image formation is cross-track pulse compression. This is also known as match filtering. This is step is necessary because using a longer transmitting pulse carries more energy than a short pulse with the same peak power which enhances the signal-to-noise ratio. The resulting cross-track resolution is not given by the duration of the transmitted pulse, but instead by its bandwidth. The task of pulse compression is done through correlation of the received data with the base-band transmitted pulse.

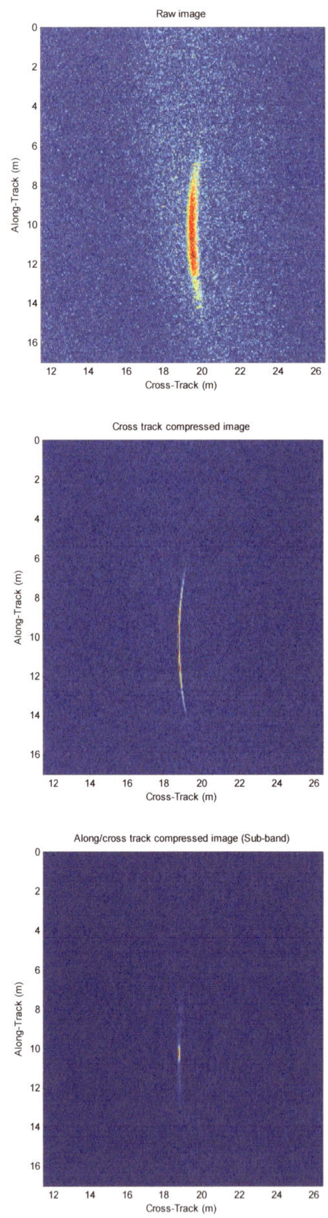

Fig. 14. Raw image, cross-track compressed image and along/cross-track compressed image.

At this stage data filtering and frequency equalization can be applied.
The next step is synthetic aperture formation that should use the available navigation data to synthesize the virtual array and form the sonar image. Fig. 14 shows these steps in succession for an image of an artificial target placed in the river bottom for a test mission. Note that the first image has low along and cross track resolution because its unprocessed, the second image has better cross-track resolution due to pulse compression and finally the last image, which is the result of synthetic aperture processing, resembles a small point.
Synthetic aperture image formation can be done through the use of several algorithms which can be classified into frequency domain algorithms, such as the wave-number algorithm, chirp scaling algorithm or the inverse scaled Fourier transform algorithm, and time domain algorithms such as the explicit matched filter or the back-projection algorithm ([Gough, P. T. (1998); Silkaitis, J.M. et al (1995)]).
The wave-number algorithm relies on inverting the effect of the imaging system by the use of a coordinate transformation (Stolt mapping) through interpolation in the spatial-frequency domain. The compressed echo data is converted to the wavenumber domain (along/cross-track Fourier transforms), matched filtering is applied supposing a target at a reference range followed by a nonlinear coordinate transformation ([Gough, P. T. (1998)]).
The chirp-scaling algorithm avoids the burdensome non-linear interpolation by using the time scaling properties of the chirps that are applied in a sequence of multiplications and convolutions. Nevertheless the chirp scaling algorithm is limited in use to processing of uncompressed echo data obtained by the transmission of chirp signals.
An approach based on the inverse scaled Fourier transform (ISFFT) previously developed for the processing of SAR data can also be followed. This algorithm interprets the raw data spectrum as a scaled and shifted replica of the scene spectrum. This scaling can then be removed during the inverse Fourier transformation if the normal IFFT is replaced by a scaled IFFT. This scaled IFFT can be implemented by chirp multiplications in the time and frequency domain (Fig. 15). The obtained algorithm is computationally efficient and phase preserving (e.g. fit for interferometric imagery). Motion compensation can be applied to the acquired data in two levels: compensation of the known trajectory deviations and fine corrections trough reflectivity displacement, auto-focus or phase-retrieval techniques. The deviations from a supposed linear path are compensated thorough phase and range shift corrections in the echo data. Velocity variations can be regarded as sampling errors in the along-track direction, and compensated through resampling of the original data ([Fornaro, G. (1999)]).
The back-projection algorithm, on the other hand, enables perfect image reconstruction for any desired path (assuming that rough estimate of the bottom topography is known), since it does not rely on the simple time gating range corrections ([Hunter, A. J. et al (2003); Shippey, G. et al (2005); Silva, S. (2007b)]). Instead, it considers that each point in one echo is the summation of the contributions of the targets in the transducer aperture span with the same range. With this algorithm one is no longer forced to use or assume a straight line for the sonar platform displacement. The platform deviations from an ideal straight line are not treated as errors, but simply as sampling positions. In the same way, different transducers array geometries are possible without the need for any type of approximation. This class of synthetic aperture imaging algorithms, although quite computational expensive in comparison with frequency domain algorithms, lends itself very well to non-linear acquisition trajectories and, therefore, to the inclusion of known motion deviations from the expected path. To reconstruct the image each echo is spread in the image at the correct

coordinates (back-projected) using the known transducer position at the time of acquisition (Fig. 16). It is also possible to use an incoherent version of this algorithm (e.g.: that does not use phase information). But the obtained along-track resolution is considerably worse ([Foo, K.Y. et al (2003)]).

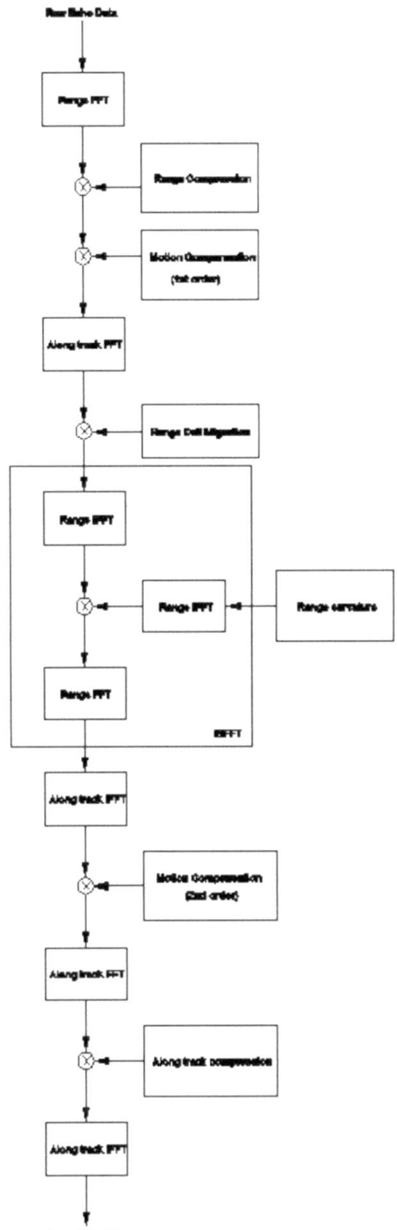

Fig. 15. ISFFT algorithm flow diagram.

The back-projection algorithm can also be implemented in matrix annotation ([Silva, S. et al (2008 a)]). The navigation information and system geometry is used to build the image formation matrix leading to the reconstructed image. The transmitting and receiving beam patterns and the corresponding swath variation with the platform oscillation is also weighted in the matrix. This makes this algorithm well suited for high resolution sonar systems with wide swaths and large bandwidths that have the assistance from high precision navigation systems. The main advantage of this algorithm is the ease of use within an iterative global contrast optimization auto-focus algorithm ([Kundur, D. et al (1996)]). The image formation is divided into two matrixes: a fixed matrix obtained from the sonar geometrical model and navigation data (corresponds to the use of a model matching algorithm, such as the explicit matched filtering); and a matrix of complex adjustable weights that is driven by the auto-focus algorithm. This is valid under the assumption that the image formation matrix is correct at pixel level and the remaining errors are at phase level (so that the complex weight matrix can correct them).

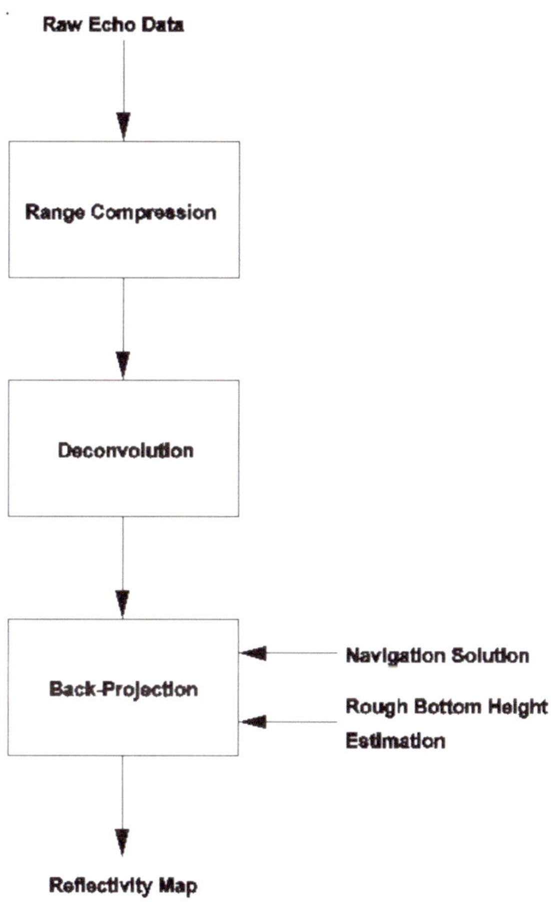

Fig. 16. Back-projection algorithm signal flow diagram.

9. Auto-focus

Since the available navigation data sources, be it DGPS or INS systems, cannot provide enough precision to enable synthetic aperture processing of high resolution (high frequency)

sonar data [Bellettini et al (2002); Wang et al (2001)], the phase errors caused by the unknown motion components and medium turbulence must be estimated to prevent image blurring.
Auto-focus algorithms exploit redundancy and or statistical properties in the echo data to estimate certain image parameters that lead to a better quality image. Therefore, the auto-focus problem can be thought as a typical system estimation problem: estimate the unknown system parameters using a random noise input. If the auto-focus algorithms estimates the real path of sonar platform they are called micronavigation algorithms [Bellettini et al (2002)] (sometimes with the aid of navigation sensors such as inertial units) otherwise they are generically designated as auto-focus algorithms. Redundant phase centre algorithm and shear average algorithm are examples of micronavigation algorithms.
Since redundancy in data is greatly explored, common auto-focusing algorithms require restrictively along-track sample rates equal or higher than the Nyquist sample rate. This imposes unpractical velocity constrains, especially for system that use few receivers (as is the case with the sonar system described here). It is not possible to obtain micro-navigation from an under-sampled swath or to perform displaced centre phase navigation with only one transducer. So, with these impairments, global auto-focus algorithms are required in sonar systems that use simple transducers arrays and under-sampled swath. The use of global auto-focus algorithm presents several advantages for synthetic aperture sonar image enhancing. They differ from other algorithm because they try to optimize a particular image metric by iteratively changing system parameters instead of trying to extract these parameters from the data. Global auto-focus algorithms can correct not only phase errors due to navigation uncertainties, but also phase errors that are due to medium fluctuations.
It is required that the synthetic aperture algorithm uses the available navigation solution to form an initial image. Starting with the available navigation solution, the errors are modelled in a suitable way. If the expected errors are small they can be modelled as phase errors for each along-track position. If the sonar platform dynamic model is known, the number of search variables can be greatly reduced by parameterizing this model ([Fortune, S. A. et al (2001)]). These parameters are weighted together with the image metric and serve as a cost function for the optimization algorithm to search the solution space (Fig. 18).
Nevertheless, these errors are hardly ever smaller than the original signal wavelength, and so create a solution surface that is difficult to search for the optimum set of parameters. However, if we have access to the raw data, by dividing the received signal bandwidth in several smaller bands and conjugate complex multiplying the pulse compressed signals obtained in each band one by the other, a new resulting signal is obtained with an effective longer wavelength corresponding to the frequency difference between the two sub-bands ([Silva, S. (2008 b)]). This longer wavelength effectively reduces the impact of phase fluctuation from the medium and platform motion uncertainties. Using this, it is possible to divide the signal bandwidth into several sub-bands and combine them in to signals with different wavelengths. At the first step, a large wavelength is used since the expected motion correction is also large. After achieving a predefined level of image quality, the auto-focus algorithm then proceeds by using a smaller wavelength and the previous estimated position parameters.
This step is repeated with decreasingly smaller wavelength and position error, until the original wavelength is used. The result is a faster progression through the solution surface, with lower probabilities of falling into local minima.

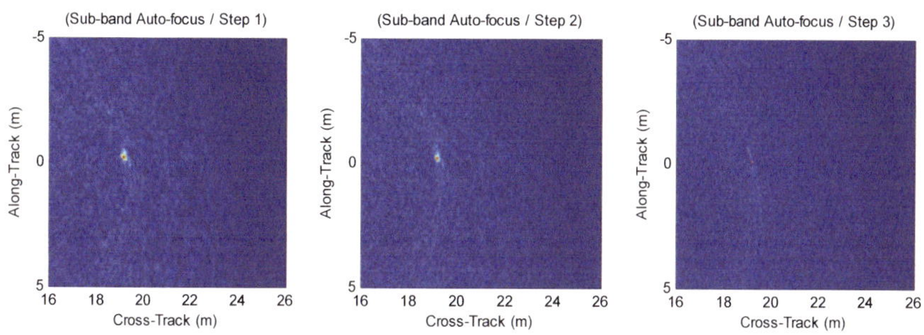

Fig. 17. Sonar image of the artificial target through the various auto-focus steps.

Fig. 17 shows an image of an artificial point target in 3 successive auto-focus steps. The algorithm starts wit a longer wavelength thus producing a low resolution image. As it progresses through the process, the target gets a sharper appearance.

For image quality metric a quadratic entropy measure can be used, which is a robust quality measure and enables fast convergence than a first order entropy measure or a simple image contrast measure. This is a measure of image sharpness. The lower the entropy measure, the sharper the image.

To calculate the quadratic entropy one needs to estimate the image information potential IP. Instead of making the assumption that the image intensity has a uniform or Gaussian distribution, the probability density function is estimated thought a Parzen window method using only the available data samples ([Liu, W. et al (2006)]):

$$IP(x) = \frac{1}{N^2} \sum_{j=1}^{N} \sum_{i=1}^{N} k_\sigma (x_j - x_i) \tag{1.14}$$

Where $k_\sigma (x - x_i)$ is the Gaussian kernel defined as:

$$k_\sigma (x - x_i) = \frac{1}{\sqrt{2\pi\sigma}} e^{-\frac{(x-x_i)^2}{2\sigma^2}} \tag{1.15}$$

Because this method of estimation requires a computational intensive calculation of the sum of Gaussians, this is implemented through the Improved Fast Gaussian Transform described in [Yang, C. et al (2003)].

This auto-focus method is suitable for systems working with an under-sampled swath and few transducers. No special image features are necessary for the algorithm to converge.

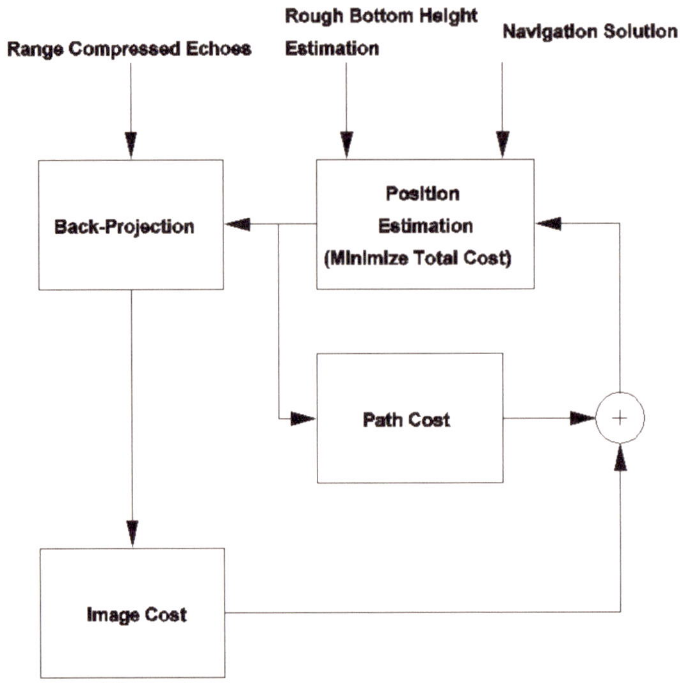

Fig. 18. Auto-focus block diagram.

Fig. 19. Artificial target used for resolution tests.

10. Results

To test the system and access its capabilities a series of test missions were performed in the Douro river, Portugal. For the first tests an artificial target was placed in the muddy river bottom and the autonomous boat programmed to make several paths through the area.

The artificial target is a half octahedral reflector structure made of aluminium (Fig. 19). It measures 20x20x20cm, but the target response seen by the sonar should be like a point after correct image synthesis.

Fig. 20 shows one image of the artificial target obtained though the matrix implementation of the back-projection algorithm as describe previously.

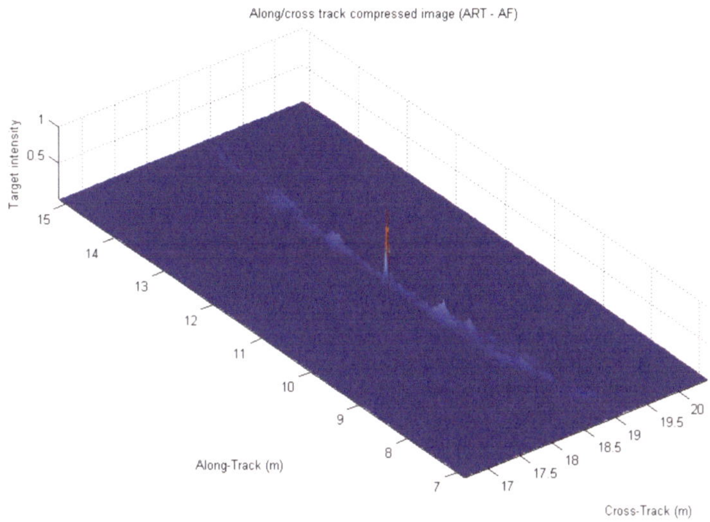

Fig. 20. Sonar image of the artificial target placed in the river bottom.

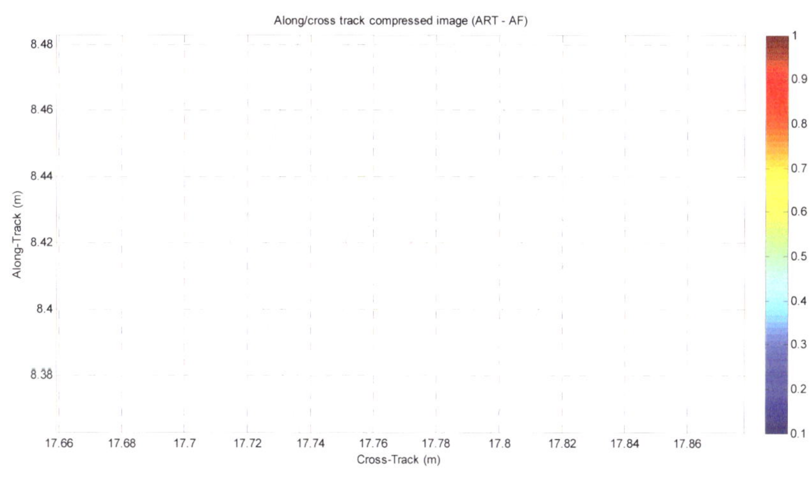

Fig. 21. Synthetic aperture sonar resolution.

As can be seen if Fig. 21, after auto-focus the image obtained from the artificial target presents sharp point like response, achieving the theoretical maximum resolution of the sonar system: 2.5x2.5 cm.

Fig. 22 shows an image obtained near the river shore before synthetic aperture processing and Fig. 23 show the same image processed using the described back-projection algorithms. It is possible to see several hyperbolic like target responses from rocks in the river bed that, after synthetic aperture image processing, assume the correct point like form.

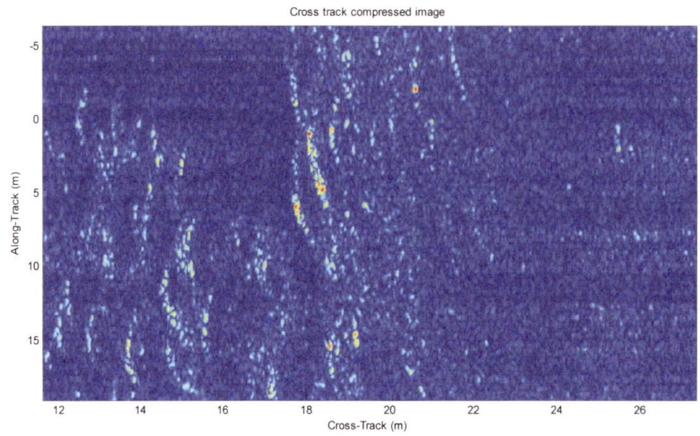

Fig. 22. Cross-track compressed reflectivity map an area near the river shore.

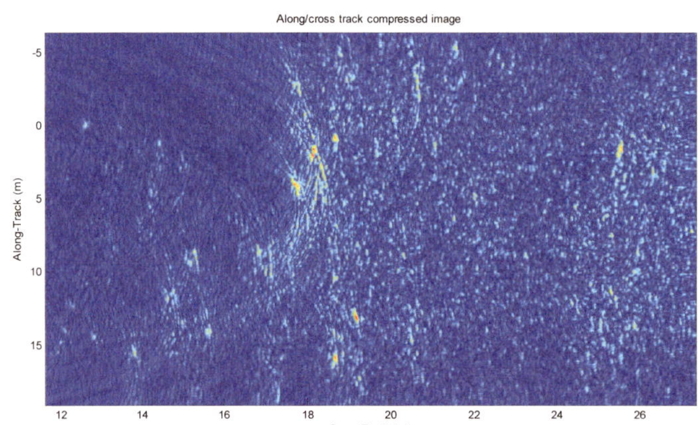

Fig. 23. Along/Cross-track compressed reflectivity map an area near the river shore.

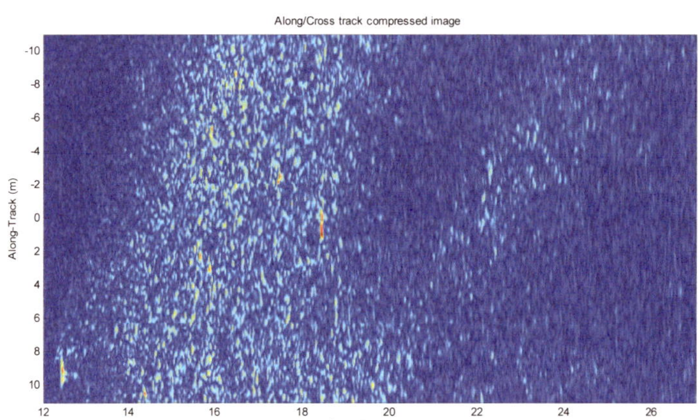

Fig. 24. Reflectivity map of harbour entrance.

The hyperboles are wavy due to the uncontrolled platform motion but, because the platform motion is known, the image nevertheless correctly synthesized.
Fig. 24 shows another sonar image obtained with the described system, this time it's a sand bank near a small harbour entrance in the Douro river.

11. Conclusion

As demonstrated, synthetic aperture sonar is a technique that enables attainment of high quality, high resolution underwater images.
Autonomous surface vehicles provides several advantages for synthetic aperture imagery. Not only it is possible to control the boat motion in this way, it is also possible to obtain navigation measurements with precisions in the order of the wavelength used in high resolution sonar systems. Furthermore unsupervised surveillance applications that combine the high quality sonar images with the effectiveness of an autonomous craft are possible.
Sonar images obtained in this way can be easily integrated in geographical information systems.
Using back-projection algorithms one is no longer restricted to linear paths, and deviations from this path are not treated as errors, but simply as sampling positions.
Phase errors due to navigation uncertainties and medium fluctuations cause blurring of the image. Nevertheless, the results can be further enhanced through auto-focus procedures that iterate the solutions until convergence to a predefined image quality parameter is achieved.
The use of high frequency signals imposes demanding restrictions in motion estimation and medium stability due to the sensibility of the image formation process to phase errors. A clever combination of the received signals enables the creation of a new one with an equivalent frequency equal to difference of the centre frequencies of the previous ones. This longer wavelength signal effectively masks phase uncertainties and enables efficient auto-focus of the synthetic aperture sonar image.
The synthetic aperture sonar thus enables enhanced imagery of underwater realms that combined with suitable platforms, such as an autonomous boat, can be obtained at low cost and high availability.

12. Future work

Synthetic aperture sonar imagery is a powerful technique for underwater imaging that is becoming widespread as the problems and difficulties inherent to it are solved.
The use of multiple receiver systems will enable a higher area coverage ratio and ease navigation system precision requirements through the use of micronavigarion techniques. With the maturing of image formation algorithms, real-time image formation will further extend the application possibilities of synthetic aperture sonar systems.
Bottom height mapping is possible through the use of a double array of transducers and also by exploring the possibility of dual-pass interferometry. In this case the combination of images of the same scene obtained from different positions of the platform will allow the construction of three dimensional maps of the analyzed surfaces.

13. References

Bellettini, A.; Pinto, M. A., (2002) . "Theoretical accuracy of the synthetic aperture sonar micronavigation using a displaced phase centre antenna", *IEEE Journal of Oceanic Engineering*, 27(4), pp. 780-789,, October 2002.

Chatillon, J.; Adams, A. E.; Lawlor, M. A.; Zakharia, M. E. (1999). "SAMI: A Low-Frequency Prototype for Mapping and Imaging of the Seabed by Means of Synthetic Aperture", *IEEE Journal of Oceanic Engineering*, vol. 24, no. 1, pp. 4-15, January 1999.

Cruz, N.; Matos, A.; Cunha, S.; Silva, S. (2007). "ZARCO - An Autonomous Craft for Underwater Surveys", *Proceedings of the 7th Geomatic Week*, Barcelona, Spain, Feb 2007.

Cutrona, L. J. (1975), "Comparison of sonar system performance achievable using synthetic-aperture techniques with the performance achievable by more conventional means", *The Journal of the Acoustical Society of America*, Volume 58, Issue 2, August 1975, pp. 336-348.

Douglas, B. L.; Lee, H. (1993) "Synthetic-aperture Sonar Imaging with Multiple-element Receiver Array", *IEEE International Conference on Acoustics*, Speech, and Signal Processing, vol. 5, pp. 445-448, April 1993.

Foo, K.Y.; Atkins, P. R.; Collins, T. (2003) "Robust Underwater Imaging With Fast Broadband Incoherent Synthetic Aperture Sonar", *Proceedings of IEEE International Conference on Acoustics*, Speech, and Signal Processing 2003, Volume 5, 2003 pp. V - 17-20 vol.5.

Fornaro, G. (1999). "Trajectory deviations in airborne SAR: analysis and compensation", *IEEE Transactions on Aerospace and Electronic Systems*, vol.35, no.3, pp.997-1009, Jul 1999.

Fortune, S. A.; Hayes, M. P.; Gough, P. T. (2001). "Statistical Autofocus of Synthetic Aperture Sonar Images using Image Contrast Optimization", *OCEANS '01 Conference Proceedings*.

Gough, P. T.; Hayes, M. P. (1989), "Measurements of acoustic phase stability in Loch Linnhe, Scotland", *The Journal of the Acoustical Society of America*, Volume 86, Issue 2, August 1989, pp.837-839.

Gough, P. T. (1998). "Unified Framework for Modern Synthetic Aperture Imaging Algorithms". *The International Journal of Imaging Systems and Technology*, Vol. 8, pp. 343-358, 1998.

Hansen, R.E.; Saebo, T.O.; Callow, H.J.; Hagen, P.E.; Hammerstad, E. (2005) "Synthetic aperture sonar processing for the HUGIN AUV", *Oceans 2005 - Europe*, vol.2, no., pp. 1090-1094 Vol. 2, 20-23 June 2005.

Hawkins, D. W.; Gough, P. T. (2004). "Temporal Doppler Effects in SAS", *Sonar Signal Processing*, Vol. 26, p. 5, 2004.

Hunter, A. J. ; Hayes, M. P. ; Gough, P. T. (2003). "A Comparison of Fast Factorised Back-Projection and Wavenumber Algorithms For SAS Image Reconstruction", *Proceedings of the World Congress on Ultrasonics*, Paris, France, September 2003.

Kundur, D.; Hatzinakos, D. (1996). "Blind image deconvolution", *Signal Processing Magazine*, IEEE, vol.13, no.3, pp.43-64, May 1996.

Liu, W.; Pokharel, P. P.; Principe, J. C.. (2006). "Correntropy: A Localized Similarity Measure", *International Joint Conference on Neural Networks*, 2006, pp. 4919 - 4924.

McHugh, R.; Shaw, S.; Taylor, N. (1998). "Spatial sampling ambiguities in synthetic aperture sonar for broadside and squint look operations", *OCEANS '98 Conference Proceedings*, vol.2, no., pp.960-964 vol.2, 28 Sep-1 Oct 1998.

Nelson, M. A. (1998). "DARPA Synthetic Aperture Sonar", *Proceedings of the Adaptive Sensor Array Processing (ASAP) Workshop*, vol. 1, pp. 141-155, 15 May 1998.

Neudorfer, M.; Luk, T.; Garrood, D.; Lehtomaki, N.; Rognstad, M. (1996) "Acoustic Detection and Classification of Burried Unexploded Ordinance (UXO) Using Synthetic Aperture Sonar", *PACON' 96*, 1996.

Putney, A.; Anderson, R.H. (2005) "Reconstruction of undersampled SAS data using the WIPE algorithm", *Proceedings of the OCEANS 2005 MTS/IEEE Conference*, vol., no., pp. 111-118 Vol. 1, 2005.

Saebo, T.O.; Callow, H.J.; Hansen, R.E.; Langli, B.; Hammerstad, E.O. (2007) "Bathymetric Capabilities of the HISAS Interferometric Synthetic Aperture Sonar", *Proceedings of the OCEANS 2007 MTS/IEEE Conference*, vol., no., pp.1-10, Sept. 29 2007-Oct. 4 2007.

Sammelmann, G. S.; Fernandez, J. E.; Christoff, J. T. (1997) "High Frequency / Low Frequency Synthetic Aperture Sonar", *Proocedings of SPIE - International Society for Optical Engineering*, vol. 2079, pp-160-171, 1997.

Shippey, G.; Banks, S.; Pihl, J. (2005). "SAS image reconstruction using Fast Polar Back Projection: comparisons with Fast Factored Back Projection and Fourier-domain imaging", *Proceedings of the OCEANS 2005 MTS/IEEE-Europe Conference*, vol.1, no., pp. 96-101 Vol. 1, 20-23 June 2005.

Silkaitis, J.M.; Douglas, B.L.; Hua Lee. (1995). "Synthetic-aperture sonar imaging: system analysis, image formation, and motion compensation", *Conference Record of the Twenty-Ninth Asilomar Conference on Signals, Systems and Computers 1995*, vol.1, no., pp.423-427 vol.1, 30 Oct-1 Nov 1995.

Silva, S.; Cunha, S.; Matos, A.; Cruz, N. (2007 a). "An In-SAS System For Shallow Water Surveying", Proceedings *of the 7th Geomatic Week, Barcelona*, Spain, Feb 2007.

Silva, S.; Cunha, S.; Matos, A.; Cruz, N. (2007 b). "An Autonomous Boat Based Synthetic Aperture Sonar," *Proceedings of the OCEANS 2007 MTS/IEEE Conference*, vol., no., pp.1-7, Sept. 29 2007-Oct. 4 2007.

Silva, S.; Cunha, S.; Matos, A.; Cruz, N. (2008 a). "An Algebraic Approach to Synthetic Aperture Sonar Image Reconstruction", *Proceedings of the OCEANS 2008 MTS/IEEE Conference*, Sept. 15 2008-Sept. 18 2008.

Silva, S.; Cunha, S.; Matos, A.; Cruz, N. (2008 b). "Sub-Band Processing of Synthetic Aperture Sonar Data", *Proceedings of the OCEANS 2008 MTS/IEEE Conference*, Sept. 15 2008-Sept. 18 2008.

Silva, S.; Cunha, S.; Matos, A.; Cruz, N. (2008 c). "Shallow Water Height Mapping With Interferometric Synthetic Aperture Sonar", *Proceedings of the OCEANS 2008 MTS/IEEE Conference*, Sept. 15 2008-Sept. 18 2008.

Tomiyasu, K. (1978). "Tutorial review of synthetic-aperture radar (SAR) with applications to imaging of the ocean surface", *Proceedings of the IEEE*, vol.66, no.5, pp. 563-583, May 1978.

Wang, L.; Belletini, A.; Fioravanti, S.; Chapman, S. ; Bugler, D. R.; Perrot, Y.; Hétet, A. (2001). "InSAS'00: interferometri SAS and INS aided SAS imaging", *Proceeding of the IEEE Oceans 2001 Conference*, volume 1, pp. 179-187, MTS/IEEE, 2001.

Yang, C.; Duraiswami, R.; Gumerov, N. A.; Davis, L. (2003). "Improved Fast Gauss Transform and Efficient Kernel Density Estimation", *Proceedings of the Ninth IEEE International Conference on Computer Vision (ICCV'03)*, 2003, pp. 664 - 671 vol.1.

3

Motion Compensation in High Resolution Synthetic Aperture Sonar (SAS) Images

R. Heremans, Y. Dupont and M. Acheroy
Signal and Image Center (SIC-RMA)
Belgium

1. Introduction

The sequence of echoes detected by an active Synthetic Aperture Sonar (SAS) is coherently added in an appropriate way to produce an image with greatly enhanced resolution in the azimuth, or along-track direction when compared with an image obtained from a standard Side Looking Sonar (SLS).

The SAS processing originates from the Synthetic Aperture Radar (SAR) concept. A complete introduction to SAR technique can be found in (Sherwin et al., 1962), (Walker, 1980), (Wiley, 1985) and (Curlander & McDonough, 1991).

Raytheon was issued in 1969 a patent for a high-resolution seafloor imaging SAS (Walsh, 1969) and 1971 analyzed a potential system in terms of its resolution and signal-to-noise ratio. Cutrona was the first well-known radar specialist to point out how the various aspects of SAR could be translated to an underwater SAS (Cutrona, 1975). Hughes (1977) compared the performance of a standard SLS to an SAS and showed that the potential mapping rate for SAS was significantly higher than for side-looking sonar. At the time there was an assumption that the instability of the oceanic environment would prevent the formation of SAS imagery. Experimental work, which was performed by Williams (1976) and Christoff et al. (1982), refuted the instability worry. The verification of this assertion performed at higher frequencies by Gough & Hawkins (1989). Later, other concerns regarding the stability of the towed platform were also raised and some rail- or wire-guided trails where set up to avoid this extra complication. Nowadays there are a multiple of systems as hull mounted SAS systems, towed SAS systems and Autonomous Underwater Vehicles (AUV) systems. For further reading one can find an extended historical background of SAS in (Gough & Hawkins, 1997).

Time and experience were needed to adapt SAR algorithms to SAS systems; the SAS systems use smaller radiating elements in proportion to the wavelength, which leads to higher radiation pattern of SAS with respect to SAR. The range migration effect on synthetic aperture processing is significant and pronounced in SAS imagery. An additional difference between SAR and SAS systems is the synthetic aperture time being greater in one order of magnitude in SAS, which leads to a phase corruption due to the medium fluctuations and platform instabilities. Typical synthetic aperture times for SAR are of the order of seconds with a medium coherence of some days, whereas for SAS the typical synthetic aperture time is of the order of several minutes with a similar medium coherence time (Marx et al. 2000).

In this chapter the theoretical background of SAS processing will be presented followed by the importance of motion compensation in high resolution imagery. To validate the accuracy of the motion estimation a simulator was developed in which different motion errors were applied. The quantification of the motion estimation was validated through the 2D Fourier space (ω,k)-reconstruction algorithm.

2. Processing requirements for SAS

In order to explain profoundly (Bruce, 1992) the aspects of the synthetic aperture sonar technique, one has to introduce first the Side-Looking Sonar (SLS). The SLS is an imaging technique that provides two-dimensional reflectance maps of acoustic backscatter energy from the ocean floor. The maps are characterized by an intensity variation that is proportional to the acoustic backscatter signal strength (Somers, 1984). The geometry of the SLS is similar to the one of SAS. The systems ensonify strips on the seafloor using an acoustic projector that moves in a nominally straight line above the ocean floor with velocity v as shown in Fig. 1. In this figure θ_E is the elevation beam width of the projector, θ_H is the azimuth beam width of the projector, R_s is the instantaneous slant range measured from the sensor to a point on the seafloor, and W is the one-sided ensonified ground-plane swath width. The reflectance map in SLS has coordinates of cross-track (measured in the slant plane) and along-track distances. The near-edge elevation[1] angle θ_e is the elevation angle at minimum range, the elevation beam width and the height of the sonar with respect to the seafloor determine the size of the ensonified swath width. As the platform moves, a projector transmits acoustic pulses and a hydrophone listens for the echoes. The time delay of each echo provides a measure of slant range, while the ping-to-ping motion of the projector gives the along-track image dimension. As the seafloor is imaged, the data are recorded and represented in the slant range plane. In Fig. 1. R_{min} and R_{max} are the minimum and maximum slant ranges respectively and d is the one-sided blind zone defined to be the no illuminated area directly below the platform (i.e. nadir).

2.1 Slant range processing for SLS and SAS

In SAS, along track and across track processing are completely disentangled. For both SLS and SAS systems, the slant range processing is identical and often also called across track processing. This processing is performed to obtain range information from the seafloor. Pulsed systems sense range by measuring the delay interval between the transmission and the reception of the pulse, as shown in the upper right figure of Fig. 1. Assuming range gates of duration τ, the two seafloor objects O1 and O2 separated by ΔR_g, will be resolved if their returns do not overlap in time. The round-trip travel time for a pulse associated with the object at range R_s is given by;

$$t = \frac{2R_s}{c} \tag{1}$$

and the incremental delay due to the proximate object O2 is;

[1] The grazing angle θ_g is given by the elevation angle θ_e as: $\pi/2 - \theta_e = \theta_g$.

$$t+\tau=\frac{2(R_S+\Delta R_s)}{c} \quad (2)$$

where c is the propagation speed in the medium. A measure of the slant plane separability is obtained by subtracting (1) from (2), and is given by

$$\tau=\frac{2\Delta Rs}{c}. \quad (3)$$

The relationship between ground-plane and slant-plane (see lower right figure of Fig. 1.) is approximated as

$$\Delta R_g=\frac{\Delta R_s}{\cos\theta_g}. \quad (4)$$

Therefore, two objects on the seafloor are fully resolvable if their ground separation satisfies

$$\Delta R_g\geq\frac{c\tau}{2\cos\theta_g}. \quad (5)$$

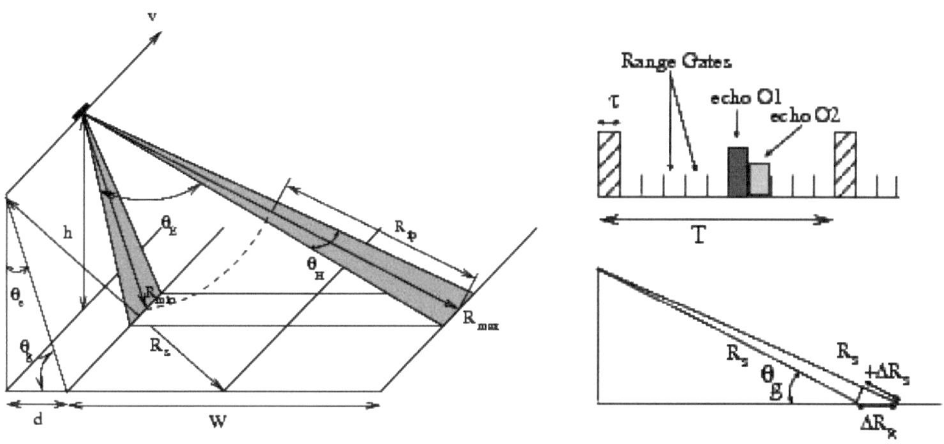

Fig. 1. Sonar geometry; (left) One-sided SLS geometry, (right) Time domain representation of a transmitted pulse and corresponding echoes.

Through equation (5) the range resolution is directly proportional to the ping duration τ and finer range resolution requires transmission of shorter pulses. However, in order to achieve very fine range resolution at long ranges, it may not be feasible to transmit very short pulses due to peak power limitation on the transmitter. In such cases, longer duration coded pulses can be transmitted with either frequency or phase modulation serving as the code. Upon arrival, the coded echo is decoded (compressed) so that the effective pulse length is significantly shorter in duration than the transmitted pulse (Stimson, 1983). The general relationship for range resolution ρ_r, is given by

$$\rho_r=\frac{c\tau_{eff}}{2}=\frac{c}{2B_{eff}} \quad (6)$$

where τ_{eff} is the effective pulse length after compression and B_{eff} represents the bandwidth of the compressed pulse.
The maximum unambiguous slant-range footprint can be determined by the effects of Pulse Repetition Interval (PRI) variations. If the PRI is set sufficiently large so that all echoes from a pulse are received prior to the transmission of the next pulse, there won't be an ambiguity. However if the PRI is decreased, the echoes from one pulse may not arrive at the receiver prior to the transmission of the following pulse. The minimum PRI value is equivalent to the requirement that the time between pulse transmission be greater than the time difference between the slant-range returns from the far- and near-edges of the slant-range footprint, R_{fp}. Thus the minimum acceptable PRI is,

$$PRI_{\min} = \frac{2R_{fp}}{c}. \tag{7}$$

2.2 Azimuth processing for SLS

Fig. 2. shows the slant-plane view of an SLS with a projector of length D and azimuthal beam width θ_H. The parameters $\delta_a^{\min}$ and $\delta_a^{\max}$ correspond to the linear azimuthal beam width at the minimum and maximum slant ranges respectively. The half-power angular beam width of a uniformly weighted rectangular aperture of length D is given in (Skolnik, 1980) by the approximate relationship

$$\theta_H \cong \frac{\lambda}{D} \tag{8}$$

where λ is the acoustic wavelength of the signal. The resolution at slant-range distance R_s of an SLS system is given by

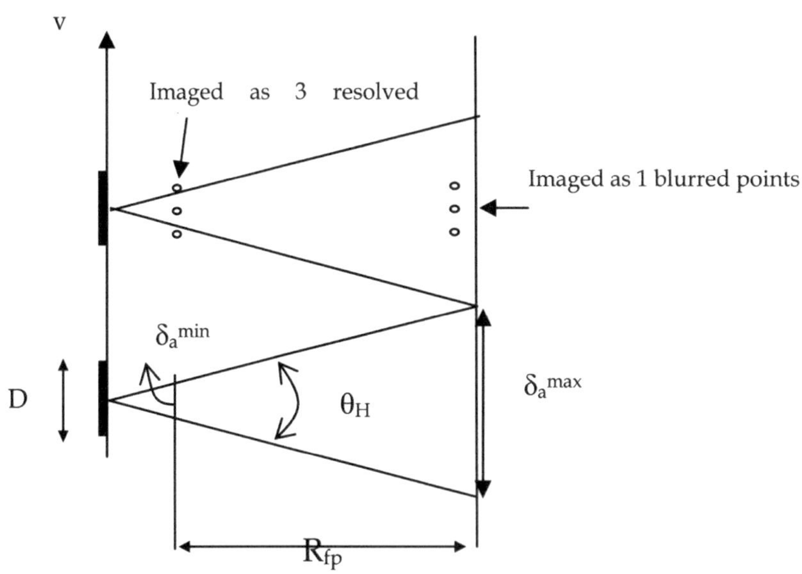

Fig. 2. Slant range, along track SASgeometry with beam spread as function of slant-range. Real-beam resolution degrading with increasing range

$$\delta_a \cong \frac{R_s \lambda}{D}. \tag{9}$$

In order to keep the resolution small as range increases, the frequency and/or the physical-aperture length D must be increased. The azimuth resolution however will ever stay dependent on slant-range.
In order to avoid along-track gaps in the image due to the beam pattern, one needs at least one sample for each along track resolution cell. Since the resolution cell width varies with range, the PRI selection will depend on the desired along-track resolution. Most side-scan systems are designed to realize the near-edge resolution

$$PRI_{\max} \leq \frac{\delta_a^{\min}}{v} = \frac{\Delta_{at}}{v} \tag{10}$$

where $\delta_a^{\min}$ corresponds to the along-track sampling spacing Δ_{at}. In some cases, the along-track sampling spacing may be chosen finer than the azimuth resolution. This situation is called over sampling. If the lowest sampling rate while satisfying the Nyquist criterion is applied (called critical sampling) the along-track sample spacing is exactly equal to the azimuth resolution.

2.3 Azimuth processing for SAS

The physical aperture of a SAS system may be regarded as one element of a linear array extending in the direction of the platform motion as shown in Fig. 3. The SAS processing can than be compared to the combination of the individual receivers from the linear array into an equivalent single receiver. L_{max} is the maximum synthetic-aperture length possible for a given azimuth beam-width θ_H, while L_{act} is the actual synthetic aperture length that may be shorter than L_{max}.
The azimuth resolution is obtained by sensing, recording, and processing the ping-to-ping phase history resulting from the variation in slant range caused by the projector's main lobe illumination pattern moving past seafloor scatterers. The maximum synthetic-array length is defined by the linear azimuth beamwidth at a given slant range R_s, $L_{\max} = R_s \theta_H$. The minimum effective horizontal beamwidth of the synthetic array is given by $\theta_{\min} = \lambda / (2L_{\max})$. The finest along track resolution that can be achieved from a focused synthetic aperture system is defined (Skolnik, 1980) as $\rho_a^{\min} = R_s \theta_{\min} = D/2$.

3. Necessity of motion compensation in SAS

Following equation (7) Synthetic Aperture Sonar, suffers the upper bound on the pulse repetition frequency (PRF) imposed by the relatively slow sound speed in water. This in turn limits the platform velocity, and consequently introduces motion errors more easily due to the ocean instabilities like waves, water currents and wind. The effect of those motion errors on the SAS reconstruction will be exhibited in section 3.1.5. Motion compensation is the main key to obtain high-resolution SAS images, which are in turn indispensable to be able to perform not only reliable small target detection but also classification and identification. The prime concept solving the micronavigation issue is called Displaced Phase Centre Array (DPCA) that exploits in a unique way the spatial and temporal coherence properties of the seafloor backscatter in a multiple receiver array configuration. The DPCA concept will be explained and illustrated on simulated data in section 5.

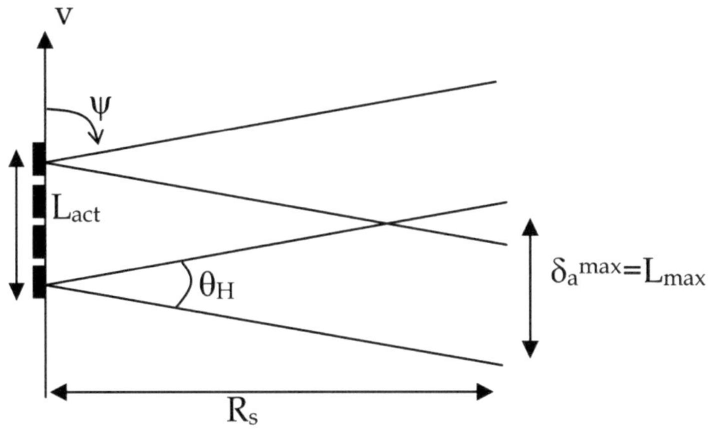

Fig. 3. Slant-plane view of an SLS with azimuthal beamspread θ_H. Real beam resolution degrades with increasing range R_s.

3.1 Simulator

The sonar simulator was designed to obtain the echo of a series of known point scatterers in a chosen scene. Let us consider first the case of a single transmitter/single receiver configuration.

3.1.1 Single transmitter/ single receiver configuration

Fig. 4 presents the 2D geometry of broadside strip-map mode synthetic aperture imaging systems. The surface imaged is substantially flat and has on top of it a collection of sub wavelength reflectors collectively described as the object reflectivity function *ff(x,y)*. Mostly *ff(x,y)* is referred to as the object and consists of a continuous 2D distribution of omni-directional (aspect independent) and frequency independent reflecting targets. This target area is illuminated by a side-looking sonar system travelling along a straight locus u with a velocity v, moving parallel to the y-axis of the target area. The origins of the delay time axis t and the range axis x have been chosen to coincide. As the sonar platforms travels along u, it transmits a wide bandwidth phase modulated waveform $p_m(t)$ of duration τ seconds which is repeated every T seconds. On reception, the coherent nature of the transmitter and receiver allows the echoes that have come from different pulses to be arranged into a 2D matrix of delay time t versus pulse number. Since the platform ideally travels a constant distance between pulses, the pulse number can be scaled to position along the aperture u in meters. Assuming the stop-start assumption, denoting that the sonar system does not move further along u between the emission of the pulse and the respective reception of the signal, the strip-map system model represents the echoes detected at the output of the receiver and is approximately described by;

$$ee_m(t,u) \approx \int_x ff(x,u) \otimes_u \left\{ a(t,x,u) \otimes_t p_m\left(t - \frac{2}{c}\sqrt{x^2 + y^2}\right) \right\} dx \tag{11}$$

where a(t,x,y) is the spatial-temporal response of the combined transmitter and receiver aperture. The output of the system given by a convolution in along-track, emphasize the two main problems faced by the inversion scheme which are:

- the system response is range variant
- there is a range curvature effect also called range migration.

Note that any function with a subscript *m* implies modulated notation, i.e. the analytic representation of the function still contains a carrier term $exp(i\omega_0 t)$, where ω_0 represents the carrier radian frequency. Demodulated functions are subscripted with a *b* to indicate a base band function. The processing of base banded data is the most efficient method, as it represents the smallest data set for both the FFT (Fast Fourier Transform) and any interpolators. Many synthetic aperture systems perform pulse compression of the received reflections in the receiver before storage. The pulse compressed strip-map echo denoted by ss_m is given by,

$$ss_m(t,u) = p_m(t) \otimes_t ee_m(t,u). \tag{12}$$

The temporal Fourier transform of equation (12) is

$$Ss_m(\omega,u) = P_m(\omega)\Im_t\{ee_m(t,u)\} = P_m(\omega)Ee_m(\omega,u) \tag{13}$$

with the Fourier transform of the raw signal,

$$Ee_m(\omega,u) = P_m(\omega)\int_x\int_y ff(x,y)A(\omega,x,y-u)\exp\left(-i2k\sqrt{x^2+(y-u)^2}\right)dx\,dy \tag{14}$$

with ω and *k* the modulated radian frequencies and wave-numbers given by $\omega = \omega_b+\omega_0$ and $k=k_b+k_0$.

Throughout this chapter, radian frequencies and wave-numbers with the subscript *0* refer to carrier terms while radian frequencies and wave-numbers without the subscript *b* refer to modulated quantities. At this point it is useful to comment on the double-functional notation and the use of the characters *e* and *s* like they are appearing in equations (11) till (14). The character *e* is used to indicate non-compressed raw echo data, while *s* is used for the pulse-compressed version (see equation (13)). Due to the fact that the echo is characterized by a 2D echo matrix (i.e. the scattering intensity as a function of azimuth and slant range) one needs a double-functional notation to indicate if a 1D Fourier transform, a 2D Fourier transform or no Fourier transform is applied on the respective variable. A capital character is used when a Fourier transform is applied. The first position of the double-functional notation refers to the slant-range direction (fast time) whereas the second position refers to the along-track direction (slow time). For example, sS_b describes the pulse-compressed echo in the range/Doppler domain since a 1D Fourier transform is taken in the along-track domain. The subscript *b* indicates that the pulse compressed echo data are also base banded. Putting the expression given in equation (14) into equation (13) leads to,

$$\begin{aligned} Ss_m(\omega,u) = {} & |P_m(\omega)|^2 . \int_x\int_y f(x,y).A(\omega,x,y-u). \\ & \exp\left(-i2k\sqrt{x^2+(y-u)^2}\right)dx\,dy. \end{aligned} \tag{15}$$

One can obtain the base banded version of equation (12) by taking the temporal Fourier transform on the base banded version of (14),

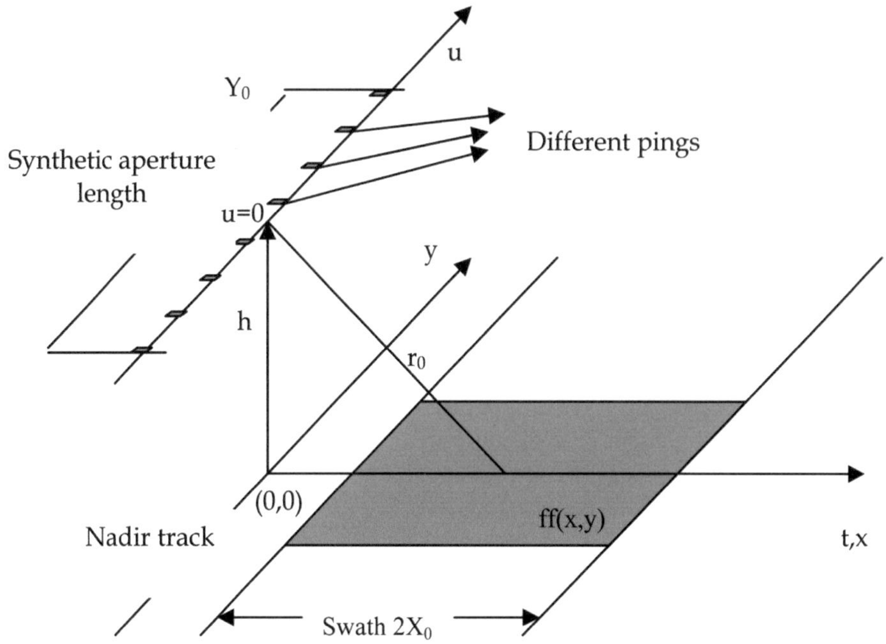

Fig. 4. Imaging geometry appropriate for a strip-map synthetic aperture system

$$\begin{aligned} Ss_b(\omega,u) &= P_b(\omega_b).Ee_b(\omega,u) \\ &= \left|P_b(\omega_b)\right|^2.\int_x \int_y f(x,y).A(\omega,x,y-u).\exp\left(-i2k\sqrt{x^2+(y-u)^2}\right). \end{aligned} \tag{16}$$

The term $2\sqrt{x^2+(y-u)^2}$ represents the travel distance from the emitter to the target and back to the receiver. In case of the start-stop assumption, the factor 2 appearing in front of the square root indicates that the travel time towards the object equals the one from the object back to the receiver. In the case the start-stop assumption is not valid anymore or in the case of multiple receivers, one has to split the term into two parts, one corresponding to the time needed to travel from the emitter to the target and one to travel from the target to the corresponding receiver. The above formulas, needed to build the simulator, will be extended in the following section towards the single transmitter multiple receiver configuration.

3.1.2 Single transmitter/multiple receiver configuration

The link with the single receiver can be made by reformulating equation (15) as follows,

$$\begin{aligned} Ee_m(\omega,u) = \sum_n P_m(\omega) \int_x \int_y f(x,y)\, A(\omega,x,y-u). \\ \exp(-ik(R_{out}(u,n)+R_{back}(u,n,h)))dx\,dy \end{aligned} \tag{17}$$

with $R_{out}(u,n)$ the distance from the transmitter to target n and $R_{back}(u,n,h)$ the distance from target n to the receiver h for a given along-track position u. In the case of a multiple receiver array R_{out} does not depend on the receiver number,

$$R_{out}(u,n) = \sqrt{x_n^2 + (y_n - u)^2}, \tag{18}$$

whereas R_{back} is dependent on the receiver number h, given by,

$$R_{back}(u,n,h) = \sqrt{x_n^2 + (y_n - u - hd_h)^2} \tag{19}$$

with d_h the along-track distance between two receivers. In the simulator the 3D echo matrix (depending on the along-track u, the return time t and the hydrophone number h) will represent only a limited return time range corresponding with the target-scene. There is no interest in simulating return times where there is no object or where there is not yet a possibility to receive back signal scattered on the seafloor from the nearest range. Therefore the corresponding multiple receiver corresponding expression of equation (16) becomes,

$$\begin{aligned} Ee(\omega,u,h) = \sum_n P(\omega).\int_x \int_y f(x,y).A(\omega,x,y-u). \\ \exp(ik\{2r_0 - [R_{out}(n,u) - R_{back}(n,u,h)]\})\, dx\, dy \end{aligned} \tag{20}$$

where the sum is performed over all N targets and r_0 is the centre of the target-scene.

3.1.3 Input signal p(t)

The echo from a scene is depending on the input signal p(t) generated by the transmitter and its Fourier transform $P(\omega)$. When a Linear Frequency Modulated (LFM) pulse p(t) is used it is expressed by,

$$p_m(t) = rect\left(\frac{t}{\tau}\right)\exp(i\omega_0 t + i\pi K t^2) \tag{21}$$

with ω_0 (rad/s) the carrier radian frequency and K (Hz/s) the LFM chirp-rate. The rect function limits the chirp length to $t \in [-\tau/2, \tau/2]$. The instantaneous frequency is obtained by differentiation of the phase of the chirp,

$$\omega_i(t) = \frac{d\phi(t)}{dt} = \omega_0 + 2\pi K t\,. \tag{22}$$

This leads to a frequency of the input signal of ranging from $\omega_0 - \pi\,\tau\, K$ till $\omega_0 + \pi\,\tau\, K$, leading to a chirp bandwidth $B = K\tau$. Using the principal of stationary phase, the approximate form of the Fourier transform of the modulated waveform is

$$P_m(\omega) = rect\left(\frac{\omega - \omega_0}{2\pi B}\right)\sqrt{\frac{i}{K}}\exp\left[-i\frac{(\omega - \omega_0)^2}{4\pi K}\right]. \tag{23}$$

The demodulated Fourier transform or pulse compressed analogue of $P_m(\omega)$,

$$P_c(\omega) = P_m(\omega).P_m^*(\omega) = \frac{1}{K} rect\left(\frac{\omega - \omega_0}{2\pi B}\right) \tag{24}$$

gives a rectangular pulse.

3.1.4 Radiation pattern

The radiation pattern or sonar footprint of a stripmap SAS system maintains the same as it moves along the track. The radiation pattern when the sonar is located at u=0 is denoted by (Soumekh, 1999)

$$h(\omega,x,y)\,. \tag{25}$$

When the sonar is moved to an arbitrary location along the track the radiation pattern will be $h(\omega,x,(y-u))$ which is a shifted version of $h(\omega,x,y)$ in the along-track direction. The radiation experienced at an arbitrary point (x,y) in the spatial domain due to the radiation from the differential element located at $(x,y)=(x_e(l),y_e(l))$ with $l\in S$, where S represents the antenna surface and where the subscript e is used to indicate that it concerns the element location, is,

$$\begin{aligned}
&\frac{1}{r}i(l)p\left[t-\frac{\sqrt{(x-x_e(l))^2+(y-y_e(l))^2}}{c}\right]dl= \\
&\frac{1}{r}i(l)\exp(i\omega t)\exp\left[-ik\sqrt{(x-x_e(l))^2+(y-y_e(l))^2}\right]dl
\end{aligned} \tag{26}$$

where $r=\sqrt{x^2+y^2}$ and $i(l)$ is an amplitude function which represents the relative strength of that element and where the transmitted signal is assumed to be a single tone sinusoid of the form $p(t)=exp(i\omega t)$. In the base banded expression the term exp(iωt) disappears and will not be considered in the following discussion. The total radiation experienced at the spatial point (x,y) , is given by the sum of the radiation from all the differential elements on the surface of the transmitter:

$$h_T(\omega,x,y)=\frac{1}{r}\int_{l\in S}dl\,i(l)\underbrace{\exp(-ik\sqrt{(x-x_e(l))^2+(y-y_e(l))^2})}_{Spherical\ PM\ signal} \tag{27}$$

Figure. 5. shows the real (blue) and absolute value (red) of $h_T(\omega,x,y)$ for a carrier frequency of f_0=50 kHz which corresponds with a radiance frequency $\omega=2\pi f_0$.
The spherical phase-modulated signal (PM) can be rewritten as the following Fourier decomposition,

$$\begin{aligned}
&\exp\left[-ik\sqrt{(x-x_e(l))^2+(y-y_e(l))^2}\right]= \\
&\int_{-k}^{k}\exp\left[-i\sqrt{k^2-k_u^2}(x-x_e(l))-ik_u(y-y_e(l))\right]dk_u.
\end{aligned} \tag{28}$$

By substituting this Fourier decomposition in the expression for h_T, and after interchanging the order of the integration over l and k_u, one obtains,

$$\begin{aligned}
h_T(\omega,x,y)=&\frac{1}{r}\int_{-k}^{k}\exp\left(-i\sqrt{k^2-k_u^2}\,x-ik_uy\right)\times \\
&\underbrace{\int_{l\in S}i(l)\exp\left[i\sqrt{k^2-k_u^2}\,x_e(l)+ik_uy_e(l)\right]dl\,dk_u}_{Amplitude\ pattern\ A_T(\omega,k_u)}
\end{aligned} \tag{29}$$

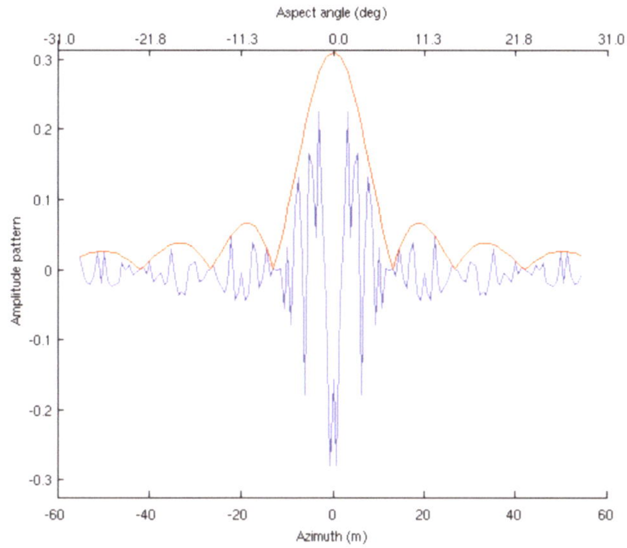

Fig. 5. Total radiation h_T experienced at a given point (x,y)=(100,[-60,60]) for a given carrier frequency f_0=50 kHz. In blue the real part of h_T is shown, in red the absolute value.

This means that the radiation pattern can be written as an amplitude-modulated (AM) spherical PM,

$$h_T(\omega,x,y)=\frac{1}{r}\int_{-k}^{k}dk_u\,A_T(\omega,k_u)\exp\left(-i\sqrt{k^2-k_u^2}\,x-ik_u y\right) \tag{30}$$

with

$$A_T(\omega,k_u)=\int_{l\in S}i(l)\exp\left[i\sqrt{k^2-k_u^2}\,x_e(l)+ik_u y_e(l)\right]dl\ . \tag{31}$$

The surface for a planar transmitter is identified via,

$$(x_e(l),y_e(l))=(0,l)\quad for\quad l\in\left[\frac{-D}{2},\frac{D}{2}\right], \tag{32}$$

where D is the diameter of the transmitter. Uniform illumination along the physical aperture is assumed to be, $i(l)=1$ for $l\in\left[\frac{-D}{2}.\frac{D}{2}\right]$ and zero elsewhere. Substituting these specifications in the model for the amplitude pattern A_T, we obtain,

$$\begin{aligned}A_T(k_u)&=\int_{-D/2}^{D/2}\exp(ik_u)dl\\&=D\sin c\left(\frac{Dk_u}{2\pi}\right)\end{aligned} \tag{33}$$

Equation (33) indicates that the transmit mode amplitude pattern of a planar transmitter in the along-track Doppler domain k_u is a sinc function that depends only on the size of the transmitter and is invariant in the across-track frequency ω.

3.1.5 Motion error implementation

In an ideal system performance, as the towfish, Autonomous Underwater Vehicle (AUV) or Hull mounted sonar system moves underwater it is assumed to travel in a straight line with a constant along-track speed. However in real environment deviations from this straight along-track are present. By having an exact notion on the motion errors implemented in the simulated data, one can validate the quality of the motion estimation process (section 5).

Since SAS uses time delays to determine the distance to targets, any change in the time delay due to unknown platform movement degrades the resulting image reconstruction. Sway and yaw are the two main motion errors that have a direct effect on the cross-track direction and will be considered here. The sway causes side to side deviations of the platform with respect to the straight path as shown in Fig. 6. This has the effect of shortening or lengthening the overall time-delay from the moment a ping is transmitted to the echo from a target being received. Since, in the case of a multiple receiver system, sway affects all of the receivers equally, the extra time-delay is identical for each receiver. A positive sway makes targets appear closer than they in reality are. In general a combination of two sway errors exist. Firstly the sway at the time of the transmission of the ping and secondly any subsequent sway that occurs before the echo is finally received. Since the sway is measured as a distance with units of meters, we can easily calculate the extra time delay, $\Delta_{sway}(u)$ given the velocity of sound through water c. The extra time delay for any ping u is,

$$\Delta t_{sway}(u) = \frac{X_{TX}(u) + X_{RX}(u)}{c} \tag{34}$$

where $X_{TX}(u)$ represents the sway at the time of the transmission of the ping under consideration and where $X_{RX}(u)$ represents the sway at the time of the reception of the same ping. Both quantities are expressed in meter. One assumes often that the sway is sufficiently correlated (i.e. slowly changing) so that it is approximately equal in both transmitting and receiving case,

$$\Delta t_{sway}(u) = \frac{2\overline{X}_{sway}(u)}{c}. \tag{35}$$

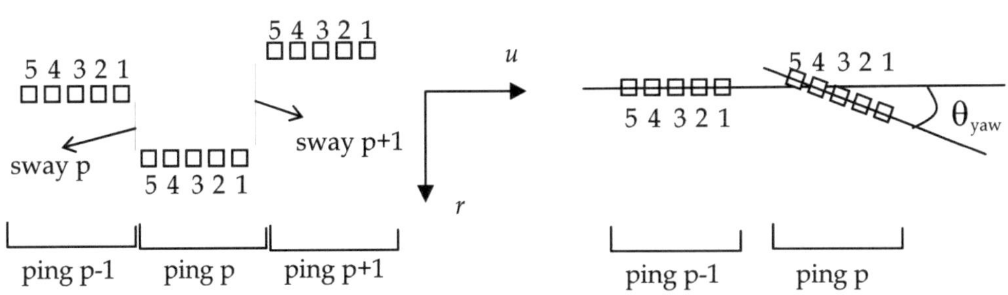

Fig. 6. The effect of sway (left) and yaw (right) on the position of the multiple receivers indicated by the numbers 1 till 5. The coordinate reference is mentioned between the two representations.

In Fig. 7. one sees the effect on the reconstruction of an image with a non-corrected sway error (middle) and with a corrected sway error in the navigation (right).

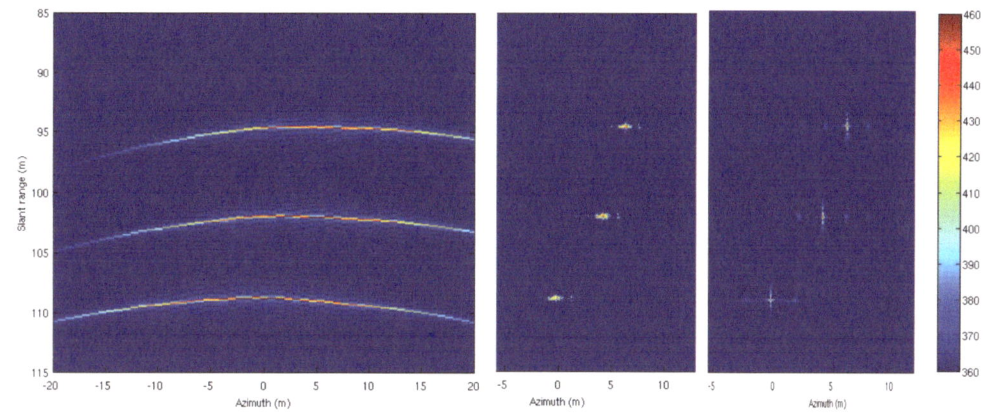

Fig. 7. Echo simulation of 3 point targets with sway error in the motion (left), (ω,k)-image reconstruction without sway motion compensation (middle) and with sway motion compensation (right).

For sway one has thus the problem of the array horizontally shifting from the straight path but still being parallel to it. With yaw, its effect is a rotated array around the z-axis such that the receivers are no longer parallel to the straight path followed by the platform as illustrated on the right in Fig. 6. Generally speaking there are two yaw errors; firstly when a ping is transmitted and secondly when an echo is received. The examination of those two gives the following; for the case where the transmitter is located at the centre of the rotation of the array, any yaw does not alter the path length. It can safely be ignored, as it does not introduce any timing errors. When the transmitter is positioned in any other location a change in the overall time delay occurs at the presence of yaw. However this change in time delay is common to all the receivers and can be thought of as a fixed residual sway error. This means that the centre of rotation can be considered as collocated with the position of the transmitter.

Yaw changes the position of each hydrophone uniquely. The hydrophones closest to the centre of rotation will move a much smaller distance than those that are further away. The change in the hydrophone position can be calculated through trigonometry with respect to the towfish's centre of rotation. The new position $\overrightarrow{x_h}'$ for each hydrophone h is given by,

$$\overrightarrow{x_h}' = \begin{pmatrix} \cos\vartheta_y & \sin\vartheta_y \\ -\sin\vartheta_y & \cos\vartheta_y \end{pmatrix} \cdot \overrightarrow{x_h} \tag{36}$$

where $\overrightarrow{x_h}$ =(x,y) indicates the position of hydrophone h relative to the centre of rotation and $\overrightarrow{x_h}'$ =(x',y') indicates the new position of hydrophone h relative to the centre of rotation after rotating around the z-axis due to yaw. θ_y represents the angle that the array is rotated around the z-axis. For small yaw angles the change in the azimuth u is small and can be ignored. Equation (36) becomes

$$x_h' = x_h + u_h \sin\vartheta_y \,. \tag{37}$$

Knowing the velocity c of the sound through the medium, one can use equation (37) to determine the change in the time delay $\Delta t_{yaw\{h\}}(u)$ for each hydrophone h

$$\Delta t_{yaw\{h\}} = \frac{\Delta x_h'}{c} \tag{38}$$

where $\Delta x_h'$ represents x_h-x_h' being the cross-track change in position of hydrophone h. Fig. 8. shows the image reconstruction of a prominent point target that has no motion errors in the data compared to one that has been corrupted by a typical yaw.
Once the surge, sway and yaw error vectors are chosen as a function of the ping number, they can be implemented in the simulator as follows;

$$TX_r^{off} = tx_r^o \cos\vartheta_y^p + tx_{az}^0 \sin\vartheta_y^p$$
$$TX_{az}^{off} = -tx_r^o \sin\vartheta_y^p + tx_{az}^0 \cos\vartheta_y^p$$

$$\begin{aligned} R_{out} &= \sqrt{\left(x_n - TX_r^{off} - sway(p)\right)^2 + \left(y_n - u(p) - TX_{az}^{off}\right)^2} \\ R_{back} &= \sqrt{\left(x_n - RX_r^{off} - sway(p)\right)^2 + \left(y_n - u(p) - RX_{az}^{off}\right)^2} \end{aligned} \tag{39}$$

Here for a transmitter situated at the centre of the array one can choose the reference system in a way that tx_r^0 and tx_{az}^0 are situated at the origin, where the subscript r refers to slant range and az to the azimuth or the along-track coordinate. Remark that R_{out} is a scalar whereas R_{back} is an array N_h numbers of hydrophones long.

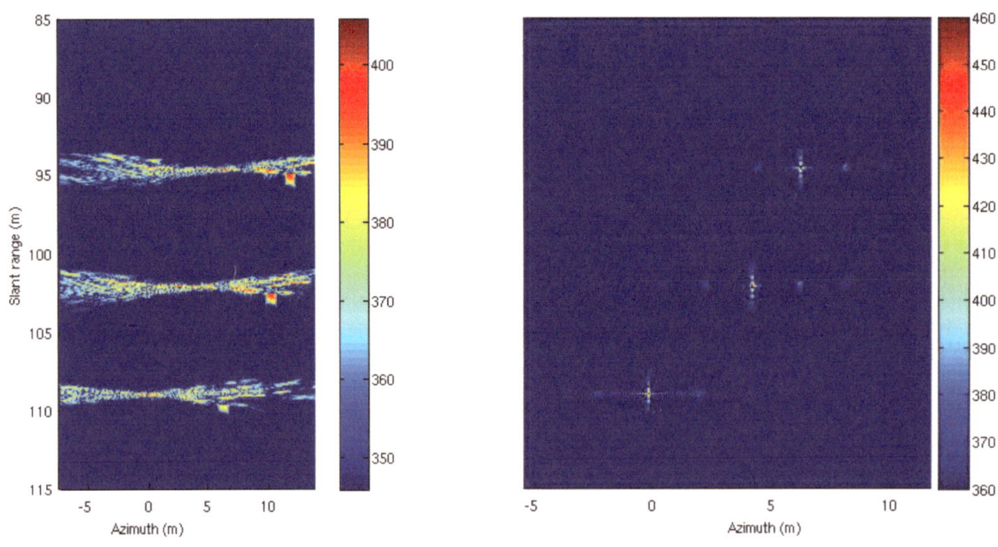

Fig. 8. (w,k)-image reconstruction without yaw motion compensation (left) and with yaw motion compensation (right).

4. (ω,k)- synthetic aperture sonar reconstruction algorithm

Form section 3 one studies that a reconstructed SAS image is very sensitive to the motion position and it is necessary to know the position of the sonar at the order of approximately 1/10th of the wavelength (a common term to express this is micro navigation). In the following sections a brief overview will be given on one particular SAS reconstruction algorithm, i.e. the *(ω,k)*-reconstruction algorithm (Callow et al. 2001), (Groen, 2006). Afterwards the motion estimation will be explained and finally the motion compensation is illustrated on the *(ω,k)*-algorithm.

The wave number algorithm, appearing under different names in the literature: *seismic migration algorithm, range migration algorithm or (ω,k)-algorithm,* and is performed in the two-dimensional Fourier transform on either the raw $EE(\omega,k_u)$ or pulse compressed data $SS(\omega,k_u)$. The key to the development of the wave number processor was the derivation of the two-dimensional Fourier transform of the system model without needing to make a quadratic approximation. The method of stationary phase leads to,

$$\Im\left\{\exp\left(-i2k\sqrt{x^2+(y-u)^2}\right)\right\} \cong \sqrt{\frac{\pi x}{ik}}\exp\left(-i\sqrt{4k^2-k_u^2}.x-ik_u.y\right). \tag{40}$$

The most efficient way to implement the wave number should be performed on the complex valued base banded data as it represents the smallest data set for both the FFT and the stolt interpolator. It is also recommended that the spectral data stored during the conversion from modulated to base banded is padded with zeros to the next power of two to take advantage of the fast radix-2 FFT. A coordinate transformation also represented by the following *stolt mapping operator* $S_b^{-1}\{.\}$,

$$\begin{aligned} k_x(\omega,k_u) &= \sqrt{4k^2-k_u^2}-2k_0 \\ k_y(\omega,k_u) &= k_u \end{aligned} \tag{41}$$

The wave number inversion scheme, realizable via a digital processor is than given by,

$$FF_b'(k_x,k_y) = S_b^{-1}\left\{\sqrt{\frac{k}{k_0}}.\exp\left[i\left(\sqrt{4k^2-k_u^2}-2k\right).r_0\right].P_b^*(\omega_b).EE_b'(\omega_b,k_u)\right\}. \tag{42}$$

The inverse Stolt mapping of the measured (ω_b,k_u)-domain data onto the (k_x,k_y)-domain is shown in Fig. 9.

The sampled raw data is seen to lie along radii of length $2k$ in the (k_x,k_y)-wave number space. The radial extent of this data is controlled by the bandwidth of the transmitted pulse and the along-track extent is controlled by the overall radiation pattern of the real apertures. The inverse Stolt mapping takes these raw radial samples and re-maps them onto a uniform baseband grid in (k_x,k_y) appropriate for inverse Fourier transformation via the inverse FFT. This mapping operation is carried out using an interpolation process. The final step is to invert the Fourier domain with a windowing function $WW(k_x,k_y)$ to reduce the side lobes in the final image,

$$\hat{ff}(x,y) = \left|\Im^{-1}_{k_x,k_y}\left\{WW(k_x,k_y).\hat{FF}(k_x,k_y)\right\}\right|. \tag{43}$$

This windowing operation can be split into two parts; data extraction and data weighting. In data extraction the operation first extracts from the curved spectral data a rectangular area of the wave number data. The choice of the 2-D weighting function to be applied to the

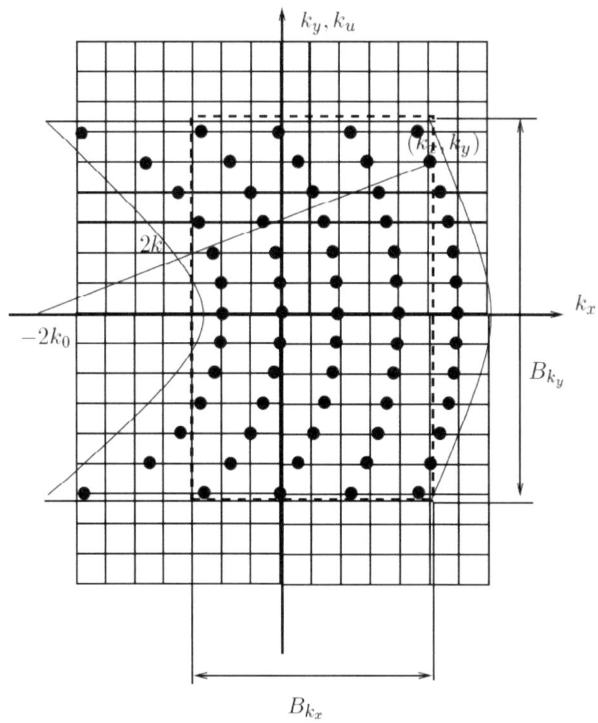

Fig. 9. The 2D collection surface of the wave number data. The black dots indicate the locations of the raw data samples along radii 2k at height k_u. The underlying rectangular grid shows the format of the samples after mapping (interpolating) to a Cartesian grid (k_x,k_y). the spatial bandwidths B_{kx} and B_{ky} outline the rectangular section of the wave number data that is extracted, windowed and inverse Fourier transformed to produce the image estimate.

extracted data is arbitrary. In the presented case a rectangular window and a 2-D Hamming window is used. Before applying the k_y weighting across the processed 3dB radiation bandwidth, the amplitude effect of the radiation pattern is deconvoluted as,

$$WW(k_x,k_y).rect\left(\frac{k_y}{B_{k_y}}\right).\frac{1}{A(k_y)}.W_h\left(\frac{k_x}{B_{k_x}}\right).W_h\left(\frac{k_y}{B_{k_y}}\right) \tag{44}$$

where $W_h(\alpha)$ is a 1D Hamming window defined over $\alpha \in [-1/2, 1/2]$ and the wave number bandwidths of the extracted data shown in Fig. 9. are

$$\begin{aligned} B_{k_x} &= \sqrt{4k_{\max}^2 - \left(\frac{2\pi}{D}\right)^2} - 2k_{\min} \approx \frac{4\pi B_c}{c} - \frac{2\pi^2}{k_{\max}D^2} \\ B_{k_y} &= \frac{4\pi}{D} \end{aligned} \tag{45}$$

here k_{min} and k_{max} are the minimum and maximum wave numbers in the transmitted pulse, B_c is the pulse bandwidth (Hz) and D is the effective aperture length. The definition of the x-

and y-axes depends on the number of samples obtained within each wave number bandwidth, the sampling spacing chosen, and the amount of zero padding used. The Hamming window over a length $N+1$ is given by its coefficients, which are calculated by,

$$w_h(n) = 0.5386 - 0.46164.\cos\left(\frac{2\pi n}{N}\right), \quad with\ 0 \leq n \leq N. \tag{46}$$

The resolution in the final image is given by,

$$\begin{aligned} \delta x_{3dB} &= \frac{2\pi}{B_{k_x}} = \alpha_w.\frac{c}{2B_{eff}} \\ \delta y_{3dB} &= \frac{2\pi}{B_{k_y}} = \alpha_w.\frac{D}{2} \end{aligned} \tag{47}$$

where α_w=1.30 for the Hamming window.

5. Platform motion estimation

In order to be able to explain properly the functioning of the Displaced Phase Centre Array (DPCA) algorithm (Bellettinit & Pinto 2000 and 2002) he simulation echoes on 9 targets arranged around the central target were generated. Some a priori known sway and yaw errors were included in the straight line navigated track. The positions of the 9 targets can be seen in Fig. 10 (left) and are given relative to the target that is situated at the center of the scene (0,0). At the right side the corresponding echo is shown. The simulator used an array consisting of 15 hydrophones separated by ΔR_x=21.88 cm. The slant range covered goes

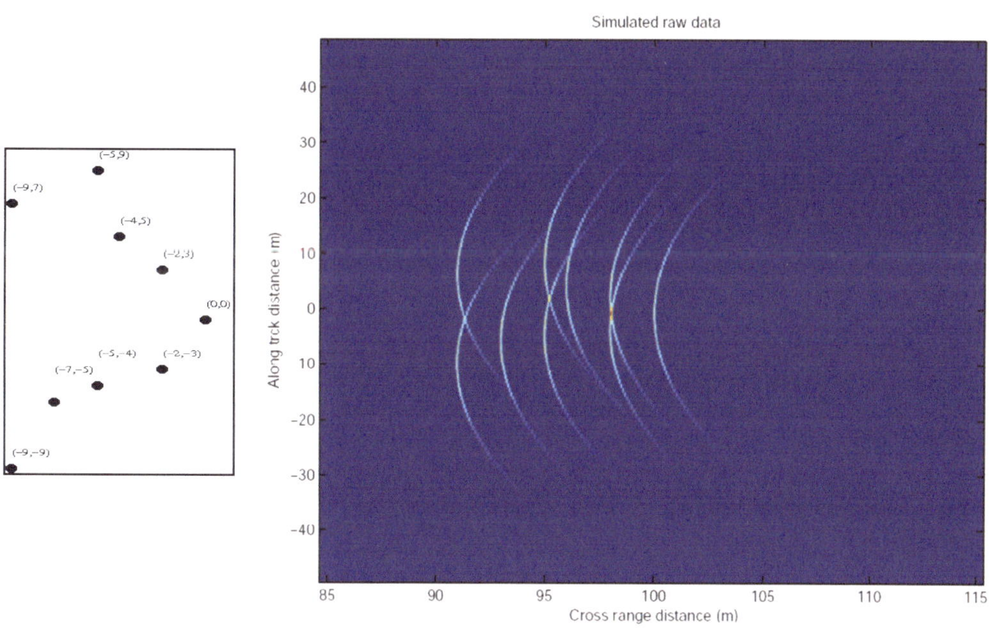

Fig. 10. Target position (r,u) given relative to the center of the scene at r_0=100 m.

from 84.64 m till 115.33 m. The data shown at the right side of Fig.10 shows only the return observed in receiver number 1. The echo data for a fixed ping number are presented in Fig. 11. for ping=61 (left) and for ping=62 (right) as function of the 15 hydrophones (x-axis) and the range (y-axis).

One can clearly see around the smallest ranges that the distance to the target is function of the receiver itself. The platform speed was chosen in a way (v=1.2763 m/s) that the echoes for receiver 1 till 8 for ping number n should coincide with those of 8 till 15 for ping number $n+1$. This set of 2 times 8 receivers serves as the input data for the dpca-code. The aim of the DPCA-motion compensation method is to estimate from the maximal correlation between the two pings data the surge, sway and yaw.

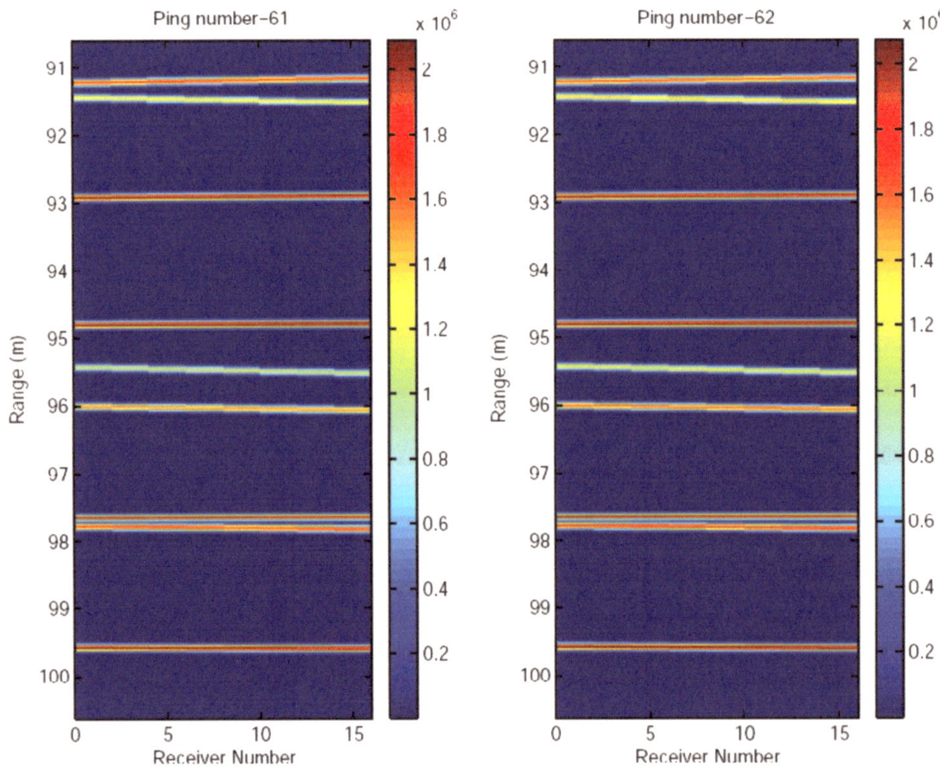

Fig. 11. The echoes are shown for the 15 different receivers as a function of the slant range for (left) ping number 61 and (right) for ping number 62. In this simulation no motion errors were introduced and the speed of the platform (v=1.2763 m/s) was chosen in a way that 8 receivers are overlapping between two consecutive pings.

5.1 Phase Center Array (PCA) approximation

In the PCA approach one approximates a true bistatic sonar as a monostatic sonar. On other words one will treat the bistatic transmitter/receiver pair as if it were a single co-located transducer located midway between the two. The error caused in making the phase-center approximation is the difference between the two-way bistatic path and the two-way equivalent monostatic path (Fig. 12). Writing out the approximation error, ε, gives;

$$\varepsilon = \sqrt{x^2 + (u-y)^2} + \sqrt{x^2 + (u+\sigma-y)^2} - 2\sqrt{x^2 + \left(u + \frac{\sigma}{2} - y\right)^2} \quad (48)$$

where σ is the relative position of the hydrophone with respect to the transmitter. The series expansion for $\sigma/y << 1$ one obtains

$$\varepsilon \approx \frac{\sigma^2}{4r}\cos^2\vartheta + \frac{\sigma^4}{64r^3}\cos^2\vartheta\left(4 - 5\cos^2\vartheta\right) + \ldots \quad (49)$$

where ϑ is the bearing of the target with respect to the position of the co-located transducer and r the distance between the co-located transducer and the target. For a far field condition, PCA holds when $\frac{\sigma^2}{4r} << \lambda_0$.

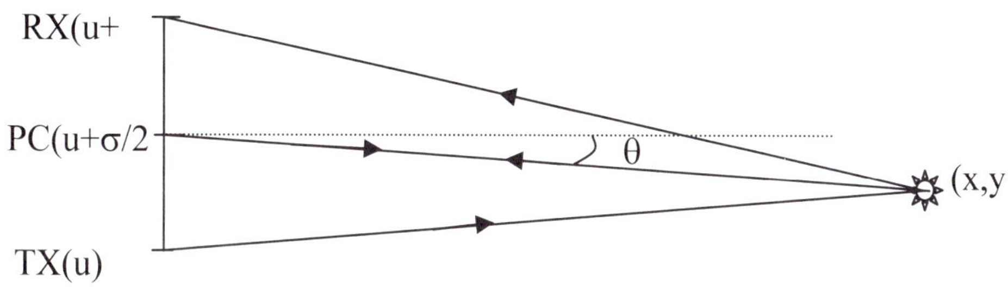

Fig. 12. Geometry of the phase centre approximation. Tx, Rx, PC are respectively the position of the transmitter, the receiver and the phase centre position. The target is located at position (x,y).

5.2 Array geometry and platform displacement

A 1D array consisting of 15 receivers displaced at ΔR_x=21.88 cm is used in the simulator. The sonar speed has been chosen to be v_S=1.2763 m/s and the pulse repetition time (PRT) to be T=0.6 s. The transmitter situates in the middle of the receiver array leading to a phase centre array as indicated in Fig. 13. When a higher platform speed is chosen, less overlapping phase centres will exist. This means that for being able to perform a DPCA motion estimation the platform speed will be rather moderate.

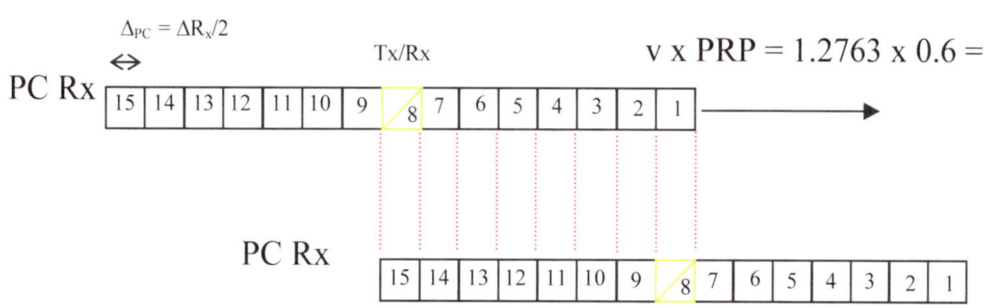

Fig. 13. The number of overlapping phase centers for a 15 hydrophone array with a displacement of $7\Delta_{PC}$ leading to 8 overlapping PC's between ping n and ping n+1. The

position of the 15 receivers for ping n+1 was shifted for better presentation. In reality PC Rx(1) of ping n overlaps with PC Rx(8) of ping n+1 etc. The diagonal in Rx(8) indicates that at this position there is also a transmitter element present.

The following formula can be used to define the platform speed for a given PRT in order to have $n \leq H$, with H the total number of receivers in the array, overlapping phase centres;

$$v = \frac{(H - n)\Delta_{PC}}{PRT}. \quad (50)$$

Due to these overlapping phase centres between two consecutive pings, a correlation analysis can be performed to get an idea about the motion errors. Therefore the notion of x-lag (cross-range direction) and t-lag (range direction) has to be introduced.

5.3 The notion of x-lag

When the platform displacement is chosen like mentioned in section 5.2, eight receivers will overlap for two consecutive pings like shown in Fig. 13. However in real situations the platform motion will deviate from this ideal trajectory. When the speed of the platform will be lower than v_8, R_{x1}(n) will no longer collocate with R_{x8}(n+1). Here, to indicate the receiver number a subscripted index is used, and the ping number is given between brackets. Assuming that the platform speed was only v=1.0940 m/s instead of v8, than R_{x1}(n) will collocate with R_{x7}(n+1), R_{x2}(n) with R_{x8}(n+1), .. and R_{x9}(n) with R_{x15}(n+1). Therefore the x-lag will be used to determine the along track motion of the platform. Due to the fact that the data volume on which the DPCA will be applied has to be kept as small as possible, one stocks only a limited amount of receiver returns. First of all one makes a mean speed estimation of the platform (most reliable is the autopilot speed of the vessel). Corresponding this speed one knows the approximate overlapping phase centres by the use of equation (50). Only those receiver returns will be considered in the DPCA analysis and corresponds to an x-lag = 0. An x-lag=+1 will be executed on $R_{x2\text{-}8}$(n) and $R_{x8\text{-}14}$(n+1) as shown in Fig. 14. In this analysis 5 x-lags are tested going from -2, -1, 0, 1, 2, being sensitive in detecting speed deviations between v=1.641 m/s for an x-lag=-2 and v=0.9116 m/s for an x-lag=+2.
From Fig. 14 one sees that the correlation for an x-lag =-2 or +2 is performed on 6 receivers for each ping, for an x-lag=-1 or +1 it is performed on 7 receivers and for an x-lag=0 it is performed on the eight receivers that were memory stacked.

5.4 The notion of t-lag

As mentioned before, the x-lag notion refers to the receiver array, while the t-lag notion refers to the range. Therefore, the t-lag will be used to determine the time delay leading to a estimation of the sway and the yaw. In the DPCA analysis, for each x-lag[♦], a cross correlation is taken between the two consecutive ping data with a chosen t-lag of 8 based on following expression:

$$c_{fg}^{m} = \sum_{n-\infty}^{\infty} f[n+m] \times g^{*}[n] \quad (51)$$

♦ Here 5 different x-lags are considered, leading to 5 cross correlation plots as shown in Fig. 15.

where m indicates the value of the t-lag (i.e. shift in the slant range between the two vectors f and g), f and g are the overlapping Rx couple for respectively ping n and ping $n+1$ and g^* represents the complex conjugate of g. Therefore, for each overlapping Rx couple, one obtains 17 cross correlations if a t-lag of 8 is chosen (Fig. 15). For a t-lag of 8 m runs from -8, -7, …,0, 1, …7, 8. The cross correlation is normalized in a way that the autocorrelations at zero lag are identically 1.0.

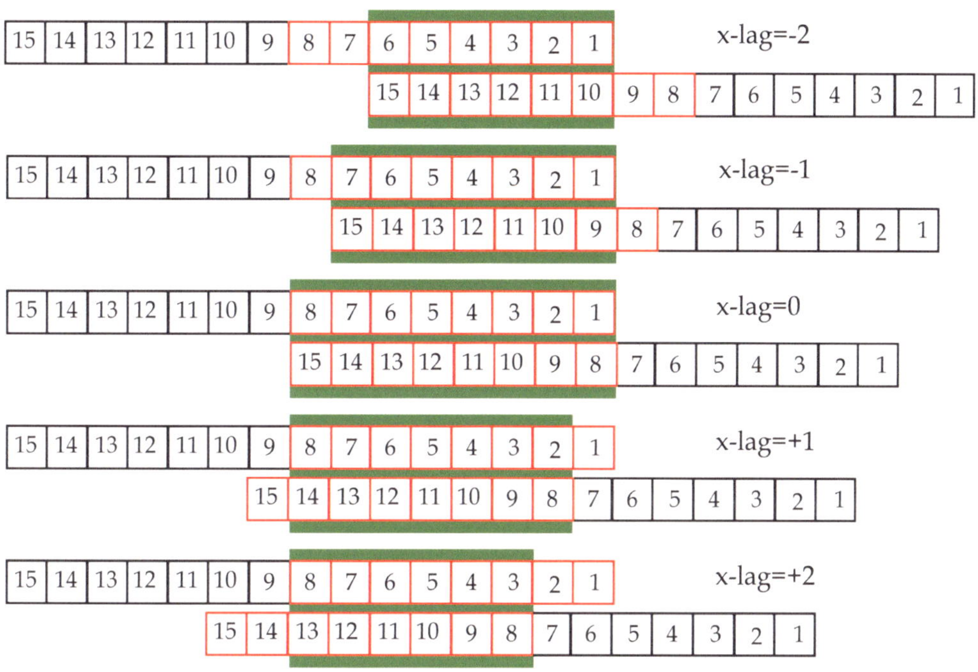

Fig. 14. The overlapping receivers between ping n and n+1 for the different x-lags under consideration (-2, -1, 0, 1, 2). The receivers indicated with the green bar are the ones that will be cross correlated. The receivers surrounded in red are the ones that are memory stacked during the DPCA processing. Each receiver contains the time series containing the echo information.

Thus for an *x-lag=-1* the cross correlation takes place between 7 phase centre couples (see Fig. 16.), i.e. $(R_{x1}(p),R_{x9}(p+1))$, $(R_{x2}(p),R_{x10}(p+1))$, …, $(R_{x7}(p),R_{x15}(p+1))$. The x-axis in the cross correlation pots represents those couples and is called *Rx couple id*. Remark that an *Rx couple id=1* for an *x-lag=-1* corresponds to $(R_{x1}(p),R_{x9}(p+1))$ whereas an *Rx couple id=1* for an *x-lag=+1* corresponds to $(R_{x2}(p),R_{x8}(p+1))$. Remark also that the number of *Rx couple ids* depends on the *x-lag* that is considered. There are 6 couples for *x-lag=+2* or *–2*, 7 couples for *x-lag=+1* or *–1* and 8 couples for *x-lag=0*. Once the Rx couple id is chosen, (lets take for instance $(R_{x4}(p),R_{x12}(p+1))$) the slant range vectors are correlated with a y-ordinate given by the chosen t-lag=8. Fig. 17 is an illustration of the t-lag=-4, 0 and +4 for an x-lag=-1 and for the first receiver couple id.

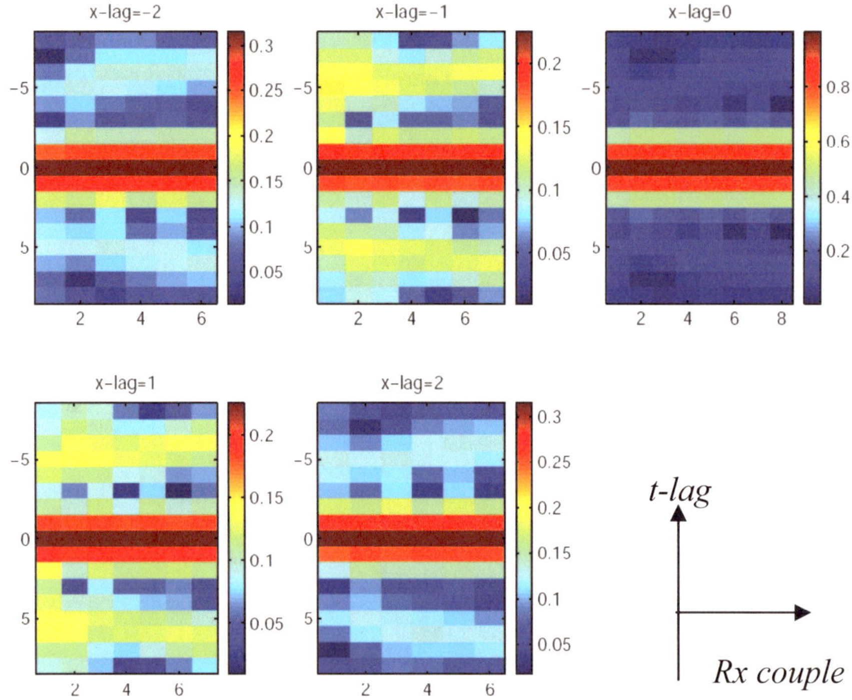

Fig. 15. Cross correlation (c^{m}_{fg}) with a t-lag of 8, giving 17 y-samples and 6, 7, 8, 7, 6 x-samples for the respective -2, -1, 0 , 1, 2 x-lags.

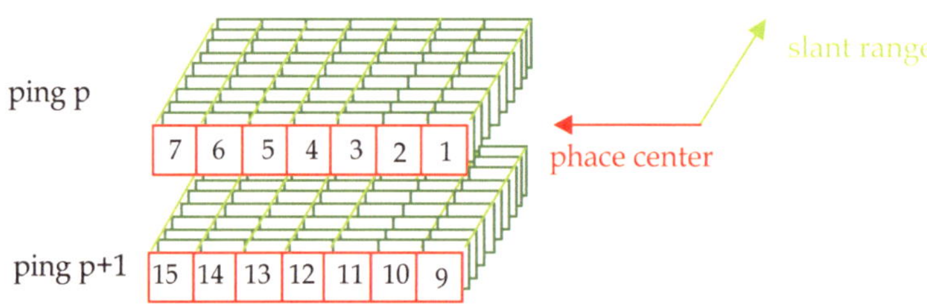

Fig. 16. (Slant range - phase center)-representation for an x-lag= -1 between ping p and p+1. The receiver (i.e. more precisely the phase center corresponding a particular Rx) couples are ($R_{x1}(p)$,$R_{x9}(p+1)$), ($R_{x2}(p)$,$R_{x10}(p+1)$), …, ($R_{x7}(p)$,$R_{x15}(p+1)$).

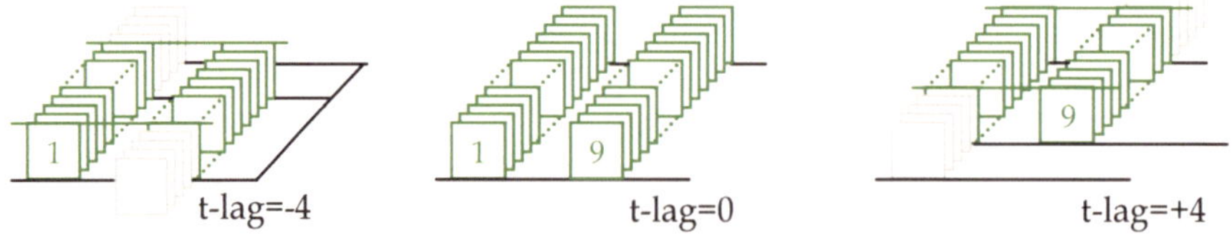

Fig. 17. Respective t-lag= -4, 0 and +4 representation of the Rx couple id=1 (i.e. ($R_{x1}(p)$,$R_{x9}(p+1)$)) for the x-lag= -1. The gray slant range bins will not contribute in the calculation of the cross correlation given in equation (51).

5.5 Beam forming on the cross correlation

To estimate the amplitude and angle of arrival one can perform beam forming on the cross correlation plots. But first, a general theory on beam forming is introduced in order to be able to explain the extension of this theory onto the cross correlation plots. This extension will form the core of the DPCA motion estimation idea.

5.5.1 Spatial time delay

Fig. 18. shows a linear array of equally spaced hydrophone elements being intercepted by a propagating wave front at beam angle θ. To make the far field approximation of the top figure one makes the following assumptions;

$$\begin{aligned} &\theta_i \approx \theta \\ &r_i \approx r \quad (\text{for amplitude variations}) \\ &r_i \approx r + x_n \quad (\text{for phase variations}) \end{aligned} \tag{52}$$

with

$$x_n = \left(n - \frac{H+1}{2}\right) d \sin\theta \,. \tag{53}$$

The corresponding spatial time delay associated with the distance x_n is given by

$$t_n = \frac{x_n}{c} \tag{54}$$

where c is the velocity of the wave and n the receiver under consideration. The corresponding spatial phase delay is given by

$$\phi_{x_n} = \frac{2\pi}{\lambda} x_n \tag{55}$$

Fig. 18. Equally spaced linear array of hydrophone elements being intercepted by a propagating wave front at beam angles θ_j for the near field configuration (left) with beam angle θ for the far field (right) approximation.

It is usually more meaningful when discussing array performance to express the phase delays in terms of the carrier frequency f_0 and the array frequency f_a. The array frequency is the frequency whose half-wavelength is equal to the inter hydrophone spacing d, i.e.

$$d = \frac{\lambda_a}{2} = \frac{c}{2 f_a} \,. \tag{56}$$

Substituting this and (53) into (55) yields

$$\phi_{x_n} = 2\pi\left(n - \frac{H+1}{2}\right)\frac{\lambda_a}{2\lambda}\sin\theta\,. \tag{57}$$

5.5.2 Beam steering

An array can be electronically steered by introducing processing phase or time delays into the hydrophone outputs. The processing delay inserted in series with the nth element output in order to steer the array at angle θ_0 is given by

$$\phi_{p_n} = 2\pi\frac{\Delta r_n}{\lambda} \quad \text{with } \Delta r_n = n d \sin\theta_0 \tag{58}$$

where Δr_n is the path correction due to the steering of the beam (Fig. 19.). If the steering angle is around the central receiver,

$$n \in \left[\frac{-H+1}{2}, \frac{-H+1}{2}+1, \ldots, \frac{H-1}{2}\right]. \tag{59}$$

The beam forming process sums the delayed outputs from the hydrophone elements to generate a beam output voltage. This beam voltage can be written in a normalized form

$$V_b = \frac{1}{N_h}\sum_{n=0}^{N_h-1} V_n \exp\left(-i\phi_{p_n}\right) \tag{60}$$

$$= \frac{1}{N_h}\sum_{n=0}^{N_h-1} V_n \exp\left(-in\Delta\phi\right) \tag{61}$$

where $\Delta\phi = \frac{\pi f}{f_a}\sin\theta_0$ is the processing phase increment. Let $\theta_0(k)$ represent the kth beam-steering angle, than

$$\Delta\phi(k) = \frac{\pi f}{f_a}\sin\theta_0(k) \tag{62}$$

represents the processing phase increment associated with the *k*[th] beam-steering angle. Restricting the beam angles $\theta_0(k)$ such that

$$\Delta\phi(k) = \frac{2\pi}{N_h}k \tag{63}$$

Substituting the value of $\Delta\phi(k)$ into equation (61) yields

$$V_b(k) = \frac{1}{N_h}\sum_{n=0}^{N_h-1} V_n \exp\left(-in\frac{2\pi}{N_h}k\right) \tag{64}$$

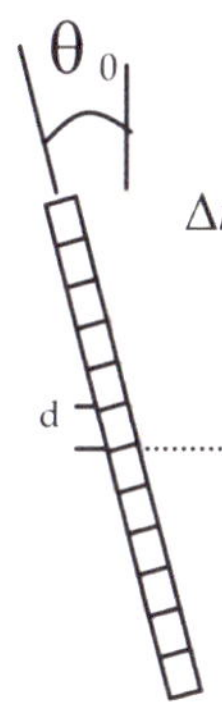

$$\Delta r_{N_h} = \frac{N_h - 1}{2} d \sin\theta_0$$

$$\Delta r_j = \left(j - \frac{N_h + 1}{2} \right) d \sin\theta_0$$

$$\Delta r_1 = -\frac{N_h - 1}{2} d \sin\theta_0$$

Fig. 19. Steering the multiple hydrophone array at angle θ_0.

From equation (62) and (63) one obtains,

$$\theta_0(k) = \sin^{-1}\left(\frac{2k}{N_h(f / f_a)} \right) = \sin^{-1}\left(\frac{kc}{L_h f} \right) \tag{65}$$

with

$$0 \le k \le \frac{N_h(f / f_a)}{2} = \frac{L_h f}{c} \tag{66}$$

with L_h the synthetic aperture over the overlapping h receivers. Equation (64) is used to beamform the cross correlation plots, which leads to a correlation beam. The y-axis indicates still the different t-lags, whereas the x-axis represents the look angle.

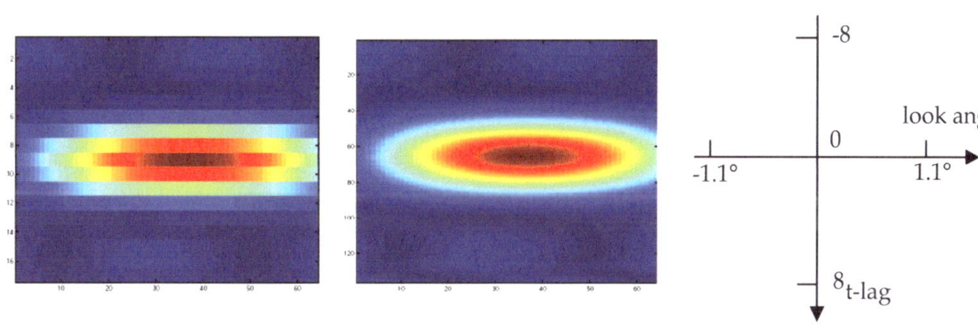

Fig. 20. The beam formed cross correlation matrix before and after interpolation. The interpolation is done along the t-lag with an over sampling factor of 8, meaning that 17 t-lags will become 136.

A restriction on the look angle is chosen between θ=-1.1 degree till θ=1.1 degree corresponding $\pm \frac{c}{f_0 L} \frac{180}{\pi}$. The result of the beam forming on one particular cross correlation plot is shown in Fig. 20. In order to find the correlation peaks one has first to smooth the correlation beams using for example a linear interpolation.

5.5.3 Correlation peaks and temporal delays

To find the correlation peaks and the temporal delays as a function of the viewing angle a parabolic maximum finder was designed. For a parabol given by $f(x) = a_0 + a_1 x + a_2 x^2$, the refined analytical maximum is given by

$$f(x = -a_1/(2a_2)) = \frac{4a_0 a_2 - a_1}{4a_2} \tag{67}$$

Since the 3 point parabolic fit is performed on a normalized dataset, one has to convert the maximum found between -1 and 1 back to its initial scale. For that a simple interpolation is performed. The representation of the 3 points, on which the parabolic fit is performed, is shown by the blue crosses on Fig. 21., the result of the parabolic fit is shown by the green line. The maximum of the parabol is shown by the blue square. One sees a small correction to the integral maximum value towards the new refined maximum.

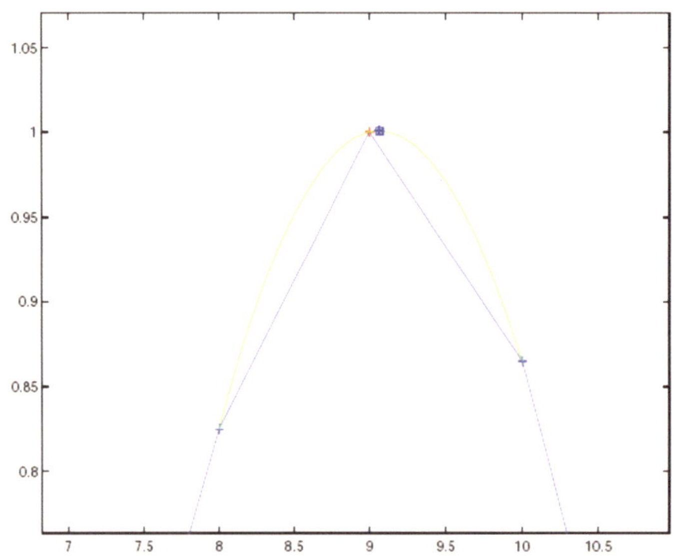

Fig. 21. Illustration of the parabolic maximum finder.

For each x-lag (i.e. -2, -1, 0, 1, 2) the maximum for each look angle is determined on the cross correlation plots after beam forming. Those maxima as function of the look angle are shown in the second line of Fig. 22. The corresponding phase delays are shown in the bottom line of Fig. 22. So for each x-lag the best beam is found containing the best beam angle for which the envelope reaches a maximum. Further, the delay that corresponds with the best beam angle is considered the best delay. When the best beam angles are set out as function of the x-lags, the parabolic maximum finder is used to find the corresponding best x-lag. In general this best spatial lag $B_{lag}(x)$ will not be an integer value but a real number and corresponds to a measure for the surge estimation via,

$$Surge = B_{lag}(x)\, d/2\,. \tag{68}$$

The best lag delay is a measure for the sway estimation via,

$$Sway = B_{lag}(delay)\, C/2\,. \tag{69}$$

And the best look angle θ corresponding this best lag is a measure for the yaw estimation via,

$$Yaw = B_{lag}(\theta)\, \lambda/L\,. \tag{70}$$

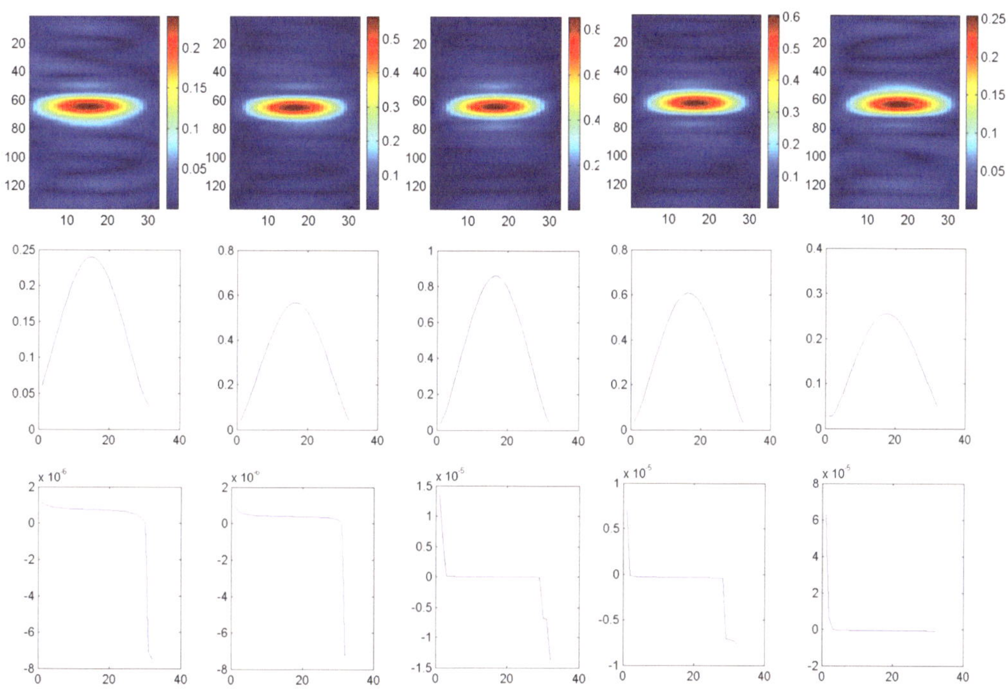

Fig. 22. Each column represents one x-lag, going from -2 (utmost left) to +2 (utmost right). The first line represents the cross correlation plots after beam forming. The x-axis represents the look angle going from -1.1 till 1.1 degree. For each look angle the 3 point maximum is defined and is shown in the figures at the second line. The third line represents the corresponding phase delays.

For each successive ping-pair the surge, sway and yaw can thus be extracted as is shown in equation (68), (69) and (70). The result of the sway estimation compared to the actual sway that was generated in the simulator is shown on the left-hand side in Fig. 23. The red crosses represent the DPCA sway estimations between a set of different ping-pairs. The line represents the actual generated sway or true sway. On the right-hand side of Fig. 23 the difference is shown between the estimated sway and the true sway expressed in mm. The highest difference between the true and estimated sway is 2 mm, which is well within the 1/10th of the applied wavelength (λ=3 cm for a carrier frequency f_0=50 kHz).
Fig. 24 shows the result of the yaw estimation compared to the actual yaw as a function of the ping number (left). The yaw values are expressed in radians. The absolute error between the true and the estimated yaw is of the order of 10^{-4} radians (right).

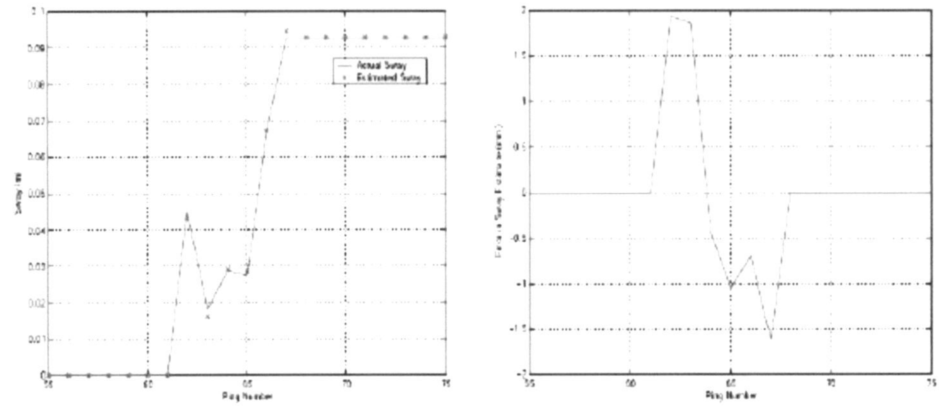

Fig. 23. Result of the sway estimation (red crosses) compared to the simulated sway or actual sway (full line) as a function of the ping number (left). The difference between the actual sway and the estimated sway expressed in mm (right).

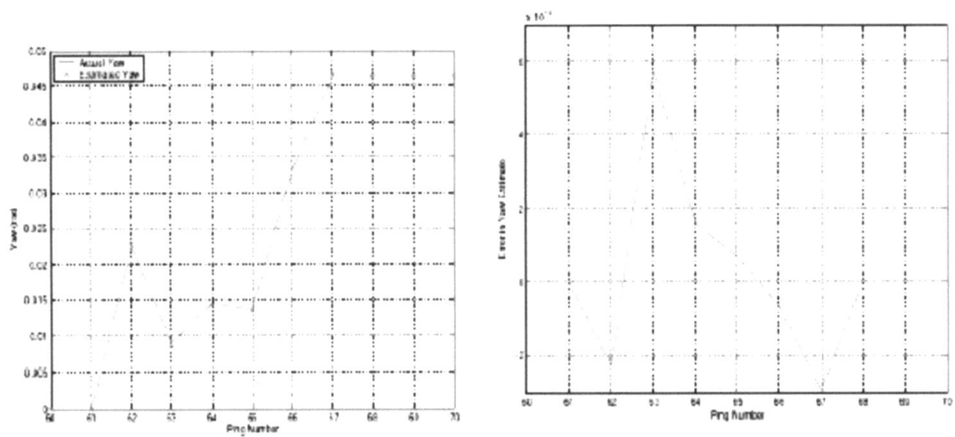

Fig. 24. Result of the yaw estimation (red crosses) compared to the simulated yaw or actual yaw (full line) as a function of the ping number (left). The difference between the actual y and the estimated yaw expressed in deg (right).

6. Motion correction

The correction of the surge, sway and yaw motions are done following the estimation of the x-and t-lag analysis obtained in Section 5.

Let (O,x,y) be the slant range plane (Fig. 23.), with Ox the along-track, Oy the across-track and (x_p,y_p) the coordinates of $C_p=(T_p+R_p)/2$. T_p and R_p are respectively the centres of the real transmitter and receiver position at ping p and θ_p is the angle between Oy and the bore-sight to the physical aperture. Than the relative position of the sonar platform can be expressed as,

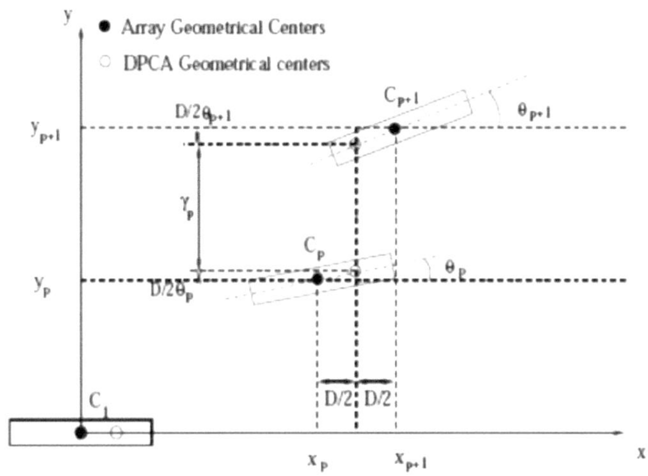

Fig. 23. SAS trajectory representation in the slant range plan

$$\begin{cases} x_{p+1} = x_p + D \\ y_{p+1} = y_p + \gamma_p + \dfrac{D}{2}\theta_p + \dfrac{D}{2}\theta_{p+1} \\ \theta_{p+1} = \theta_p + \xi_p \end{cases} \quad (71)$$

where γ_p and ξ_p are respectively the DPCA sway and yaw between pings p and p+1. The angles θ_p have been assumed small (i.e. $\sin\theta \approx \theta$). The quantity $(y_{p+1} - y_p)$, which can be interpreted as the physical sway between successive pings, is the sum of three terms. The first is the DPCA sway and the other two result from the heading of the physical reception antenna at ping p and p+1. The geometrical centre of the DPCA and the one of the physical array are separated by D/2. This leads to a difference between the real cross-track position and the cross track position of the associated phase centres $(D(\theta_p+\theta_{p+1})/2)$. The estimated trajectory can be expressed as:

$$\begin{cases} x_p = (p-1)D \\ y_p = \sum_{l=1}^{p-1}\gamma_l + D\sum_{l=1}^{p-1}(p-l-\frac{1}{2})\xi_p \\ \theta_p = \sum_{l=1}^{p-1}\xi_l \end{cases} \quad (72)$$

The accumulated errors δy_p and $\delta\xi_p$ on the DPCA are given by

$$\begin{cases} \delta y_p = \sum_{l=1}^{p-1}\delta\gamma_l + D\sum_{l=1}^{p-1}\left(p-l-\frac{1}{2}\right)\delta\xi_l \\ \delta\theta_p = \sum_{l=1}^{p-1}\delta\xi_l \end{cases} \quad (73)$$

The most important effect on SAS processing is the cross track errors. One can see in equation (73) that the along track error depends on the accumulated errors of the DPCA's sway and yaw. In a case where there is only DPCA sway errors ($\delta\xi_p$ =0) they accumulate like a random walk. In a case where there is only DPCA yaw errors ($\delta\gamma_p$ =0) they accumulate like an integrated random walk. In the last case the errors accumulate much faster and lead to a high correlated pattern of phase errors along the SAS.
The differences in cross track as well as in along track positions are leading to a time delay which can be removed by convolving the measured echo with the appropriate delta function δ(t-Δτ),

$$ee_h(t,u) = ee_h^{raw}(t,u) \otimes \delta(t - \Delta t) \quad \text{with} \quad \Delta t = \Delta t_{sway}(u) + \Delta t_{yaw\{h\}}(u) \tag{74}$$

where ee_h^{raw} represents the raw data registered at hydrophone h as a function of the delay time t and the azimuth position u. ee_h represents the motion compensated signal. In practice, instead of performing a convolution, one goes to the frequency domain (ω,k) using the fast Fourier transform in two dimensions, to perform a simple multiplication,

$$EE_h(\omega, k_u) = EE_h^{'}(\omega, k_u).\exp(-i\omega\Delta t). \tag{75}$$

8. Summary

Synthetic Aperture Sonar (SAS) is a revolutionary underwater imaging technique providing imagery and bathymetry at high spatial resolution with large area coverage. The implementation of synthetic aperture sonar utilising multiple pings to create a virtual long array for range-independent resolution was inadequate due to lack of coherence in the ocean medium, precise platform navigation and high computation rates. Moreover, SAS is far more susceptible to image degradation caused by the actual sensor trajectory deviating from a straight line. Unwanted motion is virtually unavoidable in the sea due to the influence of currents and wave action. In order to construct a perfectly-focused SAS image the motion must either be constrained to within one-tenth of a wavelength over the synthetic aperture or it must be measured with the same degree of accuracy.
The technique known as Displaced phase centers array (DPCA) has proven to be adequate technique in solving the problem of SAS motion compensation. In essence, DPCA refers to the practice of overlapping a portion of the receiver array from one ping (transmission and reception) to the next. The signals observed by this overlapping portion will be identical except for a long track and time shifts proportional to the relative motion between pings. Both shifts estimated by the DPCA are scalars representing the projection of the array receiver locations onto the image slant plane and can be used to compensate for the unwanted platform motion. Thus, the delays observed in the image slant plane can be used to refine the surge, sway and yaw motions.
With advances in innovative motion-compensation, synthetic aperture sonar is now being used in commercial survey and military surveillance systems. Emerging applications for SAS systems include economic exclusion zone mapping (EEZ), mine detection and the development of long range imaging sonar for anti-submarine warfare.

Although the development of precise navigation sensors and of stable submerged autonomous platforms the motion compensation processing is still a crucial element in the image reconstruction, pre- and/or post-processing.

9. References

Bellettini, A. and Pinto, M. A. (2000). *Experimental investigation of synthetic aperture sonar Micronavigation*, Proceedings of the Fifth ECUA 2000, Lyon, France, (445–450)

Bellettini, A. and Pinto, M. A. (2002). Accuracy of sas micronavigation using a displaced phase centre antenna: theory and experimental validation, Saclantcen report, SR-355, 24 p.

Bruce, M. P. (1992) A Processing Requirement and Resolution Capability Comparison of Side-Scan and Synthetic-Aperture SOnars, IEEE Journal of Oceanic Engineering, vol. 17, No. 1

Callow, H. J., Hayes, M. P. and Gough, P. T. (2001). Advanced wavenumber domain processing for reconstruction of broad-beam multiple-receiver sa imagery, IVCNZ, (51–56)

Castella, F. R. (1971). Application of one-dimensional holographic techniques to a mapping sonar system. Acoustic Holography, *Vol* 3.

Christoff, J. T., Loggins, C. D. & Pipkin, E. L. (1982). Measurement of the temporal phase stability of the medium. J. Acoust. Soc. Am., *Vol* 71., (1606-1607)

Curlander, J. C. & McDonough, R. N. (1991). *Synthetic Aperture Radar: Systems and Signal Processing*, Wiley, ISBN 0-471-85770-X, New York

Cutrona, L. J. (1975). Comparison of sonar system performance achievable using synthetic aperture techniques with the performance achievable by more conventional means. J. Acoust. Soc. Am., *Vol* 58., (336-348)

Gough, P. T. & Hayes, M. P. (1989). Measurement of the acoustic phase stability in Loch Linnhe, Scotland. J. Acoust. Soc. Am., *Vol* 86., (837-839)

Gough, P. T. & Hawkins, D. W. (1997). Imaging algorithms for synthetic aperture sonar: Minimising the effects of aperture errors and aperture undersampling, *IEEE J. Oceanic Eng., Vol* 22., (27-39)

Groen, J. (2006). Adaptive motion compensation in sonar array processing, *PhD thesis, Technical University Delft (TUDelft)*, Netherlands, 247 p.

Hughes, R. C. (1977). Sonar imaging with the synthetic aperture method, *Proceedings of the IEEE Oceans, Vol.* 9., (102-106)

Marx, D. ; Nelson, M. ; Chang, E. ; Gillespie, W. ; Putney, A. ; Warman, K. (2000). An introduction to synthetic aperture sonar, *Proceedings of the Tenth IEEE Workshop on Statistical Signal and Array Processing*, pp. 717-721, ISBN: 0-7803-5988-7, Pocono Manor, PA, USA, August 2000

Sherwin, C. W.; Ruina, J. P. & Rawcliffe, R. D. (1962). Some early developments in syntheti aperture radar systems. IRE Trans. Military Electronics, *Vol* 6., (111-115)

Skolnik, M. I. (1980). *Introduction to Radar Systems*, Mc-Graw-Hill, New York

Somers, M. L.; Stubbs A. R. (1984). Sidescan sonar. IEE Proceedings, vol. 131, Part F, no. 3: (243-256).

Stimson, G. W. (1983). *Introduction to Airborn Radar*, SciTech, ISBN 1-891121-01-4, New Jersey

Walker, J. L. (1980). Range-doppler imaging of rotating objects. IEEE Trans. Aerospace Electronic Syst, *Vol* 16., (23-52)

Walsh, G. M. (1969). Acoustic mapping apparatus. J. Acoust. Soc. Am., *Vol* 47., (1205)
Wang, L., Bellettini, A., Hollett, R. D., Tesei, A. and Pinto, M.A. (2001). *InSAS'00: Interferometric SAS and INS aided SAS imaging*, Proc. Oceans'01, Hawaii.
Wiley, C. A. (1985). Synthetic aperture radars. IEEE Trans. Aerospace Electronic Syst, *Vol* 21., (440-443)
Williams, R. E. (1976). Creating an acoustic synthetic aperture in the ocean. J. Acoust. Soc. Am., *Vol* 60., (60-73)

Sonar Image Enhancement

4

Ensemble Averaging and Resolution Enhancement of Digital Radar and Sonar Signals

Leiv Øyehaug[1*] and Roar Skartlien[2*]
[1]*Centre of Integrative Genetics and Department of Mathematical Sciences and Technology, Norwegian University of Life Sciences,*
[2]*Institute of Energy Technology, Department of Process and Fluid Flow Technology, Norway*

1. Introduction

In radar and sonar signal processing it is of interest to achieve accurate estimation of signal characteristics. Recorded pulse data have uncertainties due to emitter and receiver noise, and due to digital sampling and quantization in the receiver system. It is therefore important to quantify these effects through theory and experiment in order to construct "smart" pulse processing algorithms which minimize the uncertainties in estimated pulse shapes. Averaging reduces noise variance and thus more accurate signal estimates can be achieved. Considering a signal processing system that involves sampling, A/D-conversion, IQ-demodulation and ensemble averaging, this chapter forms a theoretical basis for the statistics of ensemble averaged signals, and summarizes the basic dependencies on bit-resolution, ensemble size and signal-to-noise ratio.

Repetitive signals occur in radar and sonar processing, but also in other fields such as medicine (Jane et al., 1991; Schijvenaars et al, 1994; Laguna & Sornmo, 2000) and environment monitoring (Viciani et al., 2008). Practical ensemble averaging is subject to alignment error (jitter) (Meste & Rix, 1996), but we will neglect this effect. The effective bit-resolution of the system can be increased by ensemble averaging of repetitive, A/D-converted signals, provided that the signal contains noise (Belchamber & Horlick, 1981; Ai & Guoxiang, 1991; Koeck, 2001; Skartlien & Øyehaug, 2005).

Due to varying radar and sonar cross section for scattering objects, or varying antenna gain of a sweeping emitter or receiver, the pulses exhibit variation in scaling. In the case of radar or sonar, the cross section of the target may then vary from pulse to pulse, but not appreciably over the pulse width. The scanning motion of the radar antenna may also affect the pulse scaling regardless of the target model, but we can safely neglect the time variation of the scaling due to this effect. In the case of a passive sensor, the signal propagates from an unknown radar emitter to the sensor antenna, and there is no radar target involved. Only

* The present study was conceived of and initiated during the authors' employment with the Norwegian Defence Research Establishment, P.O. Box 25, 2027 Kjeller, Norway.

the scanning motion of the emitter antenna (and possibly the sensor antenna) may in this case influence the scaling. In general, we assume that the scaling can be treated as a random variable accounted for by a given distribution function (Øyehaug & Skartlien, 2006).
In the present chapter we briefly review some of the theory of ensemble averaging of quantized signals in absence of random scaling (Sect. 2) and summarize results on ensemble averaging of randomly scaled pulses (in absence of quantization) modulated into amplitude and phase (Sect. 3). Aided by numerical simulations, we subsequently extend the results of the preceding sections to amplitude and phase modulations of scaled, quantized pulses (Sect. 4). In Sect. 5 we discuss how the theoretical results can be implemented in practical signal processing scenarios and outline some of the issues that still require clarification. Finally, in Sect. 6 we draw conclusions.

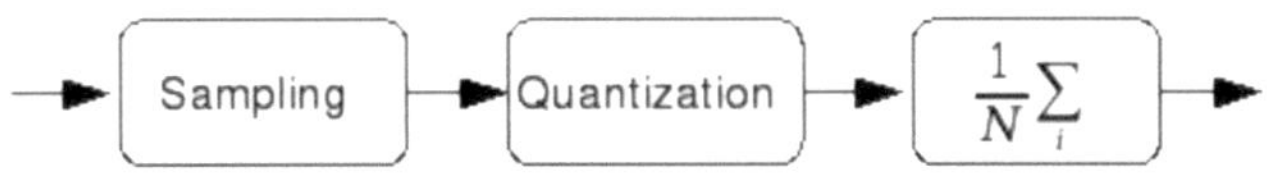

Fig. 1. The signal chain considered in Sect. 2. After sampling, the signal is quantized (A/D-converted) followed by ensemble averaging.

2. Ensemble averaging of quantized signals; benefitting from noise

This section considers the statistical properties of ensemble averages of quantized, sampled signals (Fig. 1), and demonstrates that the expectation of the quantization error diminishes with increasing noise, at the cost of a larger error variance. As the ensemble average approximates the expectation, it follows that the quantization error (in the ensemble average) can be made much smaller than what corresponds to the bit resolution of the system. We will also demonstrate that there is an optimum noise level that minimizes the combined effect of quantization error and noise. First, consider a basic analog signal with additive noise;

$$y_i(t) = s(t) + n_i(t), \tag{1}$$

where t is time, and n is random noise. We observe N realizations of y, and the index i denotes one particular realization i (or sonar or radar pulse i). We assume that s is repetitive (independent of i), while n varies with i. We assume a general noise distribution function with zero mean and variance σ^2. The recorded signal is sampled at discrete t_j giving $y_{i,j}$, and these samples are subsequently quantized through a function Q to obtain the sampled and A/D-converted digital signal $x_{i,j} = Q(y_{i,j})$. We consider the quantization to be uniform, i.e. the separation between any two neighboring quantization levels is constant and equal to Δ. The probability distribution function (pdf) of $x_{i,j}$ is discrete and generally asymmetric even if the pdf of n is continuous and symmetric.

2.1 Error statistics

We define the error in the quantized signal as $e_{i,j} = x_{i,j} - s_j$ accounting for both noise and quantization effects. The noise in different samples is uncorrelated and we assume that the correlation time is sufficiently small such that the noise between different time-samples is also uncorrelated.

To illuminate the effect of ensemble averaging, we consider the one-bit case for which the quantizer has two levels: 0 and 1. If the input is larger than 1/2 the output will be 1, otherwise the output is 0. Consider a constant "signal" $s_j = 1/2$. For zero noise the output is always 1, giving an error of 1/2. If we introduce noise with a symmetric pdf with zero mean, the quantizer output "flips" between 0 and 1 randomly. The expectation value of the output is then 1/2, since we expect an equal number of zeroes and ones on the output, hence the expectation value of the error is zero. If the input signal is larger or smaller than 1/2, there will be an error such that the expected error of the output is nonzero.

For Gaussian noise combined with a uniform quantizer with many levels, Carbone and Petri (1994) derived

$$E[e_{i,j};s_j] = \frac{\Delta}{\pi}\sum_{k=1}^{\infty}\frac{(-1)^k}{k}\exp[-2\pi^2 k^2(\sigma/\Delta)^2]\sin(2\pi k s_j/\Delta). \tag{2}$$

It is easy to see that the expected error is reduced and goes to zero for increasing noise. The reason for this is that for increasing noise, the discrete pdf of the quantized signal becomes an increasingly more accurate representation of the continuous pdf of $y_{i,j}$ (with expectation s_j). The pdf of $y_{i,j}$ gets "broader" and is thus better resolved on the fixed grid defined by the quantizer cells. The expected error attains its largest values for zero noise where it becomes a "sawtooth" function of s_j and, for intermediate noise, it becomes roughly sinusoidal as function of s_j (Fig. 2, left), since only the first few terms in the series expansion are significant.

Along with the expectation value of the error, there is an error variance, defined by $\mathrm{Var}[e_{i,j};s_j] = E[e_{i,j}^2;s_j] - E^2[e_{i,j};s_j]$, which can also be derived in terms of a trigonometric series. For Gaussian noise,

$$E[e_{i,j}^2;s_j] = \frac{\Delta^2}{12} + \sigma^2 + \left(\frac{\Delta}{\pi}\right)^2 \sum_{k=1}^{\infty}(-1)^k\left(\frac{1}{k^2} + 4\pi(\sigma/\Delta)^2\right)\exp[-2\pi^2 k^2(\sigma/\Delta)^2]\cos(2\pi k s_j/\Delta). \tag{3}$$

It is apparent from the exponential factors that in the large noise limit the error variance goes to $\Delta^2/12 + \sigma^2$ (Fig. 2, right), i.e. the variance is signal-independent. Both a vanishing error expectation, and a variance of σ^2 in the large noise limit, are exactly the properties of the analog signal before quantization.

2.2 Expectation of the sample mean

In the following we consider the ensemble average of $x_{i,j}$,

$$\bar{x}_j = \frac{1}{N}\sum_{i=1}^{N} x_{i,j}. \tag{4}$$

Ensemble averaging has the beneficial effect of reducing the variance, as we expect from basic statistics. Using the asymptotical relations above, one can show that the variance follows the usual $1/N$-law in the large noise limit, i.e. the ensemble average variance goes to

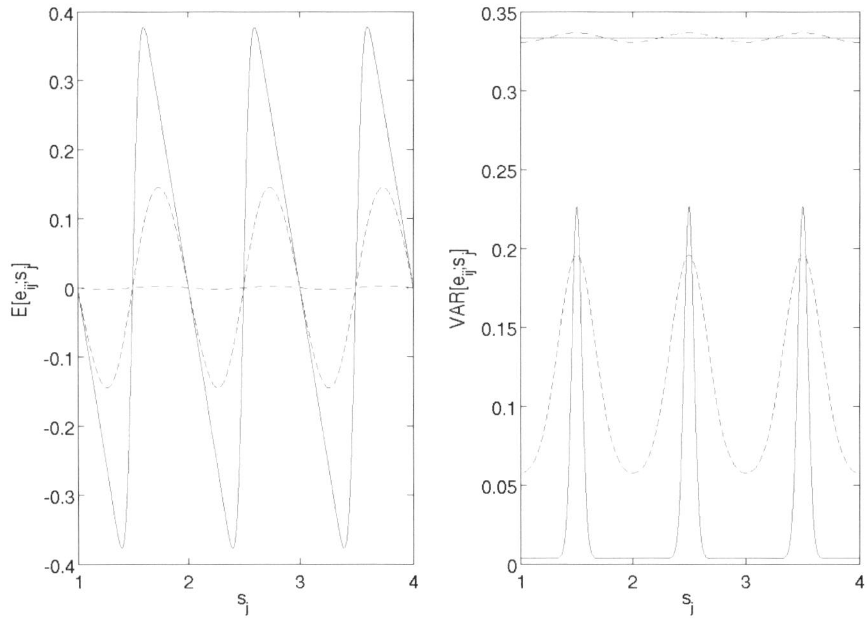

Fig. 2. Signal-dependent expected error (left) and variance (right) for $\Delta = 1$ and three noise levels; $\sigma = 0.05$ (full line) $\sigma = 0.2$ (dashed) and $\sigma = 0.5$ (dash-dotted). In (B), the straight line $\Delta^2/12+\sigma^2$ is plotted to indicate the convergence towards this value with increasing noise.

$\left(\Delta^2/12+\sigma^2\right)/N$ with increasing noise. Thus, the variance of the ensemble average can be made arbitrarily small for increasing ensemble size N. It is important to note that the expectation of the ensemble average converges to the input signal for increasing noise level. Noise is therefore beneficial in this respect, at the cost of a larger variance that can of course be compensated by increasing N.

Furthermore, for small noise, the expectation of the ensemble average differs from the input signal. Ensemble averaging will not remove this difference, since it originates from the deterministic property of the quantizer and not from the noise. We illustrate the effect of noise in Fig. 3, where we compute the average of a sinusoidal with Gaussian noise with variance σ^2 and use a simple roundoff to integer numbers as the quantizer function (i.e. Δ=1). With no noise we obtain a staircase function as expected (Fig. 3, upper panel). With increasing noise, the staircase function is smoothed out to resemble the sine-wave (Fig. 3, lower panels).

2.3 Mean square error (MSE) of the sample mean

With the sample mean as the measured quantity, the associated error is $\bar{e}_j = \bar{x}_j - s_j$. We note that as the sample mean tends to the expectation value as N tends to infinity, $\bar{e}_j$ tends to $E[e_{i,j}]$. For Gaussian noise and for sufficiently large σ/Δ, the mean square error MSE of the sample mean, obtained by averaging $\left(\bar{e}_j\right)^2$ over all discrete samples $j \in [1,...,M]$, is (Skartlien & Øyehaug, 2005),

$$\mathrm{MSE}(\sigma,N) \simeq \frac{1}{N}\left(\sigma^2 + \Delta^2/12\right) + \left(1-\frac{1}{N}\right)\frac{\Delta^2}{2\pi^2}\exp\left(-4\pi^2(\sigma/\Delta)^2\right). \tag{5}$$

When N is given, one can ask what are the bounds on σ to obtain super-resolution? For fixed $N \geq 2$, the requirement $\mathrm{MSE} < \Delta^2/12$ defines the super-resolution interval $(0, \sigma_{\max}(N))$. In the case of Gaussian noise, the upper bound $\sigma_{\max}(N)$ is given implicitly by

$$(N-1)\Delta^2\left(\frac{1}{12} - \frac{1}{2\pi^2}\exp[-4\pi^2(\sigma_{\max}/\Delta)^2]\right) = \sigma_{\max}^2, \tag{6}$$

for large enough σ/Δ. A remarkable property is that there is an optimal noise σ_{opt} in the super-resolution interval $(0,\sigma_{\max}(N))$ which minimizes the MSE. This optimal noise is $\sigma_{\mathrm{opt}} = \Delta\sqrt{\log(2(N-1))}/2\pi$ provided that σ/Δ is sufficiently large and that the noise is Gaussian. For N=100, for example, the optimum value for σ/Δ is close to 0.366, which explains the good averaging performance associated with this value in Fig. 3.

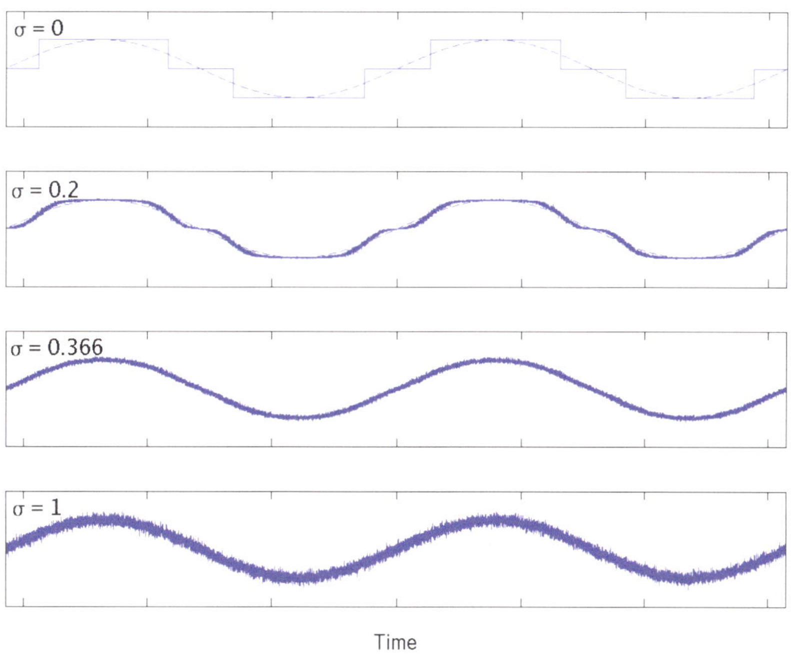

Fig. 3. An example of the effect of ensemble averaging of quantized signals. A sinusoidal with unit amplitude (dashed) plus Gaussian white noise is quantized by a simple roundoff to the nearest integer (i.e. Δ=1) and then averaged over an ensemble of 100 realizations (ensemble average: solid line). In absence of noise we obtain a staircase function (upper panel) and, with increasing noise, the staircase function is smoothed out to resemble the sine-wave. For the particular value σ/Δ=0.366 the mean square error is a minimum.

In summary, the existence of a minimum MSE is a consequence of the balancing between quantizer and noise effects. For small noise ($\sigma < \sigma_{\text{opt}}$), the noise tends to remove the effect from the quantization error in the sample mean and the MSE decreases with increasing noise. For large noise ($\sigma > \sigma_{\text{opt}}$), quantization is roughly negligible compared to the effect of the noise itself, and the MSE increases with increasing noise.

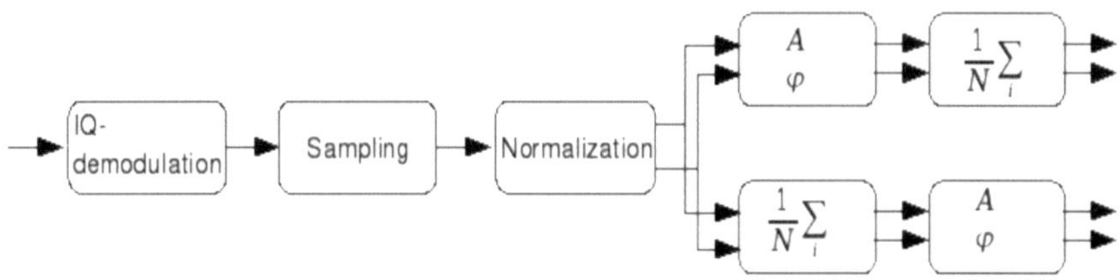

Fig. 4. The signal processing chain considered in Sect. 3. After IQ-demodulation, the signal is sampled and normalized followed by either (i) ensemble averaging of I and Q and then amplitude/phase modulation (in Sect. 3 referred to as Method I, lower branch) or (ii) amplitude/phase modulation and then ensemble averaging of amplitude and phase (Method II, upper branch).

3. Ensemble averaging of modulated signals that are randomly scaled

In this section our focus is on the statistical properties of the ensemble average of amplitude and phase of a sequence of randomly scaled, IQ-demodulated pulses. Here we ignore quantization effects. We give the pdf of amplitude and phase modulation of complex signals in Gaussian noise, then discuss in which order ensemble averaging of IQ-demodulated and normalized signals should proceed (whether amplitude and phase should be computed for each individual pulse and then averaged or the average of I and Q should be used to compute phase and amplitude averages. See Fig. 4 for the two alternative methods). Then, we review the theory behind the optimal scaling threshold, which involves discarding pulses that have amplitudes below a certain threshold.

Consider the complex signal $Z(t)$ in terms of an IQ-decomposition; $Z(t) = I(t) + iQ(t)$. An IQ-demodulator provides a signal on this form in a sonar/radar or a radio. Alternatively, one may generate the quadrature signal by Hilbert transformation. To include random noise and scaling, we adopt the signal model

$$Z_k(t) = a_k Z_0(t) + n_k(t), \tag{7}$$

where t is time delay from pulse start and k is pulse number. The scaling a_k is accounted for by a general distribution $p(a)$, where a is a positive, real number. The noise is complex and Gaussian with variance σ^2. We assume a certain noise correlation function with a characteristic correlation time that is sufficiently short such that noise in different pulses is uncorrelated. We will in the following consider the phase and amplitude modulations, which are

$$\phi_k(t) = \mathrm{Arg}\left(Z_k(t)\right), \quad A_k(t) = \mathrm{Mod}(Z_k(t)), \tag{8}$$

respectively. Both these modulations have a random component due to noise and scaling.

3.1 Phase and amplitude statistics in general terms

The starting point to obtain the pdf's in question, is to consider how the scaled and subsequently normalised complex numbers

$$\widehat{Z}_k(t) = Z_k(t)/a_k = Z_0(t) + \frac{n_k(t)}{a_k} \tag{9}$$

are distributed in the complex plane. It is obvious that the phase of $\widehat{Z}_k$ is unaffected, and that the normalized amplitude is accounted for, by the scaling. The associated amplitude and phase distributions are obtained by considering the conditional distributions for given a_k, and then integrating over the scaling distribution $p(a)$.
We obtain the conditional distribution by using the scaled variance $(\sigma/a)^2$ in place of σ^2 in the phase distribution for a Gaussian complex variable of variance σ^2 (see Davenport and Root, 1958; Papoulis, 1965),

$$p(\phi;q,a) = \frac{\exp\left(-a^2q\right)}{2\pi} + \frac{\sqrt{a^2q}\cos(\phi)\exp\left(-a^2q\sin^2(\phi)\right)}{\sqrt{4\pi}}\left[1 + 2\mathrm{Erf}\left(\sqrt{2a^2q}\cos(\phi)\right)\right], \tag{10}$$

where $q = A_0^2/(2\sigma^2)$, and Erf denotes the usual error function. With given a, the normalised amplitude $\widehat{A} = A/a$ obeys a Rician distribution (Davenport and Root, 1958; Papoulis, 1965) with variance $(\sigma/a)^2$ and amplitude parameter $A_0 = \mathrm{Mod}(Z_0)$,

$$p(\widehat{A};A_0,\sigma,a) = a^2(\widehat{A}/\sigma^2)\exp\left(-a^2(\widehat{A}^2 + A_0^2)/2\sigma^2\right)I_0(a^2\widehat{A}A_0/\sigma). \tag{11}$$

Here, I_0 is the modified Bessel function of zero order.
Integration over $p(a)$ yields the phase and amplitude distributions

$$p(\phi_k;q) = \int_0^\infty p(\phi_k;q,a)p(a)\,da, \tag{12}$$

$$p(\widehat{A}_k;A_0,\sigma) = \int_0^\infty p(\widehat{A}_k;A_0,\sigma,a)p(a)\,da. \tag{13}$$

The variances σ_A^2 and σ_ϕ^2 can now be obtained by calculation of the second moments of $p(\widehat{A}_k;A_0,\sigma)$ and $p(\phi_k;q)$.

3.2 Order of ensemble averaging

Averaging methods: There are two different ways of generating accurate phase and amplitude estimates by ensemble averaging (see Fig. 4):

- **Method I**, which refers to calculating phase and amplitude of ensemble averaged *I/a* and *Q/a*.
- **Method II,** which refers to calculating phase and amplitude of each individual realization of the pair (*I/a*,*Q/a*) before ensemble averaging.

In radar- or sonar-terms, Method I can be regarded as "coherent integration" and Method II as "incoherent integration", where "integration" is to be understood as ensemble averaging.
Method I: For sufficiently large ensemble *N*, the averages of the output of the IQ-demodulator $\overline{I}$ and $\overline{Q}$ tend to normal distributions (Gaussian random variables) by invoking the Central Limit Theorem from basic statistics. For normalized averaging, the joint distribution of $\overline{I}$ and $\overline{Q}$ is also symmetric. One can then immediately use the classical "Rician" probability distributions (10) and (11) for the amplitude and the phase, respectively, which apply to Gaussian and symmetric joint distributions. Then, for Method I, the pdf of the amplitude is of the form (11) with σ replaced by σ_N, where

$$\sigma_N^2 = \frac{\sigma^2}{N}\int_0^\infty \frac{p(a)}{a^2}\,da, \tag{14}$$

and the pdf of the phase is of the form (10) with q replaced by $q_N = A_0^2/(2\sigma_N^2)$.
Method II: In Method II we calculate the phase and the amplitude of each individual realization of the pair (*I/a*,*Q/a*), before ensemble averaging. The phase and amplitude modulations are estimated by performing an ensemble average (pulse to pulse average) over all available pulses. The ensemble averaged phase and amplitude are

$$\langle\phi(t)\rangle = \frac{1}{N}\sum_{k=1}^{N}\phi_k(t), \tag{15}$$

$$\langle A(t)\rangle = \frac{1}{N}\sum_{k=1}^{N}\frac{A_k(t)}{\overline{A_k}}, \tag{16}$$

where the amplitude ensemble average is normalized and $\overline{A_k}$ is the time averaged pulse amplitude for pulse *k*. We assume that $\overline{A_k}$ estimates a_k with sufficient accuracy such that we can neglect the stochastic component of $\overline{A_k}$ in the analysis. We note that $\overline{A_k}$ is the average instantaneous pulse amplitude over a single pulse *k* only. The pdfs of $\phi_k(t)$ and $A_k(t)/\overline{A_k}$ for fixed delay give the variance of the individual terms in the sum. The variance of the ensemble average is found by scaling this variance with $1/N$, since the individual terms are uncorrelated.
One can show that the joint pdf of (*I/a*,*Q/a*) is non-Gaussian. This can be handled by treating the Rician distributions as conditional distributions for given *a*, and then integrating over *p*(*a*) to obtain the non-Rician amplitude and phase distributions (12, 13). The resulting variances can then be calculated numerically. Finally, the variances for the ensemble averages are obtained by scaling with $1/N$, using the assumption of uncorrelated realizations.
Large SNR: In the large SNR case, the phase variance for both methods tend to the same value for $A_0 \gg \sigma$: In this limit $a^2 q \gg 1$ (for not too small *a*), such that the phase pdf (12) for Method II can be replaced by a normal distribution (see Appendix A of Øyehaug & Skartlien, 2006) with variance

$$\sigma^2_{\phi,\mathrm{M2}} = \frac{(\sigma/A_0)^2}{N}\int_0^\infty \frac{p(a)}{a^2}da. \tag{17}$$

Similarly, the phase distribution for Method I has the variance $1/(2q_N)$ in the same limit, and it follows that $\sigma^2_{\phi,\mathrm{M1}} = \sigma^2_{\phi,\mathrm{M2}}$. We conclude that the two methods give different phase variance only for moderate signal to noise ratios, which means a low amplitude radar or sonar pulse, or on the rising and falling edges of the pulse in general.
Also the amplitude variance for both methods tend to the same value for $A_0 \gg \sigma$: In this limit, the amplitude distribution tends to a Gaussian near A_0. The integrand of the amplitude distribution in (13) is then a Gaussian with expectation $E[A;a] = A_0 + (\sigma/a)^2/(2A_0)$ and it can be shown that

$$\sigma^2_{A,\mathrm{M2}} = \frac{\sigma^2}{N}\int_0^\infty \frac{p(a)}{a^2}da. \tag{18}$$

Similarly, the amplitude distribution (11) for Method I has variance σ_N^2 in the same limit. It then follows that $\sigma^2_{A,\mathrm{M1}} \to \sigma^2_{A,\mathrm{M2}}$ for $A_0/\sigma \to \infty$. We conclude that the two methods give different amplitude variances only for moderate signal to noise ratios.
Comparison of methods: Which of the two methods gives the smallest amplitude and phase variance for moderate signal to noise ratios? The answer is non-trivial, since the computation of amplitude and phase is nonlinear. We need to express how the variances depend on the noise, the signal amplitude and *N*. With the phase and amplitude pdfs (10) and (11), we obtain for Method I:

$$\begin{aligned}\sigma^2_{A,\mathrm{M1}} &= \sigma^2_{A,\mathrm{M1}}(A_0,[\sigma/A_0]^2/N),\\ \sigma^2_{\phi,\mathrm{M1}} &= \sigma^2_{\phi,\mathrm{M1}}([\sigma/A_0]^2/N).\end{aligned} \tag{19}$$

For Method II, we can assume that the terms in the averaging sum are uncorrelated, and obtain the usual $1/N$-law,

$$\begin{aligned}\sigma^2_{A,\mathrm{M2}} &= \sigma^2_{A,\mathrm{M2}}(A_0,(\sigma/A_0)^2)/N,\\ \sigma^2_{\phi,\mathrm{M2}} &= \sigma^2_{\phi,\mathrm{M2}}([\sigma/A_0]^2)/N.\end{aligned} \tag{20}$$

The difference between the methods then arises when we scale the *argument* with $1/N$ in contrast to scaling the *variances* with $1/N$. Either way, the amplitude as well as the phase variance decrease with increasing *N*. Comparing the performance of Methods I and II then comes down to establishing which is the smallest of the functions $f_{\mathrm{M1}}(x/N)$ and $f_{\mathrm{M2}}(x)/N$, where $x = (\sigma/A_0)^2$ and the *f*'s express the phase variance or the amplitude variance. Thus, given the value of *N* we should be able to establish for which signal-to-noise ratios Method I is favourable over Method II and vice versa. One can expect that the differences between the variances of Method I and II vary as function of *N* and σ in general. Below, we quantify these differences for given signal strength and noise level by integrating over the pdf's, when an explicit form is not available.

Amplitude: There are two independent parameters in the amplitude pdf; A_0 and σ. We plot the output NSR as function of *N* (Fig. 5A) for σ = 1 and σ = 0.1. In the former case Method II performs best, in the latter the methods are virtually indistinguishable. The plot of output NSR as function of SNR (Fig. 5B) gives the same conclusion; Method II is the best choice, this time by a clear margin for both values of σ. For high SNR, however, the methods are indistinguishable; the variances for the two methods coalesce near SNR=10.

Phase: The variance of the phase depends only on the input SNR via $q = A_0^2 / (2\sigma^2)$. Fig. 5C displays the standard deviation of the phase as function of *N* (measured in degrees) for σ = 1 and σ = 0.1. For the former and for low values of *N*, Method II is the best, otherwise the methods have close to indistinguishable variances. Considering variation in input SNR (Fig. 5D), for low input SNR, Method I is the best, for moderate SNR, Method II is the best. The difference between the two methods converges rapidly to zero with increasing input SNR. In summary, Method II appears to achieve the smallest variances, the only exception being at low-to-moderate input SNR for the phase variance.

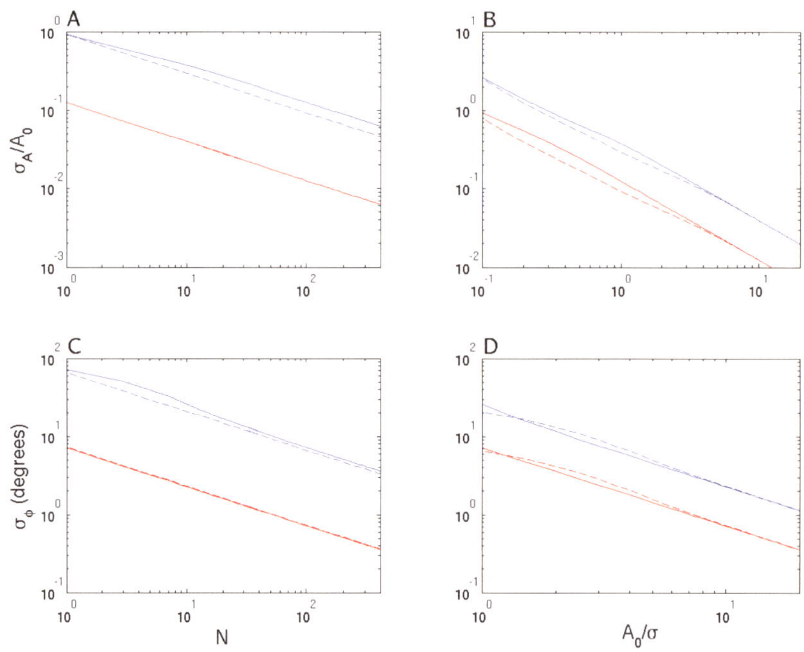

Fig. 5. Comparison of Method I and II. Output amplitude NSR σ_A/A_0 as function of (A) *N* and of (B) input SNR A_0/σ using Method I (solid line) and Method II (dashed) for σ = 1 (blue) and σ = 0.1 (red). Plots (C) and (D) display the same for the phase standard deviation (in degrees).

3.3 Optimum thresholding

In the preceding subsection we saw that when one considers moderate to good signal to noise ratios it is possible to obtain analytic, approximate expressions for the phase and amplitude variances (given in eqs. (17) and (18), respectively). Both these variances are proportional to the quantity

$$R = \frac{1}{N}\int_0^\infty \frac{p(a)}{a^2}\,da. \tag{21}$$

One can see the possibility of minimising R by rejecting pulses below a certain threshold a_0. A new truncated distribution $p(a;a_0)$ governs the remaining data and we obtain

$$R(a_0) = \frac{1}{n(a_0)}\int_{a_0}^\infty \frac{p(a)}{a^2}\,da, \tag{22}$$

where $n(a_0) = N\int_{a_0}^\infty p(a)\,da$ is the reduced ensemble size. A minimum point exists provided that $\int_{a_0}^\infty p(a)/a^2\,da$ decreases faster than $n(a_0)$ for small a_0, and slower than $n(a_0)$ for larger a_0. The existence and location for the optimal threshold depends entirely on the properties of $p(a)$. We find that a necessary condition for the existence of a minimum is (Øyehaug & Skartlien, 2006),

$$\int_{a_0}^\infty \left(\frac{a_{\min}}{a}\right)^2 p(a)\,da < \frac{1}{2}, \tag{23}$$

where $p(a)$ is the original distribution. Optimum thresholding is further investigated below in Sect. 4, where the signals are also assumed to be quantized.

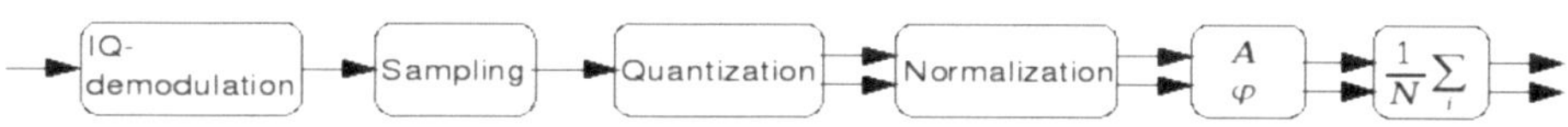

Fig. 6. The signal processing chain under consideration in Sect. 4. The input signal is demodulated into I and Q, sampled and quantized before it is normalized. The two components are then used to estimate amplitude and phase and finally the ensemble averages are computed using amplitude and phase from each individual pulse (Method II of Sect. 3).

4. Ensemble averaged randomly scaled amplitude and phase in quantized signals

This section considers the combination of the signal models that we looked at in Sects. 2 and 3, i.e. the signal under study has undergone IQ-demodulation, sampling, quantization, normalization, modulation into amplitude and phase and, finally, ensemble averaging (Fig. 6). The complex signal to be considered before modulation is

$$U_k(t) = \frac{1}{a_k}Q[Z_k(t)] = \frac{1}{a_k}Q[a_k Z_0(t) + n_k(t)]. \tag{24}$$

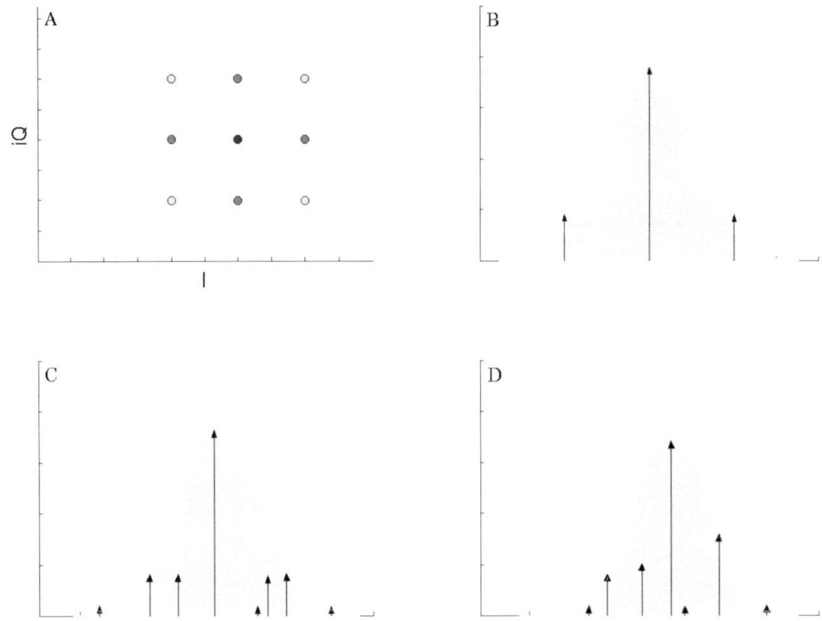

Fig. 7. (A) Typical distribution of the quantization of IQ-demodulated noisy signals in the complex (I,Q)-plane. The degree of shading of the markers indicates the frequency of each quantization level. (B) Histograms of the distribution (shaded) of noisy signals for I (the histogram for Q is similar), amplitude and phase (C, D). The associated discrete distributions of the quantized signal are depicted as arrows (not correctly scaled compared to the continuous distributions).

4.1 Statistical properties of randomly scaled and quantized complex signals

Due to quantization, the complex numbers in (24) follow a discrete pdf depicted as arrows in Fig. 7B where the normalized histogram of the Gaussian noise (corresponding to the pdf) is drawn for comparison. Deriving a general pdf for amplitude and phase that accounts for quantization as well as stochastic noise poses an extremely difficult mathematical problem that we do not attempt to solve. Instead we employ a mixture of graphical arguments and simulations to shed light on the statistics of these quantities.

In Fig. 7A there are nine possible complex values for the given noise and quantization levels, of which two have identical amplitude and two pairs of points have identical angle, which explains that there are eight different attainable values of the amplitude (Fig. 7C) and seven different for angle (Fig. 7D). We note that despite the low SNR in this example the underlying amplitude and phase pdfs (represented by shaded histograms in Figs. 7C, D) resemble very much Gaussian distributions.

We observe that the amplitude and phase pdfs when quantization is included, differ both qualitatively and quantitatively from the pdfs obtained when quantization is neglected. Despite this, the differences between the corresponding *variances* need not be as large as one might expect when comparing the pdfs. For Gaussian noise and σ/Δ sufficiently large, eqs. (2) and (3) give the following approximate expression for the variance of a signal s_j with a given scaling a (not random);

$$\begin{aligned}\mathrm{Var}[e_{i,j};s_j,a] \simeq \frac{1}{a^2}\Bigg(\frac{\Delta^2}{12}+\sigma^2-\left(\frac{\Delta}{\pi}\right)^2 \exp[-2\pi^2a^2(\sigma/\Delta)^2]\times \\ \left[\left(1+4\pi a^2(\sigma/\Delta)^2\right)\cos(2\pi s_j a/\Delta)+\exp[-2\pi^2a^2(\sigma/\Delta)^2]\sin^2(2\pi s_j a/\Delta)\right]\Bigg).\end{aligned} \tag{25}$$

With increasing σ/Δ the variance in (25) goes rapidly to $\left(\Delta^2/12+\sigma^2\right)/a^2$ such that the error variance becomes signal-independent. In Sect. 3.2 we argued that, in the large SNR limit and without taking into account quantization (i.e. Δ=0), this estimate also holds for the amplitude variance σ_A^2 and the phase variance multiplied by the squared signal amplitude $A_0^2\sigma_\phi^2$. Thus we expect, at least for small Δ, that eqs. (17) and (18) remain valid with σ^2 replaced by $\Delta^2/12+\sigma^2$. Among other things we examine this validity numerically in Sect. 4.2 below.

Consider random scaling with a uniform scaling pdf; $p(a)=1/(1-a_{\min})$ on $(a_{\min},1)$. The corresponding truncated pdf is $p(a;a_0)=1/(1-a_0)$ on $(a_0,1)$. Straightforward calculus applied to eqs. (15) and (16) establishes that, in the large SNR and σ/Δ limits,

$$\begin{aligned}\sigma_{\langle\phi\rangle}^2 &= \left(\frac{\sigma^2+\Delta^2/12}{A_0^2N}\right)\frac{1-a_{\min}}{a_0(1-a_0)},\\ \sigma_{\langle A\rangle}^2 &= \left(\frac{\sigma^2+\Delta^2/12}{N}\right)\frac{1-a_{\min}}{a_0(1-a_0)}.\end{aligned} \tag{26}$$

These estimates are subject to numerical investigation below in Sect. 4.2.

4.2 Numerical results

Numerical experiments were performed to demonstrate the validity of the asymptotical estimates (26) and to examine the effect of quantization on thresholding. We estimated the variances numerically with a uniform $p(a)$, and compared these to the asymptotic values obtained analytically. The numerical results estimate the exact variances for all SNR, whereas the analytical results are valid only asymptotically for large SNR and σ/Δ.

The numerical variance estimates are based on a series of realizations of (24). We conducted the experiments as follows. Let $a_k, k=1,\ldots,N$ be a random sequence where the elements are uniformly distributed on $(a_{\min},1)$, where $a_{\min}$ = 0.01. For a randomly selected Z_0 (see below), the complex numbers $Z_{k\ell}=a_{k\ell}Z_0+n_{k\ell}$ (where $n_{k\ell}$ is complex and Gaussian and the real and imaginary parts are independent) are computed for $k=1,\ldots,N, \ell=1,\ldots,M$, where k is the pulse index, while ℓ is a realization index. Different realizations are necessary for estimating the variances numerically. For each ℓ, we estimated the mean values $\langle A/\overline{A}\rangle$ and $\langle\phi\rangle$ by summing over k. The variances of these averages were estimated by summing over ℓ.

For convenience, the sequence in a_k is sorted according to increasing scaling to easily handle the thresholding. Each k then corresponds to a scaling threshold a_k. Only data with scaling $a \geq a_k$ were retained and used for signal estimation; for each value of k the mean values $\langle A/\overline{A}\rangle_{k\ell}$ and $\langle\phi\rangle_{k\ell}$ were computed including a_k for indices $k, k+1, \ldots, N$. Subsequently, amplitude and phase variance estimates were obtained by averaging over all realizations $\ell = 1, \ldots, M$;

$$\hat{\sigma}^2_{\langle\phi\rangle} = \frac{1}{M}\sum_{\ell=1}^{M}\left(\langle\phi\rangle_{k\ell} - \arg(Z_0)\right)^2,$$
$$\hat{\sigma}^2_{\langle A\rangle} = \frac{1}{M}\sum_{\ell=1}^{M}\left(\left\langle A/\overline{A}\right\rangle_{k\ell} - \mathrm{mod}(Z_0)\right)^2. \tag{27}$$

The simulations were performed for three values of the quantization separation Δ. To avoid signal-dependent estimates, which is generally the case (see eq. 25), for each Δ we repeated the protocol described above 100 times with Z_0 selected at random on the circle in the complex plane with modulus 4 and thereafter calculated the mean variance estimate. Comparing the asymptotical expressions in (26) with the numerical results in Fig. 8, we observe that there is a reasonable agreement between numerical and theoretical estimates, with two notable exceptions: (i) for small values of a_0 and for large noise the numerical variances deviate markedly from the theoretical estimates and (ii) for large Δ and small noise (in particular for the phase variance), the numerical variances are clearly larger than the theoretical estimate.

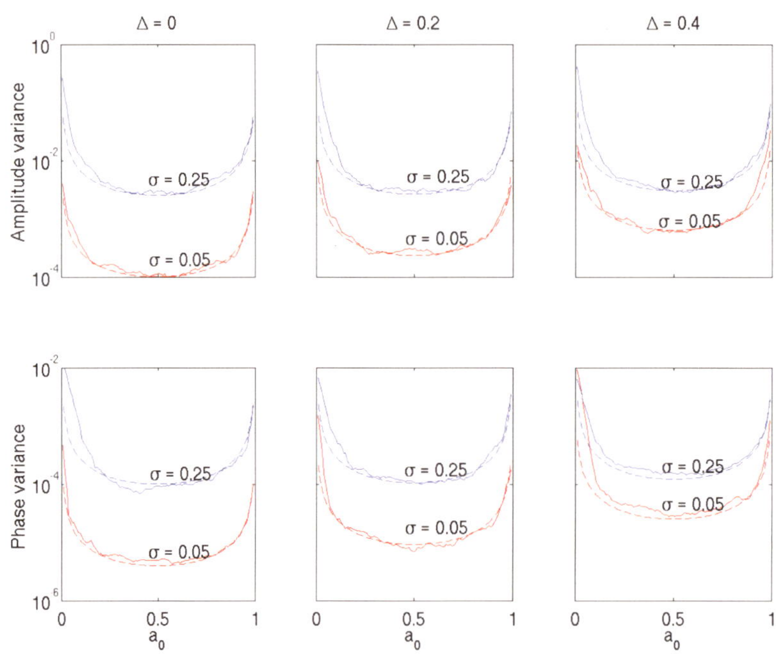

Fig. 8. Amplitude and phase variances as function of scaling threshold a_0 for the specified values of σ and Δ obtained by performing the computations described in the main text (solid lines) and corresponding asymptotical estimates (eqs. 26, dashed).

5. Discussion

As the test case in Sect. 4.2 shows, it was justified to apply the asymptotic estimates in Sect. 3 for both phase and amplitude averaging for sufficient levels of SNR ranging from roughly 10. Although this SNR is reasonable for many practical purposes, the instantaneous signal to noise ratio varies throughout the radar/sonar pulse with the instantaneous amplitude. Parts of the rising and falling flanks of the pulses will then correspond to short time intervals in which the theory should not be applied.

We adopted a smooth scaling distribution $p(a)$ in our analysis. In a practical situation, only the scaling histogram is available. The normalised histogram approximates $p(a;a_0)$ and the optimal scaling threshold can be obtained by the discrete analog to eq. (23). On the other hand, the optimum scaling threshold can of course be computed by brute force, i.e. by straightforward estimation of the variance based on available pulse signals and rejecting those pulses that contribute to a degraded ensemble average. One interesting possible future investigation is to evaluate the brute force and theoretically driven approaches in practical situations and compare them in terms of efficiency and reliability.

In Sect. 2.3 we defined and obtained a mathematical expression for the mean square error (MSE) of the ensemble average of a quantized, noisy signal. The MSE is a signal-independent measure of the average signal variance. When the signals over which we average are randomly scaled, there is no obvious way of defining the MSE. One way of circumventing this problem is to, as we did in Sect. 4.2 above, calculate variances of a large number of randomly selected points and then taking the average in order to achieve variances that are roughly signal-independent (Fig. 8). In the future, more sophisticated definitions of average variance that account for random scaling as well as quantization and stochastic noise should be developed.

Direct averaging with subsequent amplitude and phase calculation (Method I) provides the same results as Method II in the large SNR limit. Method I is potentially a more efficient averaging method, since amplitude and phase need not be computed for each pulse. However, signal degradation is more sensitive to alignment errors of the pulses before averaging; the sensitivity to precise alignment increases for increased carrier frequency due to larger phase errors for the same time lag error. This problem is much reduced when one performs averaging on amplitude and phase modulations directly (Method II).

6. Conclusion

We have reviewed the statistics of (i) averaged quantized pulses and (ii) averaged amplitude and phase modulated pulses that are randomly scaled, but not quantized. We showed that ensemble averaging should be performed on the amplitude and phase modulations rather than on I and Q. In the final point (iii), we analyzed the asymptotic statistics for ensemble averaged amplitude and phase modulated pulses that are both randomly scaled and quantized after IQ-demodulation. We studied the effect of thresholding (rejecting pulses below a certain amplitude) and found that theoretical estimates of the variance as function of threshold, closely agree with numerical estimates.

We believe that our analysis is applicable to radar and sonar systems that rely on accurate estimation of pulse characteristics. We have covered three key aspects of the problem, with the goal of reducing statistical errors in amplitude and phase modulations. Extensions or modifications of our work may be necessary to account for the signal chain in a specific digital system.

7. References

Ai, C. & Guoxiang, A. (1991). Removing the quantization error by repeated observation. *IEEE Trans. Signal Processing*, vol. 39, no. 10, (oct 1991) 2317-2320, ISSN: 1053-587X.

Belchamber, R.M. & Horlick, G. (1981). Use of added random noise to improve bit-resolution in digital signal averaging. *Talanta*, vol. 28, no 7, (1981) 547-549, ISSN: 0039-9140.

Carbone, P. & Petri, D. (1994). Effect of additive dither on the resolution of ideal quantizers. *IEEE Trans. Instrum. Meas.*, vol. 43, no 3, (jun 1994) 389-396, ISSN: 0018-5456.

Davenport, W.B. Jr. & Root, W.L. (1958), *An Introduction to the Theory of Random Signals and Noise*, McGraw-Hill Book Company, Inc., New York.

Jane, R. H., Rix, P., Caminal, P. & Laguna, P. (1991). Alignment methods for averaging of high-resolution cardiac signals - a comparative study of performance. *IEEE Trans. Biomed. Eng*, vol. 38, no. 6 (jun 1991) 571-579, ISSN: 0018-9294.

Koeck, P.J.B. (2001). Quantization errors in averaged digitized data. *Signal Processing*, vol. 81, no. 2, (feb 2001) 345-356, ISSN: 0165-1684.

Laguna, P. & Sornmo, L. (2000). Sampling rate and the estimation of ensemble variability for repetitive signals. *Med. Biol. Eng. Comp*, vol. 38, no. 5, (sep 2000) 540-546, ISSN: 0140-0118.

Meste, O. & Rix, H. (1996). Jitter statistics estimation in alignment processes. *Signal Processing*, vol. 51, no. 1, (may 1996) 41-53, ISSN: 0165-1684.

Øyehaug, L. & Skartlien, R. (2006). Reducing the noise variance in ensemble-averaged randomly scaled sonar or radar signals. *IEE Proc. Radar Sonar Nav.*, vol. 153, no. 5, (oct 2006) 438-444, ISSN: 1350-2395.

Papoulis, A. (1965). *Probability, Random Variables, and Stochastic Processes*, McGraw-Hill Book Company, Inc., New York, ISBN: 0-07-048448-1.

Schijvenaars, R.J.A., Kors, J.A. & Vanbemmel, J.H. (1994). Reconstruction of repetitive signals. *Meth. Inf. Med.*, vol. 33, no. 1, (mar 1994) 41-45, ISSN: 0026-1270.

Skartlien, R. & Øyehaug, L. (2005). Quantization error and resolution in ensemble averaged data with noise. *IEEE Trans. Instrum. Meas.*, vol. 53, no. 3, (jun 2005) 1303-1312, ISSN: 0018-5456.

Viciani, S., D'Amato, F., Mazzinghi, P., Castagnoli, F., Toci, G. & Werle, P. (2008). A cryogenically operated laser diode spectrometer for airborne measurement of stratospheric trace gases. *Appl. Phys. B*, vol. 90, no. 3-4, (mar 2008), 581-592, ISSN: 0946-2171.

Sonar Detection and Analysis

5

Independent Component Analysis for Passive Sonar Signal Processing

Natanael Nunes de Moura, Eduardo Simas Filho and José Manoel de Seixas
Federal University of Rio de Janeiro – Signal Processing Laboratory/COPPE – Poli
Brazil

1. Introduction

Systems employing the sound in underwater environments are known as sonar systems. SONAR (Sound Navigation and Ranging) systems have been used since the Second World War (Waite, 2003), (Nielsen, 1991). These systems have the purpose of examining the underwater acoustic waves received from different directions by the sensors and determine whether an important target is within the reach of the system in order to classify it. This gives extremely important information for pratical naval operations in different conditions. Fig. 1 shows a possible scenario for a sonar operation, in which two targets: the ship that is a surface contact and another submarine. In this case, the submarine's hydrophones are receiving the signals from the two targets and the purpose is to identify both targets.

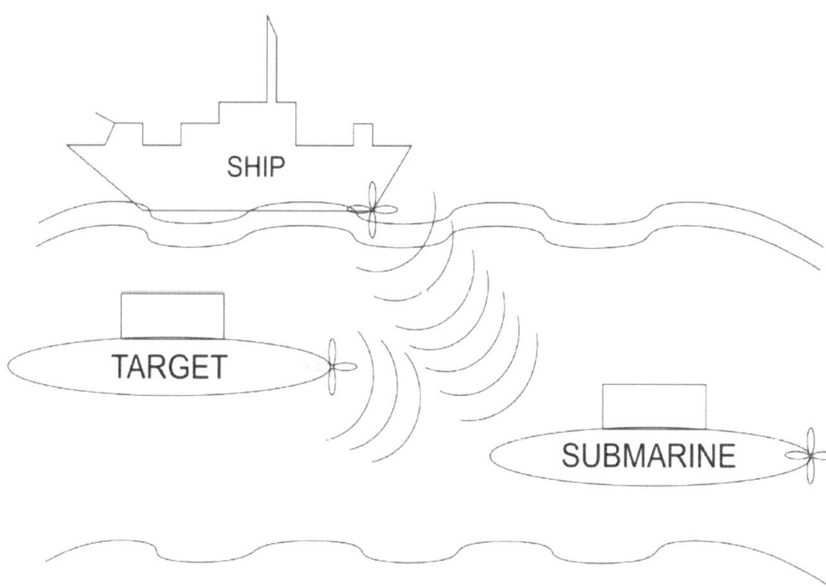

Fig. 1. Possible scenario for sonar operation

Depending on the sonar type, it may be, passive or active. The active sonar system transmits an acoustic wave that may be reflected by the target and signal detection, parameter estimation and localization can be obtained through the corresponding echoes (Nielsen,

1991), (Waite, 2003). A passive sonar system performs detection and estimation using the noise irradiated by the target itself (Nielsen, 1991) (Clay & Medwin, 1998), (Jeffsers et al., 2000). The major difficulty in passive sonar systems is to detect the target in huge background noise environments. As much in active and passive mode, the sonar operator, (SO) listens to the received signal from one given direction, selected during the beamforming, envisaging target identification. This chapter focus on passive sonar systems and how the received noise is analysed that may arise. In particular, the signal interference in neighbour directions is discussed. Envisaging interference removal, Independent Component Analysis (ICA) (Hyvärinen, 2000) is introduce and recent results obtained from experimental data are described. The chapter is organised as it follows. In next Section, the analysis performed by passive sonar systems is detailed described. Section 3 introduces ICA principles and algorithms. Section 4 shows how ICA may be applied for interference removal. Finally, a chapter summary and perspectives of passive sonar signal processing are addressed in Section 5.

2. Passive sonar analysis

A passive sonar system is typically made from a number of building blocks (see Fig. 2); described in terms of its aim and specific signal processing techniques that have been applied for signal analysis.

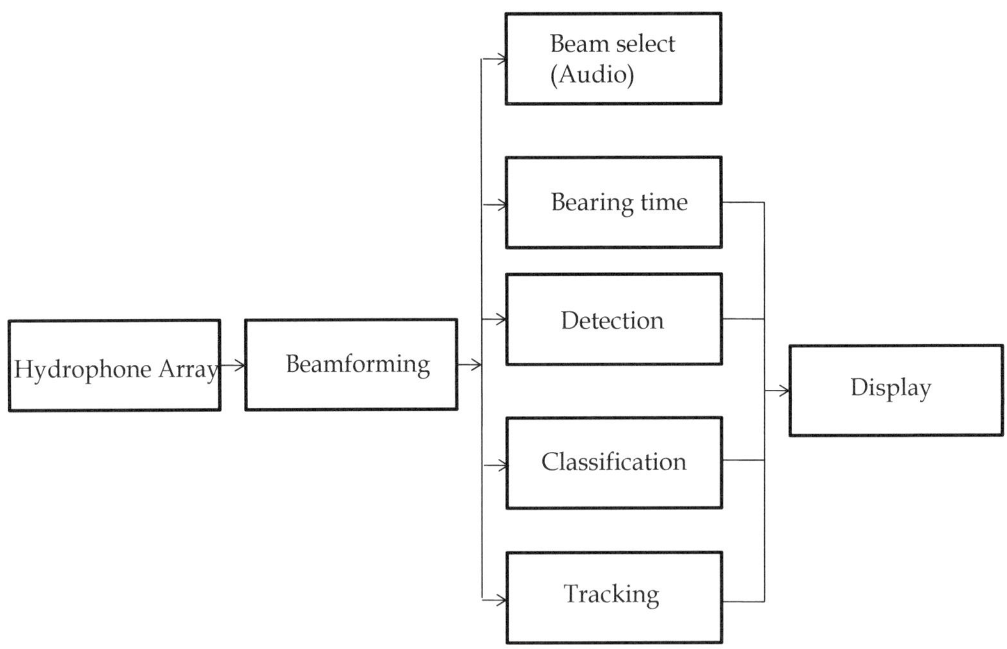

Fig. 2. Blocks diagram for passive sonar system

2.1 Sensors array

The passive sonar systems rely very much on the ability of their sensors in capturing the noise signals arriving in different directions. Typically, sensors (hydrophones) are arranged in arrays for fully coverage of detection directions The hydrophone array may be linear, planar, circular or cylindrical. For the experimental results in Section 4, signals, were acquired through a cylindrical hydrophone array (CHA) while realizing an omnidirectional surveillance. This type of array comprises a number of sensor elements, which are distributed along staves. Therefore, the design performance depends on the number of staves, the number of hydrophones and the number of vertical elements in a given stave. For instance, the CHA from which the experimental tests were derived has 96 staves.

2.2 Beamforming

The beamforming operation aims at looking at a given direction of arrival (DOA) with the purpose of observing the target energy of a given direction through a bearing time display (Krim & Viberg, 1996). The signals are acquired employing the delay and sum (ds) technique to realize the DOA, allowing omnidirectional surveillance (Knight et al., 1981). In case of the experimental results to be described in Section 4, the directional beam is implemented using 32 adjacent sensors as it is shown in Fig. 3. A total of 32 adjacent staves were used to compute the direction of interest which gives an angular resolution of 3.75º.

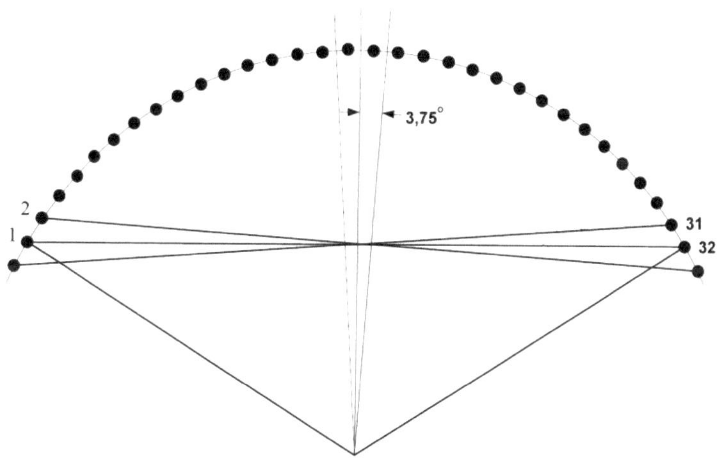

Fig. 3. Arrange of hydrophones for beamforming on a determined direction

Fig. 4 shows a bearing time display. In this figure, the horizontal axis represents the bearing position (full coverage, -180 to 180 degrees) and the vertical axis represents time, considering one second long acquisition window. This corresponds to waterfall display. The energy

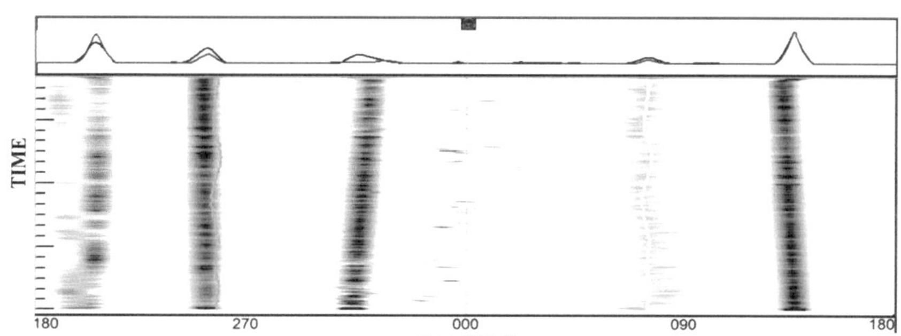

Fig. 4. A bearing time display

measurement for each bearing at each time window has a gray scale representation. The sonar operator relies very much on the bearing time display, the sonar operator relies very much on the in the time display for possible target observation. An audio output permits the operator to listen to the target noise from a specific direction of interest.

2.3 Signal processing core

After beamforming, passive sonar signal processing comprises detection, classification and, in some situations, target tracking. For detection, two main analysis are performed; LOFAR (LOw Frequency Analysis and Recording) and DEMON (Demodulation of Envelope Modulation On Noise). The LOFAR analysis is also used for target classification.

2.3.1 LOFAR analysis

The LOFAR is a broadband spectral analysis (Nielsen, 1991) that covers the expected frequency range of the target noise as, for instance, machinery noise. The basic LOFAR block diagram is shown in Fig. 5.

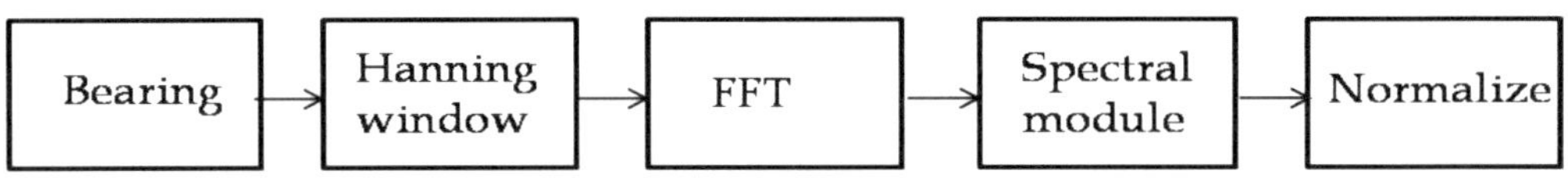

Fig. 5. Block diagram of the LOFAR analysis

As it can be depicted from Fig. 5, at a given direction of interest (bearing), the incoming signal is firstly multiplied by a Hanning window (Diniz et al., 2002), In the sequence, short-time Fast Fourier Transform (FFT) (Brigham, 1988) is applied to obtain signal representation in the frequency-domain (Spectral module). The signal normalization follows typically employing the TPSW (Two-Pass Split Window) algorithm (Nielsen, 1991) for estimating the background noise (see Fig. 6).

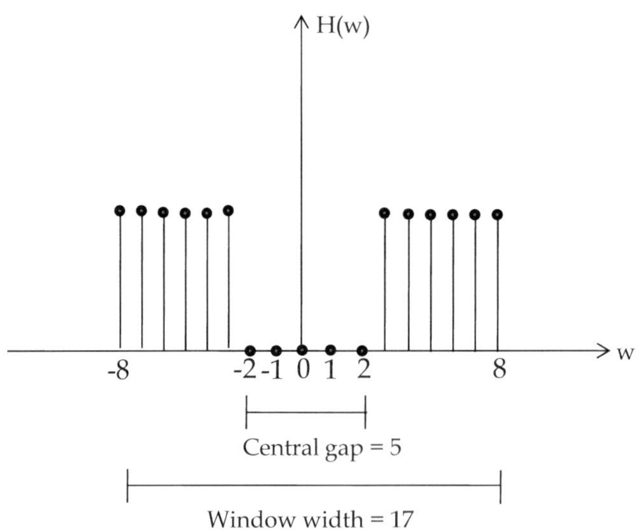

Fig. 6. TPSW window.

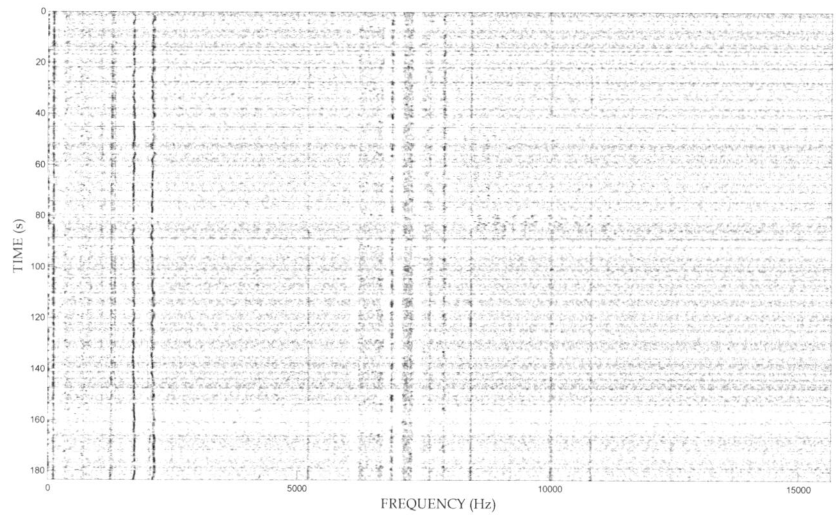

Fig. 7. Typical LOFAR display.

This window will slide along the signal and performing a local average to achieve the removal of background noise and making a sign of normalization. This TPSW normalization aims at estimation a mean spectrum by computing a local mean for each sample. This makes it possible to remove the bias and perform peak equalization, so that the amplitudes in all spectrums present similar values.
Fig. 7 shows a typical display from LOFAR analysis. The horizontal axis corresponds to frequency, in this case covering range of 0 to 15.625 Hz, and the vertical axis represents time. In this case, 200 acquisition windows (one second long each were accumulated). As can be seen in Fig. 7, some rays of often persist over time, thus characterizing the type of target being identified.

2.3.2 DEMON analysis

DEMON is a narrowband analysis that operates over the cavitation noise of the target propeller with the purpose of identifying the number of shafts, shaft rotation frequency and the blade rate (Nielsen, 1999), (Trees, 2001). As these parameters provide a detailed knowledge about the target propellers and normally the propeller noise is characteristic of each contact, this analysis shows good detection capabilities. Fig. 8 shows the block diagram of classical DEMON analysis.

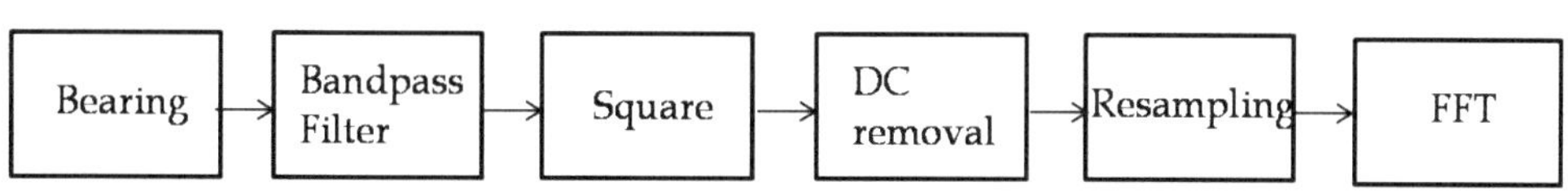

Fig. 8. Block diagram of the DEMON analysis

Given a direction (bearing) of interest, noise signal is bandpass filtered to limit the cavitation frequency range. The cavitation frequency goes from hundreds until thousands of Hz. Therefore, it is important to select the cavitation band and obtain the maximum information for ship identification. In sequence, the signal is squared as in traditional demodulation (Yang et al., 2007) (Trees, 2001) and the TPSW algorithm is used to reduce the background noise (Nielsen, 1991). Using TPSW, it is possible to emphasize target signal peaks. In most cases, the signal sampling rate is relatively high, so that the band of interest is sampled with coarse resolution with respect to observation needs. Thus, it is necessary to resample the signal for better observation in the range. Finally, a short-time Fast Fourier Transform algorithm is applied to observe the peaks in frequency domain. Fig. 9 shows a typical DEMON plot. The horizontal axis represents the rotation scale (in RPM) while the vertical axis correspond to signal amplitude (in dB). This allows target identification, as shaft rotation and the number of blades may be obtained. The largest amplitude reveals the speed of shaft rotation, while the subsequent harmonics indicate the number of blades. In this example, the shaft rotation is about 148 RPM and next hamonics are, 293.6, 441.8, 587.1 and 735.3 RPM, from which the number of blades can be obtained.

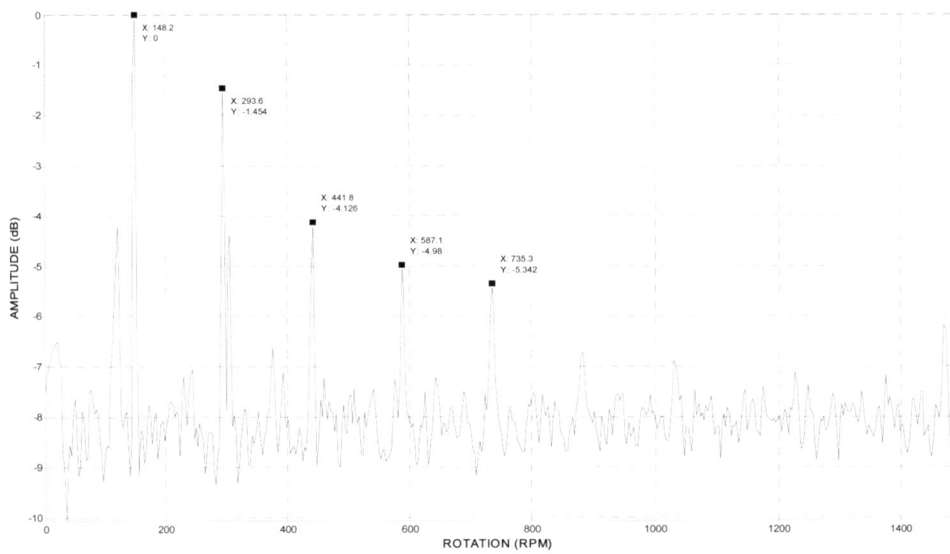

Fig. 9. typical DEMON display.

2.3.3 Classification

Another important task for passive sonar systems is target classification. Usually classification is based on extracting relevant features that characterize target classes and using such features to decide whether a detected target belongs to a given class. As already mentioned, features are typically extracted in frequency domain using the LOFAR analysis. But the stress, many directions of interest and high number of classes, automatic classification often uses computational intelligence algorithms to obtain the target class. Neural networks (Haykin, 2001) have successfully been used for passive sonar signal classification. (Moura, 2007), (Torres., 2004);(Seixas, 2001) and (Soares Filho, 2001). Other signal processing techniques have been applied to realize the classification task.(Peyvandi, 1998) used a hidden Markov model with Hausdorff similarity measurement to detect and classify targets. Another way to perform the detection and classification of targets is to use the Prony's method (Marple, 1991), which provides an alternative time-frequency mapping (signals are modelled through a sum of damped sinusoidal components) suitable to acoustic signals.

2.3.4 Tracking

Eventually, tracking a target over time may be important. Usually this is perfomed after target detection at a specific direction. In some situations, the sonar operator performs tracking manually, but modern sonars have an automatic system to support this task. Although Kalman filters (Lee, 2004) have often been used to implement passive tracking (Rao, 2006), other techniques, (Mellema, 2006) have also been obtaining good results in target tracking application.

2.3.5 Interference

As it may be depicted from Fig. 9, interference from neighbour bins, as it is the case for bearings 190° and 205°, and the self-noise produced by the submarine in which the sonar system is installed may mask the original target features. Thus, when such is the case, a preprocessing scheme may be developed aiming at reducing signal interference, facilitating target identification.This procedure is addopted in Section 4 using the ICA (Hyvärinen, 2001).

3. Independent component analysis

The Independent Component Analysis (ICA) considers that a set of N observed signals $\mathbf{x}(t) = [x_1(t), ..., x_N(t)]^T$ is originally generated from a linear combination of signal sources $\mathbf{s}(t) = [s_1(t), ..., s_N(t)]^T$:

$$\mathbf{x}(t) = \mathbf{A}\mathbf{s}(t) \quad (1)$$

where, $\mathbf{A}$ is the NxN mixing matrix (Hyvärinen et al., 2001). Formulated this way, ICA is also referred to as Blind Source Separation (BSS) (Cardoso, 1998) and its purpose is to estimate the original sources $\mathbf{s}(t)$ using only observed data, $\mathbf{x}(t)$. A solution can be obtained if we find the inverse of the mixing matrix $\mathbf{B} = \mathbf{A}^{-1}$ and apply this inverse transformation on the observed signals to obtain the original sources.

$$\mathbf{s}(t) = \mathbf{B}\mathbf{x}(t) \quad (2)$$

A general principle for estimating the matrix $\mathbf{B}$ can be found by considering that the original source signals are statistically independent (or as independent as possible). High-order statistics (HOS) information is required during the search for independent components. There are many mathematical methods for calculating the coefficients of matrix $\mathbf{B}$. The nonlinear decorrelation and the maximally nongaussianity are among the most applied ones (Hyvärinen & Oja, 2000). There are some indeterminacies in the ICA model, the order of extraction of the independent components can change and scalar multipliers (positive or negative) may be modifying the estimated components. Fortunately these limitations are insignificant in most applications (Hyvärinen et al., 2001).

3.1 Statistical independence

Considering two random variables $\mathbf{x}$ and $\mathbf{y}$, they are statistically independent if and only if (Papoulis, 1991):

$$p_{x,y}(\mathbf{x},\mathbf{y}) = p_x(\mathbf{x})p_y(\mathbf{y}) \tag{3}$$

where $p_{x,y}$ (**x**, **y**), p_x (**x**) and p_y (**y**) are, respectively, the joint and marginal probability density functions (pdf) of **x** and **y**. Equivalent condition is obtained if for all absolutely integrable functions g(**x**) and h(**y**) the expression on Eq. 4 holds:

$$E\{g(\mathbf{x})h(\mathbf{y})\} = E\{g(\mathbf{x})\}E\{h(\mathbf{y})\} \tag{4}$$

where E{.} is the expectation operator (Hyvärinen et al., 2001).
In typical blind signal processing problems, there is very little information on the source signals and so direct estimation of the pdfs is a very difficult task. Eq. 4 gives an alternative independence measure and is the origin of a class of ICA algorithms that searches for nonlinear decorrelation.
Independent variables are uncorrelated, although, the reciprocal is not always true. Linear correlation is verified by second order statistics, while independence needs higher order information. In the nonlinear decorrelation methods, nonlinear functions introduce high-order statistics, making it possible the search for independent components.
As from Eq. 4, two random variables are statistical independent if they are nonlinearly uncorrelated. As it is not possible to check all integrable functions g(.) and h(.), estimates of the independent components are obtained while guaranteeing nonlinear decorrelation between a finite set of nonlinear functions (Hyvärinen et al., 2001).
For example, a well known linear ICA algorithm, proposed by Cichocki and Unbehauen in (Hyvärinen & Oja, 2000), searches for independent components while providing decorrelation between a hyperbolic tangent and a polynomial function, both applied to the input signals (observations).

3.1.1 Non-gaussianity and independence

The ICA/BSS model described in Eq. 1 can be re-written as:

$$x_i = \sum_{j=1}^{N} a_{ij} s_j \qquad i = 1,\ldots,N \tag{5}$$

Considering the central limit theorem (Spiegel et al., 2000): "The sum of two (independent) random variables is always closest to a Gaussian distribution than the original variables distributions". As described in Eq. 5, the observed signals x_i are formed by an averaged summation of the sources s_i. Thus, x_i is closer to Gaussian-distributed variables than s_i. In other words, the independent components can be obtained through maximization of non-gaussianity (Hyvärinen et al., 2001).
The gaussianity (and consequently the statistical dependence) of a random variable can be measured through higher order cumulants. Considering a random vector **x**, the moment α_k and central moment μ_k of order k are defined by (Spiegel et al., 2000):

$$\alpha_k = E\{\mathbf{x}^k\} = \int_{-\infty}^{\infty} \mathbf{x}^k p_x(\mathbf{x})dx \tag{6}$$

$$\mu_k = E\{(\mathbf{x}-\alpha_1)^k\} = \int_{-\infty}^{\infty} (\mathbf{x}-\alpha_1)^k p_x(\mathbf{x})dx \tag{7}$$

where $\alpha_1 = m_x$ is the mean of **x**. If the random variable **x** is zero mean ($m_x=0$), than for all k holds: $\alpha_k = \mu_k$.
The cumulant κ_k of order k is defined as a function of the moments (Spiegel et al., 2000). For a zero mean random variable **x**, the first four cumulants are:

$$\kappa_1 = 0; \qquad \kappa_2 = E\{\mathbf{x}^2\} = \alpha_2; \qquad \kappa_3 = E\{\mathbf{x}^3\} = \alpha_3; \tag{8}$$

$$\kappa_4 = E\{\mathbf{x}^4\} - 3[E\{\mathbf{x}^2\}]^2 = \alpha_4 - 3\alpha_2^{\ 2}$$

The third and fourth order cumulants are called respectively skewness (κ_3) and kurtosis (κ_4) (Kim & White, 2004). Cumulants of order higher than four are rarely applied in practical ICA/BSS problems. Some interesting properties of cumulants are:

$$\begin{aligned} \kappa_k(\mathrm{x}+\mathrm{y}) &= \kappa_k(\mathrm{x}) + \kappa_k(\mathrm{y}) \\ \kappa_k(\mathrm{x}) &= 0, \text{for k} > 2 \text{ if x is Gaussian} \end{aligned} \tag{9}$$

Therefore, cumulants of order higher than two may be applied to estimate data gaussianity. The skewness value, for example, is related to pdf symmetry ($\kappa_3=0$ indicates symmetry). Spanning the interval $[-2, \infty)$, kurtosis is zero for a Gaussian variable. Negative values of kurtosis indicate sub-gaussianity (pdf flatter than Gaussian) and positive values super-gaussianity (pdf sharper than Gaussian) (Spiegel et al., 2000). Examples of Gaussian, sub and super-gaussian distributions are illustrated in Figure 10. Kurtosis can be easily computed from data substituting expectations in Eq. 8 by sample means. One disadvantage is that kurtosis can be seriously influenced by outliers (observations that are numerically distant from the rest of the data), in extreme situations the kurtosis value may be dominated by a small number of points (Kim & White, 2004). Some studies are been conduced with the purpose of obtaining robust estimation of high order cumulants, specially the kurtosis (Welling, 2005).
Alternative gaussianity measures can be obtained from information theory (Cover & Thomas, 1991). These parameters are usually more robust to outliers than cumulant based ones (Hyvärinen et al., 2001).

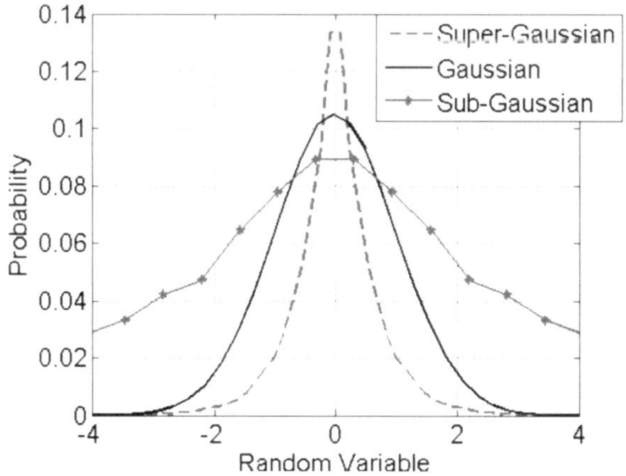

Fig. 10. Examples of Gaussian, sub and super-Gaussian distributions.

For instance, **Negentropy** of a random variable **x** is calculated through (Cover & Thomas, 1991):

$$J(\mathbf{x}) = H(\mathbf{x}_{gauss}) - H(\mathbf{x}) \tag{10}$$

where H (.) is the entropy, and $\mathbf{x}_{gauss}$ is a Gaussian random variable with the same mean and variance of **x**. Entropy is one of the basic concepts of information theory and can be interpreted as the level of information contained in a random variable. Entropy H(**x**) can also be viewed as the minimum code length needed to represent the variable **x**, considering a discrete random variable, entropy is defined as (Shannon, 1948):

$$H(\mathbf{x}) = \sum_i P(\mathbf{x} = a_i) \log P(\mathbf{x} = a_i) \tag{11}$$

where a_i are the possible values assumed by the variable **x**, and P(**x**=a_i) is the probability that **x**=a_i.
An important result is that the Gaussian variables have maximum entropy among variables of same variance (Hyvärinen et al., 2001). So both entropy and negentropy can be used as gaussianity measures. The advantage of J(**x**) is that it is always non-negative and zero when variable **x** is Gaussian. A problem with the computation of J (.) and H (.) in blind signal processing is the pdf estimation (see Eq. 10 and 11). To avoid this, approximations using high order cumulants or non-polynomial functions shall be applied (Hyvärinen et al., 2001; Hyvärinen, 1998).
Another statistical independence measure can be obtained through mutual information. The **Mutual Information** I(x_1, x_2, ..., x_m) between m random variables **x** = [x_1, x_2, ..., x_m] is obtained through (Hyvärinen et al., 2001):

$$I(x_1, \ldots, x_m) = \sum_{i=1}^{m} H(x_i) - H(\mathbf{x}) \tag{12}$$

It is proved elsewhere (Cover & Thomas, 1991) that, more efficient codes are obtained while using the set of variables **x** instead of the individual ones x_i, unless when the variables are independent ((x_1, x_2, ..., x_m)=0). So, minimization of mutual information leads to statistical independence.
The **Kullback-Leiber (KL) divergence**, defined through Eq. 13 (Hyvärinen et al., 2001):

$$C_{KL}(Q, P) = \int Q_x(\mathbf{x}) \log \frac{Q_x(\mathbf{x})}{P_x(\mathbf{x})} dx \tag{13}$$

measures the distance between the two probability densities P_x(**x**) and Q_x(**x**), as it is always nonnegative with minimum value zero when both densities are the same. If one pdf is Gaussian, maximizing C_{KL} is equivalent to maximize non-gaussianity. The KL divergence is proved to be equivalent to mutual information (Hyvärinen et al., 2001).
Using one of these statistical independence measures, several routines have been proposed to find the **B** matrix (Hyvärinen et al., 2001). Here we consider two which are among the most successful ICA algorithms.

3.2 JADE algorithm

The start point for JADE (Joint Approximate Diagonalization of Eigenmatrices) algorithms is the realization that BSS (Blind Source Separation) algorithms generally require an estimation of the distributions of independent sources or have such an assumption built into the algorithm (Cardoso, 1998). It is also noted that, optimising cumulant approximations of data implicitly perform this, leading to present a number of approximations to information theoretic algorithms that operate on second and fourth order cross cumulant.

The cumulant tensor is a linear operator defined by the cumulant of fourth order cum(x_i, x_j, x_k, x_l) (Hyvärinen et al., 2001). This linear operation generates a matrix in form of Eq. 14. In this algorithm, the eigenvalue decomposition is considered as a preprocessing.

$$\mathrm{F}_{ij}(\mathbf{M}) = \sum_{kl} m_{kl} cum(x_i, x_j, x_k, x_l) \tag{14}$$

Where, m_{kl} is an element of the matrix **M** that is transformed and **x** is an nx1 random vector. The second order cumulant is used to ensure that data are white (decorrelated) (Cardoso, 1998). A set of cumulant matrices is estimated from the whitened data, as shown in Eq. 4.Then **F(M)** is made diagonal through **W** for some $\mathbf{M}_i$.

$$\mathbf{Q} = \mathbf{W}\,\mathrm{F}(\mathbf{M}_i)\mathbf{W}^T \tag{15}$$

The minimization of the sum of the squares of the non-diagonal elements of Eq. 15 is equivalent to maximization of the sum of squares of the diagonal elements, because an orthogonal matrix **W** does not change the total sum of squares of a matrix. The maximization of JADE is a method that gives an approximate joint diagonal of $\mathbf{F}(\mathbf{M}_i)$.

$$J_{JADE}(\mathbf{W}) = \sum_i \| diag(\mathbf{W}\,\mathrm{F}(\mathbf{M}_i)\mathbf{W}^T \|^2 \tag{16}$$

3.3 FastICA algorithm

Independent components can be extracted from a mixture implementing the principles of maximization of nongaussianity, described in terms of kurtosis or negentropy (Hyvärinen et al., 2001; Hyvärinen & Oja, 2000; Shaolin & Sejnowski, 1995). Considering a mixture **x**, one defines kurtosis in Eq. 8, where **W** is the weight matrix, and **z** is a component vector. There is a whitening step as a preprocessing, and thus, **z** = **Vx**, where **V** is the whitening matrix and the correlation matrix **z** is equal to identity, E{$\mathbf{z}\mathbf{z}^T$}= **I**. So using kurtosis, it is possible to estimate the independent components from the cost function presented in Eq. 17.

$$\mathbf{x} = \mathbf{W}^{\mathrm{T}}\mathbf{z}$$

$$\frac{\partial\,|\,\mathrm{kurt}(\mathbf{W}^T\mathbf{z})\,|}{\partial\mathbf{W}} = 4\,\mathrm{sign}[\mathrm{kurt}(\mathbf{W}^T\mathbf{z})]\,\mathrm{E}\{z(\mathbf{W}^T\mathbf{z} - 3\mathbf{W}\,\|\,\mathbf{W}\,\|^2)\} \tag{17}$$

To make the algorithm faster, the gradient computation is changed to Eq. 18 a normalization was implemented to avoid a **W** overflow.

$$\Delta\mathbf{W} \propto \mathrm{sign}(\mathrm{kurt}(\,\mathbf{W}^{\mathrm{T}}\mathbf{z}))\;\mathrm{E}(\mathbf{z}(\mathbf{W}^{\mathrm{T}}\mathbf{z})^3$$

$$\mathbf{W} \leftarrow \mathbf{W}/\,\|\,\mathbf{W}\,\| \tag{18}$$

Then, the FASTICA (PEACH, 2000) optimizes Eq. 19.

$$\mathbf{W} \leftarrow E\{(\mathbf{W}^T\mathbf{z})\} - 3\mathbf{W} \tag{19}$$

Another possibility for maximizing non-gaussianity is negentropy (Hyvärinen, 1999). The classic method of approximating negentropy is using higher order cumulants and polynomial density expansions, like G($\mathbf{x}$) = log[cosh($\mathbf{x}$)] or -exp($\mathbf{x}^2/2$). Using a gradient based method, function derivatives (g) can be chosen to be applied in FASTICA.

$$\mathbf{W} \leftarrow E\{\mathbf{z}g(\mathbf{W}^T\mathrm{z})\} - E\{g'(\mathbf{W}^T\mathrm{z})\}\mathbf{W} \tag{20}$$

4. Interference removal

As already mentioned in Section 2, passive sonar signals detected at adjacent bearings may be masked by cross-channel interference. The complexity of the target identification task increases proportionally to the interference level. Considering this, blind source separation methods (Cardoso, 1998) may be useful as a preprocessing step in passive sonar signal analysis as they project the observed signals into directions of maximum independence.

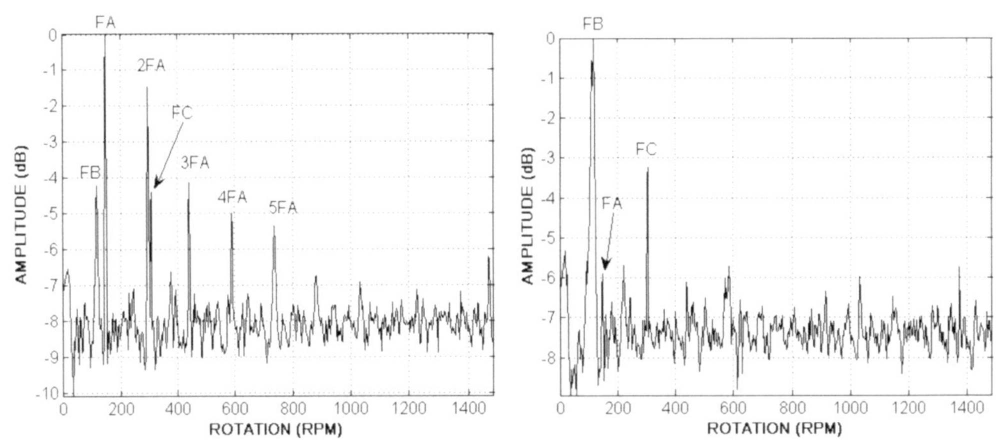

(a) (b)

Fig. 11. DEMON analysis at (a) 190° and (b) 205°.

Consider a particular problem where two targets are present at adjacent directions (190° and 205°). As illustrated in Fig. 11, the frequency components of 190° target (F_A=148 RPM and its multiples) are mixed together with information from the 205° direction (F_B=119 RPM). The same problem exists in the signal measured at bearing 205°. It was also observed that both signals (190° and 205°) are contaminated by F_C=305 RPM that is the main frequency present at direction 076°, see Fig. 12. It is known from the experimental setup that the last bearing (076°) contains information from the noise radiated by the submarine where the hydrophones array is allocated (self-noise). It can also be verified that, signal measured at direction 076° presents interference from target at 205° (F_B).

Independent component based methods are applied in the following sub-sections aiming at reducing signal interference and thus, allow contact identification through DEMON analysis performed over cleaner data. Signal processing may be performed in both time-domain and frequency-domain. The main advantage of frequency-domain methods is that, after DEMON, the signal-to-noise ratio is significantly improved, producing better separation results.

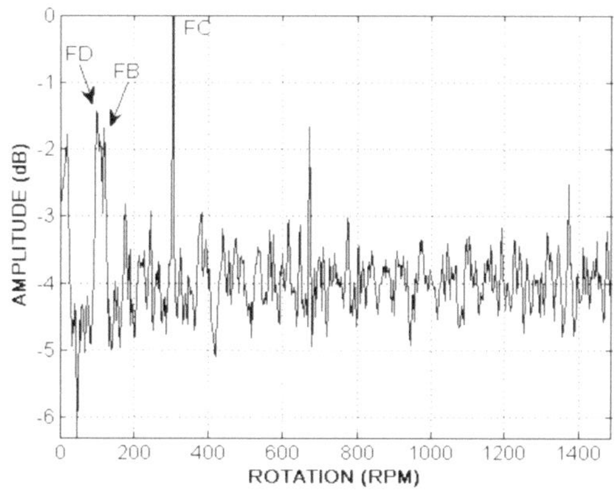

Fig. 12. DEMON analysis at 076°.

Performance comparisons between ICA algorithms applied to passive sonar signal separation were conduced in (Moura et al. 2007b) and it was observed that JADE presents slightly better performance. Considering this, the results presented in the next sections were derived through the application of JADE algorithm to perform ICA.

4.1 Time-domain BSS

A simple and straightforward implementation is to perform independent component analysis over raw-data. Signals measured at each direction (076°, 190° and 205°) are put together in order to compose a three component observation vector. An ICA algorithm (JADE) is applied to estimate three (time-domain) independent components, which will further be used as inputs to DEMON analysis block. The method is illustrated in Fig. 13.

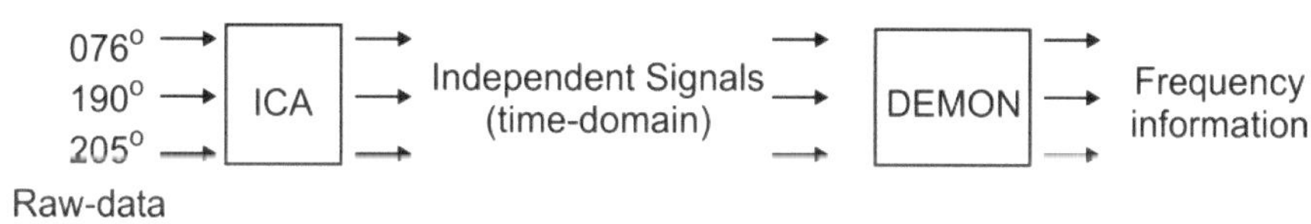

Fig. 13. Time-domain blind signal separation method.

To obtain quantitative measures of the signal separation performance, the peak amplitude values of each frequency component (after DEMON analysis) are compared for both raw-data and separated signals. Moreover, useful information is also obtained from the full-width at half of the peak value (Full-Width of Half Maximum - FWHM) of a certain frequency component *Fx*. This measure indicates whether *Fx* is accurately estimated (shorter FWHM) or not (larger FWHM). When ICA is applied, can be observed, from Fig. 14, that, considering direction 205°, the amplitudes of interfering frequencies F_A and F_C were reduced from -5.9dB and -3.2dB (in raw-data) up to, respectively, -9.1dB and -4.2dB. The background noise level (estimated from the high frequency components amplitude) was also reduced from -7dB up to -8.5dB.

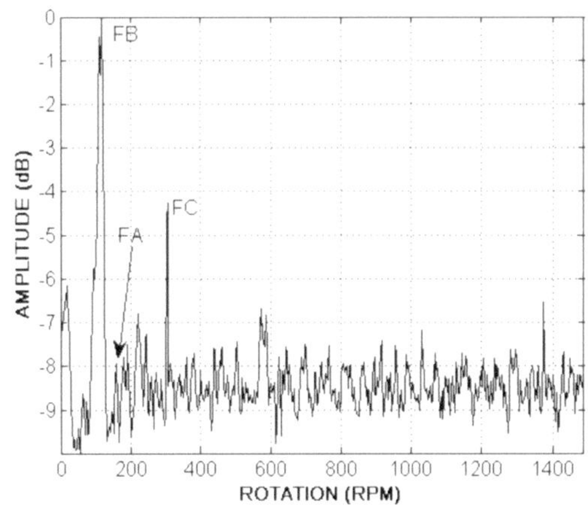

Fig. 14. DEMON analysis at 205°, for independent signal estimation.

Unfortunately, through this method, at directions 076° and 190° no significant signal separation was observed. The half-peak bandwidths were not modified either.

A main limitation of this approach is that raw-data is usually corrupted by additive underwater acoustic environment noise. It is known that, standard ICA algorithms present poor performance in the presence of noise (Hyvärinen et al., 2001). Modifications on the traditional ICA model in order to consider additive noise may increase the algorithms accuracy and thus produce better separation results (Hyvärinen, 1998b).

4.2 Frequency-domain BSS.

An alternative approach is to perform signal separation in the frequency domain. As illustrated in Fig. 15, DEMON analysis is initially performed over raw-data and frequency information from the three directions are used as inputs for an ICA algorithm, producing the independent (frequency-domain) components.

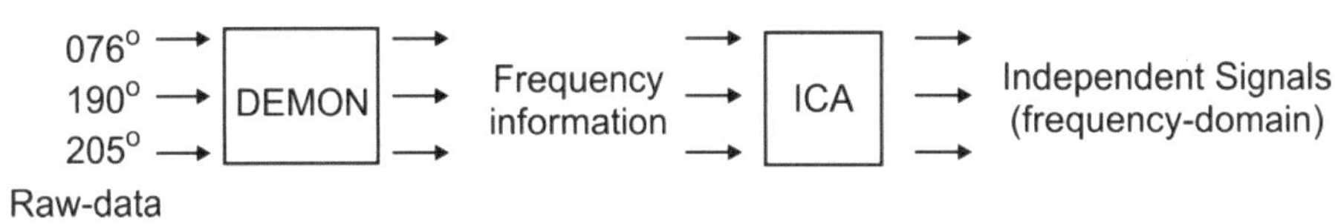

Fig. 15. Frequency-domain blind signal separation method.

As described in Section 2, DEMON analysis basically consists in performing demodulation and filtering of acoustic data in order to obtain relevant frequency information for target characterization. Most of noise and nonrelevant signals are eliminated by DEMON, allowing more accurate estimation of the independent components.

A particular characteristic is that DEMON analysis is usually performed over finite time-windows (approximate length = 250ms) and the frequency components are estimated within these windows. Aiming at reducing the random noise generated in time-frequency transformation, an average spectrum is computed using frequency information from these time slots.

In Independent Component Analysis algorithms the order and the amplitude of the estimated components are random parameters and thus different initializations may lead to

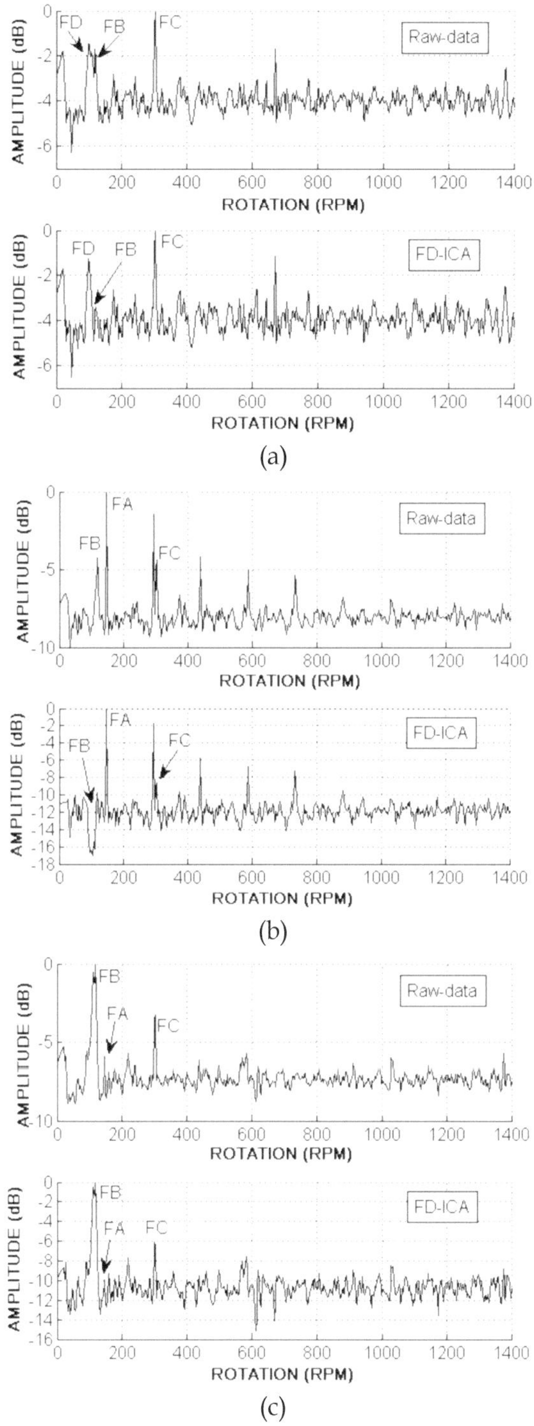

Fig. 14. DEMON analysis for both raw-data (measured acoustic signal) and frequency domain independent components (FD-ICA) at bearings (a) 076°, (b) 190° and (c) 205°.

different scaling factors and ordering (Hyvärinen et al., 2001). As in the frequency-domain BSS approach the ICA algorithms are executed after DEMON estimation at each time window, independent components from a certain direction may appear in different ordering at adjacent time-windows in this sequential procedure. Before generating the average spectrum, the independent components must be reordered (to guarantee that the averages are computed using samples from the same direction) and normalized in amplitude. The normalization is performed by converting signal amplitude into dB scale. The reordering procedure is executed by computing the correlation between independent components estimated from adjacent time slots. High correlation indicates that these components are related to the same direction.

Separation results obtained through this approach are illustrated in Fig. 14. It can be seen that, the interfering frequencies were considerably attenuated at the independent components from all three directions. The higher frequency noise levels were also reduced.

The results obtained from both time (ICA) and frequency domain (FD-ICA) methods are summarized in Table 1 (when *Fx* frequency width is not available it means that half of *Fx* peak amplitude is under the noise level). It can be observed that, for FD-ICA both the interference peaks and the width of the frequency components belonging to each direction were reduced, allowing better characterization of the target. The time domain method (ICA) produced relevant separation results only for 205° signal.

		Peak (dB)			Width (RPM)		
	Freq.	**Raw-data**	**ICA**	**FD-ICA**	**Raw-data**	**ICA**	**FD-ICA**
Direction 076	**FB**	-1,7	-0,8	-3,6	-	-	-
	FC	0	0	0	8	7,9	8,4
	FD	-1,4	-3,1	-1,3	-	-	-
Direction 190	**FA**	0	0	0	4,9	4,9	4,3
	2FA	-1,4	-1,4	-1,7	5,3	5,3	4,3
	3FA	-4,1	-4,1	-5,7	8,8	8,8	6,2
	4FA	-5	-5	-6,7	-	-	7,4
	5FA	-5,3	-5,4	-7,2	-	-	11
	FB	-4,2	-4,1	-9,8	16,6	15,8	-
	FC	-4,4	-4,4	-8,6	7,7	7,7	-
	FD	-8,7	-8,6	-16,5	-	-	-
Direction 205	**FA**	-5,9	-9,1	-9,9	-	-	-
	FB	0	0	0	16,8	16,3	15,2
	FC	-3,2	-4,2	-6,4	7	6,6	6,5
	FD	-5,6	-5,8	-9,3	-	-	-

Table 1. Separation results summary

4.4 Extensions to the basic BSS model

In order to obtain better results in signal separation and thus higher interference reduction, more realistic models may be assumed for both the propagation channel and measurement system.

For example, it is known that, signal transmission in passive sonar problems may comprise different propagation paths, and thus the measured signal may be a sum of delayed and mixed versions of the acoustic sources. This consideration leads to the so-called convolutive mixture model for the ICA (Hyvarinen et al., 2001), for which the observed signals $x_i(t)$ are described through Eq. 10:

$$x_i(t) = \sum_{j=1}^{n} \sum_{k} a_{ikj} s_j(t-k), \quad for\ i = 1,\dots,n \tag{10}$$

where s_j are the source signals. To obtain the inverse model, usually a finite impulse response (FIR) filter architecture is used to describe the measurement channel.
Another modification that may allow better performance is to consider, in signal separation model, that sensors (or propagation channel) may present some source of nonlinear behavior (which is the case in most passive sonar applications). The nonlinear ICA instantaneous mixing model (Jutten & Karhunen, 2003) is thus defined by:

$$\mathbf{x} = F(\mathbf{s}) \tag{11}$$

where $F(.)$ is a $R^N \rightarrow R^N$ nonlinear mapping (the number of sources is assumed to be equal to the number of observed signals) and the purpose is to estimate an inverse transformation G : $R^N \rightarrow R^N$:

$$\mathbf{s} = G(\mathbf{x}) \tag{12}$$

so that the components of $\boldsymbol{y}$ are statistically independent. If $G = F^{-1}$ the sources are perfectly recovered (Hyvärinen & Pajunen, 1999).
Some algorithms have been proposed for the nonlinear ICA problem (Jutten & Karhunen, 2003), a limitation inherit to this model is that, in general, there exists multiple solutions for the mapping G in a given application. If **x** and **y** are independent random variables, it is easy to prove that f(**x**) and g(**y**), where f(.) and g(.) are differentiable functions, are also independent. A complete investigation on the uniqueness of nonlinear ICA solutions can be found in (Hyvärinen & Pajunen, 1999). NLICA algorithms have been recently applied in different problems such as speech processing (Rojas et al., 2003) and image denoising (Haritopoulos et al., 2002).
Although these extensions to the basic ICA model may allow better signal separation performance, the estimation methods usually require considerable large computational requirements, as the number of parameters increases (Jutten & Karhunen, 2003) e (Hyvarinen., 2001). Thus, an online implementation (which is the case in passive sonar signal analysis) may not always be possible.

5. Summary and perspective

Sonar systems are very important for several military and civil underwater applications. Passive sonar signals are susceptible to cross-interference from acoustic sources present at different directions. The noise irradiated from the ship where the hydrophones are installed may also interfere with the target signals, producing poor performance in target identification efficiency. Independent component analysis (ICA) is a statistical signal processing method that aims at recovering source signals from their linearly mixed versions. In the framework of passive sonar measurements, ICA is useful to reduce signal interference and highlight targets acoustic features.

Extensions to the standard ICA model, such as considering the presence of noise, multiple propagation paths or nonlinearities may lead to a better description of the underwater acoustic environment and thus produce higher interference reduction. Another particular characteristic is that the underwater environment is non-stationary (Burdic, 1984). Considering this, the ICA mixing matrix becomes a function of time. To solve the non-stationary ICA problem recurrent neural networks trained using second-order statistic were used in (Choi et al., 2002) and a Markov model was assumed for the sources in (Everson & Roberts, 1999).

6. References

Brigham, E. (1988). *Fast Fourier Transform and Its Applications.* ISBN10: 0133075052, Ed. FACSIMILE. 1988.

Burdic, W. S. (1984) *Underwater Acoustic System Analysis*, Prentice-Hall, ISBN 10- 0932146635.

Cardoso, J.-F. (1998). "Blind signal separation: Statistical principles", *Proceedings of IEEE*, pp. 2009-2025, vol. 86 no. 10, October 1998.

Choi, S. ; Cichocki, A. and Amari, S-I (2002). Equivariant nonstationary source separation, *Neural Networks*, vol. 15, no. 1, pp. 121-130, January.

Clay, C. and Medwin, H. (1998). *Fundamentals of Acoustical Oceanography.* ISBN:0-12-487570-X. Academic Press.

Cover, T. M. and Thomas, J. A. (1991). *Elements of Information Theory.* Wiley.

Diniz, P. S. R., Silva, E., Lima Netto, S. (2002). *Digital Signal Processing*, ISBN: 0-521-78175-2, Cambridge University Press.

Everson, R. and Roberts, S. J. (1999) Non-stationary Independent Component Analysis *Proceedings of the IEE conference on Artificial Neural Networks*, 7 - 10 September.

Haritopoulos, H.; Yin, H. and Allinson, N. M. (2002) Image denoising using self-organizing map-based nonlinear independent component analysis, *Neural Networks*, pp. 1085-1098, 2002.

Haykin, S. (2001). *Neural Networks, Principles and Practice.* Bookman, ISBN: 9780132733502.

Hyvärinen, A. (1998a). New approximations of diferencial entropy for independent component analysis and projection pursuit, *Advances in Neural Information Signal Processing*, no. 10, pp. 273-279.

Hyvärinen, A. (1998b) Independent component analysis in the presence of Gaussian noise by maximizing joint likelihood. *Neurocomputing.* Volume 22, Issues 1-3, November , Pages 49-67.

Hyvärinen, A. and Pajunen, P. (1999). Nonlinear independent component analysis: Existence and uniqueness results, *Neural Networks*, vol. 12, no. 3, pp. 429-439.

Hyvärinen, A. and Oja, E. (2000). *Independent component analysis: Algorithms ans applications.* Helsinki University of Technology, P. O. Box 5400, FIN-02014 HUT, Filand. Neural Networks, 13 (4-5): 411-430. 2000.

Hyvärinen, A., Karhunen, J. and Oja, E. (2001). *Independent Component Analysis*, ISBN: 0-471-40540-X, John Wiley & Sons, .inc. 2001.

Jeffsers, R., Breed and B. Gallemore (2000). Passive range estimation and rate detection, *Proceedings of Sensor Array and Multichannel Signal Processing Workshop*, pp. 112-116, ISBN: 0-7803-6339-6, Cambridge, US, March 2000.

Jutten, C. and Karhunen, J. (2003) Advances in nonlinear blind source separation, *Proceedings of the 4th Int. Symp. on Independent Component Analysis and Blind Signal Separation* (ICA2003), Nara, Japan, pp. 245-256.

Kim, T.-H. and White, H. (2004) "On more robust estimation of skewness and kurtosis," *Finance Research Letters*, vol. 1, pp. 56-73.

Knight, W. C. Pridham, R. G. Kay, S. M. (1981). Digital signal processing for sonar. *Proceedings of IEEE*, ISSN: 0018-9219 vol. 69, issue-.11, pp. 1451-1506, November. 1981.

Krim, H. and Viberg, M. (1996) Two decades of array signal processing research: the parametric approach. *IEEE Signal Processing Magazine*, vol. 13, Issue: 4, pp. 67-94, ISSN: 1053-5888.

Lee, B. J. Park, J.B. Joo, Y. H. Jin, S. H. (2004). Intelligent Kalman filter for tracking a manoeuvring target. *Radar, Sonar and Navigation, IEE Proceedings*, ISSN: 1350-2395, vol. 151 issue: 6, pp. 344-350 Dec. 2004.

Marple, L., Brotherton, T. (1991). Detection and classification of short duration underwater acoustic signals by Prony's method, *International Conference on Acoustics, Speech, and Signal Processing*, pp. 1309-1312 vol.2, ISBN: 0-7803-0003-3, Toronto, Ont., Canada, May 1991.

Mellema, G. R. (2006). Reverse-Time Tracking to Enhance Passive Sonar, *International Conference on Information Fusion*, ISBN: 0-9721844-6-5, pp. 1-8, July.

Moura, N. N.; Soares Filho, W.; Seixas, J. M. de (2007a). Passive Sonar Classification based on Independent components. *Proceedings of the Brazilian congress of neural networks, 2007*, Florianópolis, Brazil, pp. 1-5. (In Portuguese).

Moura, N. N., Seixas, J. M. Soares Filho, W. and Greco, A. V. (2007b) "Independent component analysis for optimal passive sonar signal detection," *Proceedings of the 7th International Conference on Intelligent Systems Design and Applications*, Rio de Janeiro, pp. 671-678, October 2007.

Nielsen, R. O. (1991). *Sonar Signal Processing*, ISBN: 0-89006-453-9. Artech House Inc, Nortwood, MA, 1991.

Nielsen, R. O. (1999). "Cramer-Rao lower bounds for sonar broadband modulation parameters". *IEEE Journal of Ocean Engineering*, vol. 24 no. 3, pp. 285-290, July 1999.

Papoulis,A. (1991), *Probability, Random Variables, and Stochastic Processes*. McGraw-Hill.

Peyvandi, H., Fazaeefar, B., Amindavar, H (1998). Determining class of underwater vehicles in passive sonar using hidden Markov with Hausdorff similarity measure, *Proceedings of 1998 International Symposium on Underwater Technology*, pp. 258-261,ISBN: 0-7803-4273-9, Tokyo, Japan, April 1998.

Rao, S. K. (2006). Pseudo linear Kalman filter for underwater target location using intercept sonar measurements. *Symposium of Position, location and navigation*, ISBN: 0-7803-9454-2. Pp. 1036-1039, San Diego, US, April 2006.

Rojas, F.; Puntonet, C. G. and Rojas, I. (2003) Independent component analysis evolution based method for nonlinear speech processing, *Artificial Neural Nets Problem Solving Methods*, PT II, vol. 2687, pp. 679-686, 2003.

Seixas, J. M., Mamazio, D. O., Diniz, P. S. R., Soares-Filho, W. (2001). Wavelet transform as a preprocessing method for neural classification of passive sonar signals, *The 8th IEEE International Conference on Electronic, Circuits and Systems*, pp. 83-86, ISBN: 0-7803-7057-0, Malta, September 2001.

Shannon, C. E. (1948). "A mathematical theory of communication," *The Bell System Technical Journal*, pp. 379-423, July.

Shaolin Li, Sejnowski, T. J. (1995). Adaptive separation of mixed broadband sound sources with delays by a beamforming Herault-Jutten network. *IEEE Journal of Oceanic Engineering*, ISSN: 0364-9059, vol. 20, pp. 73-79, 1995.

Soares Filho, W.; Seixas, J. M. de; Calôba, L. P. (2001) Principal Component Analysis for Classifyting Passive Sonar Signals. *IEEE International Symposium on Circuits and Systems*, 2001, Sidney, Australia.

Spiegel, M. R., Schiller, J. J. and Srinivasan, R. A. (2000). *Probability and Statistics*. McGraw-Hill, 2 ed.

Torres, R. C.; Seixas, J. M. de, Soares Filho, W. (2004) Sonar Signals Classification Through Non-linear Principal Component Analysis. *Learning and nonlinear models*. Vol. I, no. 4, pp. 208-222, 2004 (In Portuguese).

Trees, H. L. V. (2001) *Detection, Estimation and Modulation Theory – Part I*, ISBN: 0-471-09517-6, John Wiley & Sons, Inc.

Waite, A. D. (2003). *Sonar for practicing Engineers, ISBN: 0-471-49750-9,* John Wiley and Sons, New York, 2003

Welling, M. (2005) Robust higher order statistics, *Proceedings of the Tenth International Workshop on Artificial Intelligence and Statistics* (AISTATS2005), pp. 405-412,Barbados, January.

Yan, D. P. and Peach, J. (2000). Comparison of blind source separation algorithms. *Advanxes in Neural networks and applications*. Pp. 18-21. 2000.

Yang, Z., Yang, L., Qing, C., Huang, D.. (2007) An improved empirical AM/FMdemodulation method. *International Conference on Wavelet Analysis and Pattern Recognition*. ISBN: 978-1-4244-1065-1, vol. 3, pp. 1049-1053.

6

From Statistical Detection to Decision Fusion: Detection of Underwater Mines in High Resolution SAS Images

Frédéric Maussang[1], Jocelyn Chanussot[2], Michèle Rombaut[2] and Maud Amate[3]
[1]Institut TELECOM; TELECOM Bretagne; UeB; CNRS UMR 3192 Lab-STICC
[2]GIPSA-Lab (CNRS UMR 5216); Grenoble INP
[3]Groupe d'Etudes Sous-Marines de l'Atlantique DGA/DET/GESMA Brest
France

1. Introduction

Among all the applications proposed by sonar systems is underwater demining. Indeed, even if the problem is less exposed than the terrestrial equivalent, the presence of underwater mines in waters near the coast and particularly the harbours provoke accidents and victims in fishing and trade activities, even a long time after conflicts.

As for terrestrial demining (Milisavljević *et al.*, 2008), detection and classification of various types of underwater mines is currently a crucial strategic task (U.S. Department of the Navy, 2000). Over the past decade, synthetic aperture sonar (SAS) has been increasingly used in seabed imaging, providing high-resolution images (Hayes & Gough, 1999). However, as with any active coherent imaging system, the speckle constructs images with a strong granular aspect that can seriously handicap the interpretation of the data (Abbot & Thurstone, 1979).

Many approaches have been proposed in underwater mine detection and classification using sonar images. Most of them use the characteristics of the shadows cast by the objects on the seabed (Mignotte *et al.*, 1997). These methods fail in case of buried objects, since no shadow is cast. That is why this last case has been less studied. In such cases, the echoes (high-intensity reflection of the wave on the objects) are the only hint suggesting the presence of the objects. Their small size, even in SAS imaging, and the similarity of their amplitude with the background make the detection more complex.

Starting from a synthetic aperture image, a complete detection and classification process would be composed of three main parts as follows:

1. *Pixel level:* the decision consists in deciding whether a pixel belongs to an object or to the background.
2. *Object level:* the decision concerns the segmented object which is "real" or not: are these objects interesting (mines) or simple rocks, wastes? Shape parameters (size,...) and position information can be used to answer this question.
3. *Classification of object:* the decision concerns the type of object and its identification (type of mine).

This chapter deals with the first step of this process. The goal is to evaluate a confidence that a pixel belongs to a sought object or to the seabed. In the following, considering the object

characteristics (size, reflectivity), we will always assume that the detected objects are actual mines. However, only the second step of the process previously described, which is not addressed in the chapter, would give the final answer.
We propose in the chapter a detection method structured as a data fusion system. This type of architecture is a smart and adaptive structure: the addition or removal of parameters is easily taken into account, without any modification of the global structure. The *inputs* of the proposed system are the parameters extracted from an SAS image (statistical in our case). The *outputs* of the system are the areas detected as potentially including an object.
The first part of the chapter presents the main principal of the SAS imaging and its use for detection and classification. The second part is on the extraction of a first set of parameters from the images based on the two first order statistical properties and the use of a mean - standard deviation representation, which allow to segment the image (Maussang *et al.*, IEEE, 2007). A third part enlarges this study to the higher order statistics (Maussang *et al.*, EURASIP, 2007) and their interest in detection. Finally, the last part proposes a fusion process of the previous parameters allowing to separate the regions potentially containing mines ("*object*") from the others ("*non object*"). This process uses the belief theory (Maussang *et al.*, 2008). In order to assess the performances of the proposed classification system, the results, obtained on real SAS data, are evaluated visually and compared to a manually labeled ground truth using a standard methodology (Receiver Operating Characteristic (ROC) curves).

2. SAS technology and underwater mines detection

SAS (Synthetic Aperture Sonar) history is closely linked to the radar one. Actually, the airborne radar imagery was the first to develop the process of synthetic aperture in the 1950's (SAR : Synthetic Aperture Radar). Then, it was applied to satellite imagery. The first satellite to use synthetic aperture radar was launched in 1978. Civilian and military applications using this technique covered enlarged areas with an improved resolution cell. Such a success made the synthetic aperture technique essential to obtain high resolution images of the earth. Following this innovation, this technique is now frequently used in sonar imagery (Gough & Hayes, 2004). The first studies in synthetic aperture sonar occurred in the 1970's with some patents (Gilmour, 1978, Walsh, 1969, Spiess & Anderson, 1983) and articles on SAS theory by Cutrona (Cutrona, 1975, 1977).

2.1 SAS principle

Synthetic aperture principle is presented on Fig. 1 and consists in the coherent integration of real aperture beam signals from successive pings along the trajectory. Thus, the synthetic aperture is longer than the real aperture. As the resolution cell is inversely proportional to the length of the aperture, longer the antenna, better the resolution. In practice, the synthetic aperture depends on the movements of the vehicle carrying the antenna. Movements like sway, roll, pitch or yaw are making the integration along the trajectory more difficult.
The synthetic aperture resolution is that of the equivalent real aperture of length L_{ERA}, given by the expression:

$$L_{\mathrm{ERA}} = 2(N-1)VT + L_{\mathrm{R}} \tag{2.1}$$

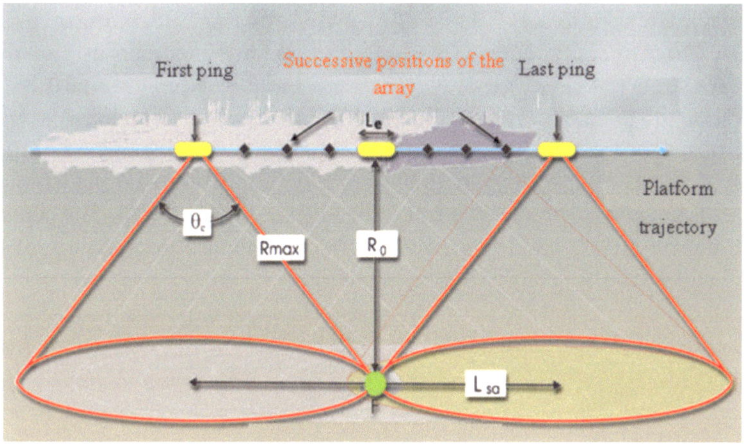

Fig. 1. SAS principle

where N is the number of pings integrated, V is the mean cross-range speed, T is the ping rate and L_R is the real aperture length.
Hence, the cross-range resolution at range R is given by:

$$\delta_S = \frac{R\lambda}{L_{ERA}} \tag{2.2}$$

The maximum travel length $(N\text{-}1)VT$ corresponds normally (but not necessarily) to the cross-range width of the insonification sector, equal to $R\lambda/L_{tr}$ when the transmitter has a uniform phase-linear aperture of length L_{tr} and operates in far field. For large N, the L_{ERA} given by (2.1) equals approximately twice this width; hence, the resolution is independent of range and frequency, and is given by the expression:

$$\delta_S = \frac{L_{tr}}{2} \tag{2.3}$$

Let us note that the cross-range resolution of the physical array $\delta_R = R\lambda/L_R$. The resolution gain g of the synthetic aperture processing is defined by the expression:

$$g = \frac{\delta_R}{\delta_S} = \frac{L_{ERA}}{L_R} \tag{2.4}$$

2.2 SAS challenges

Nowadays, SAS is a mature technology used in operational systems (MAST'08). However, some challenges remain to enhance SAS performances. For example, a precise knowledge of the motion of the antenna will permit to obtain a better motion compensation and better focused images. There are also some studies to improve beamforming algorithms, more adapted to SAS processing. Another challenge lies in the reduction of the sonar frequency. Knowing that sound absorption increases with the frequency in environments like sea water or sediment, a logical idea is to decrease imagery sonar frequency. Yet, resolution is inversely proportional to frequency and length of antenna. So for a reasonable size of array, the resolution remains quite low, especially for underwater minewarfare. SAS processing can then be used to artificially increase the length of the antenna and improve the resolution One of the purposes is the detection of objects buried in the sediment. Both civilian (pipeline detection, wreck inspection) and military (buried mines detection) applications are interested in this concept. GESMA conducted numerous sea experiments on SAS subject since the end of the 1990's. Firstly, in 1999, in cooperation with the British agency DERA, high frequency SAS was mounted on a rail in Brest area (Hétet, 2000). The central frequency was 150 kHz, the frequency band was 60 kHz and the resolution obtained was 4 cm. Fig. 2 presents two images resulting from this experiment.

Fig. 2. On the left, SAS image and picture of the associated modern mine. On the right, SAS image and picture of the associated modern mines

Then, GESMA decided to work on buried mines and conducted an experiment with a low frequency SAS mounted on a rail in 1999. It was in Brest area, the sonar frequency was between 14 and 20 kHz (Hétet, 2003). Fig. 3 presents results of this experiment. We notice the presence of a large echo coming from the cylinder.

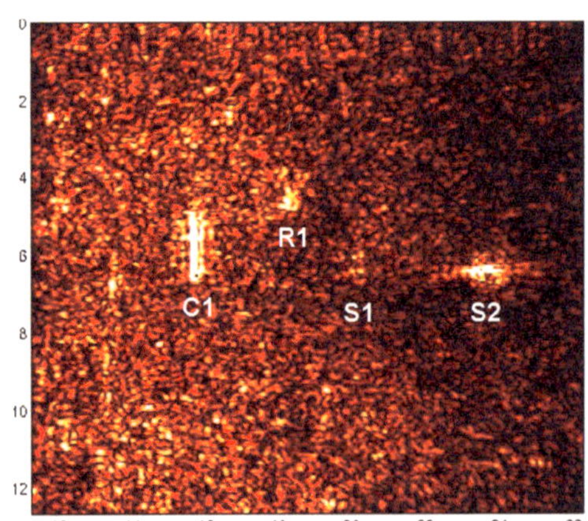

Fig. 3. SAS image of buried and proud objects at 20 m. C1 : buried cylinder ; R1 : buried rock ; S1 : buried sphere ; S2 : proud sphere

Fig. 3 shows that low frequencies allow to penetrate the sediment and to detect buried objects. Moreover, echoes are more contrasted on this image and there is a lack of the shadows for the objects, making the classification more difficult. To go a step further, in 2002, a low frequency SAS was hull mounted onboard a minehunter (Hétet *et al.*, 2004). Frequency was chosen between 15 and 25 kHz and Fig. 4 presents results of these trials conducted in cooperation with the Dutch agency TNO, Defence, Security and Safety.

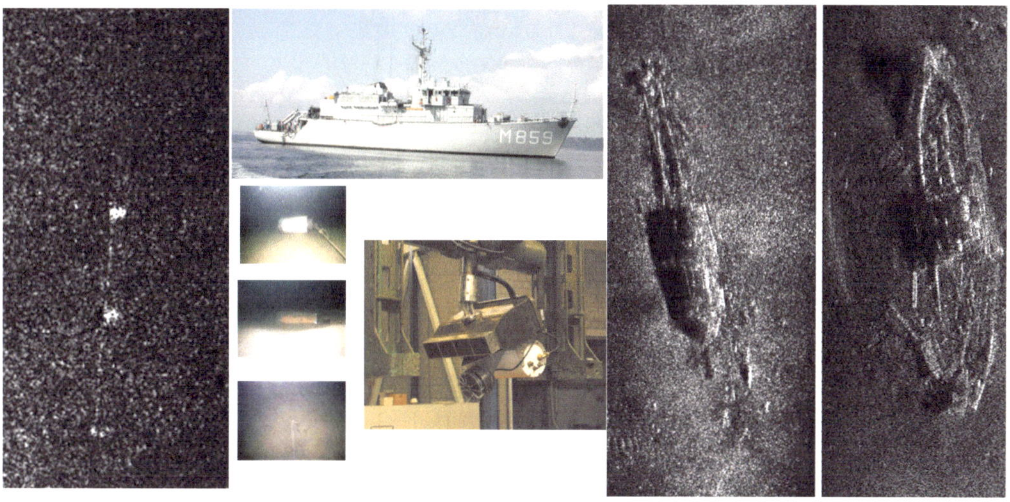

Fig. 4. Low frequency SAS images. On the left, SAS image of three cylinders. Downwards : proud cylinder, half buried cylinder, buried cylinder. In the middle : pictures of the supporting ship, the sonar and the three cylinders. On the right, SAS images of two wrecks in the bay of Brest.

Considering previous figures, low frequency and high frequency SAS images present an important difference. High frequency images allow detecting and classifying underwater objects thanks to their shadow when low frequency images present no more shadow but strong echoes. The idea is thus to use specificities of low frequency SAS to define a new approach to detect and classify buried objects.

3. Underwater mines detection using local statistical parameters

The key issue when designing a classification algorithm is to choose the right parameters discriminating the classes of interest. The two main approaches are (i) use of statistical knowledge about the process, (ii) use of expert knowledge, eventually derived from a physical model of the process. In this application, the statistical characteristics of the seabed pixels are well known and follow statistical laws (Rayleigh and Weibull distributions for instance). As a consequence, the data fusion process is based on the comparison of the statistical characteristics locally extracted for each pixel and these laws.

3.1 Statistical description of the SAS images

The sonar images, as any image formed by a coherent system (radar imagery is another example), are seriously corrupted by the speckle effect. They thus have a strong granular aspect. This noise comes from the presence of a large number of elements (sand, rocks, etc.) that are smaller than the wavelength and randomly distributed over the seabed. The sensor receives the result of the interference of all the waves reflected by these small scatterers within a resolution cell (Goodmann, 1976).

3.1.1 Speckle noise and the Rayleigh law

Sonar images provided by the sonar system are constructed by the speckle. This bottom reverberation comes from the presence of a large number of elements (sand, gravel, etc.) that are smaller than the wavelength of the used monochromatic and coherent illumination source. These elements are assumed to be randomly distributed on the seabed. As a consequence, the sensor records the result of the constructive and destructive interferences of all the waves reflected by these elementary scatterers contained in a resolution cell (Collet *et al.*, 1998, Schmitt *et al.*, 1996).

The response of a resolution cell can thus be described by the following:

$$\rho = \sum_{i=1}^{N_d} a_i \exp(j\varphi_i) = A\exp(j\phi) = X + j.Y \tag{3.1}$$

with A being the amplitude of the response on a resolution cell and ϕ representing the phase. The phases are usually considered as independent and uniformly distributed over $[-\pi, +\pi]$. With these assumptions, and if the number of elementary scatterers N_d within the resolution cell is large enough, the central limit theorem applies: X and Y can be considered as Gaussian random values. Consequently, the probability density function of amplitude $A = \sqrt{X^2 + Y^2}$ follows a Rayleigh distribution:

$$p_{R_A}(A) = \frac{A}{\alpha^2}\exp\left(-\frac{A^2}{2\alpha^2}\right), \quad A \geq 0 \tag{3.2}$$

with α being the Rayleigh's law specific parameter. This parameter is bound with the average intensity of the reflected waves.

The vth-order moment of A is given by the following:

$$\mu_{A(\nu)} = (2\alpha^2)^{\nu/2}\Gamma\left(1 + \frac{\nu}{2}\right) \tag{3.3}$$

with Γ being the Gamma function ($\Gamma(z+1) = z! = \int_0^{+\infty} e^{-t}t^z dt$). This results in an interesting property of the Rayleigh law: the standard deviation σ_A and the mean μ_A of the amplitude A are linked by a simple proportionality relation:

$$\mu_{A=} = k_R\sigma_A \quad \text{with} \quad k_R = \sqrt{\frac{\pi}{4-\pi}} \approx 1.91 \tag{3.4}$$

This property leads to modeling the speckle as a multiplicative noise. As a matter of fact, the variation of amplitude induced by the speckle and characterized by parameter σ_A is bound by the mean amplitude (μ_A) with a multiplicative coefficient k_R.

3.1.2 Non-Rayleigh models

The previous description of the speckle is the most usual and the most popular one. However, it is not satisfactory when the number of scatterers within a resolution cell (noted

N_d in the previous paragraph) significantly decreases. The central limit theorem does not hold and the Rayleigh approximation is no longer valid. This case is frequently observed in high-resolution images (Collet *et al.*, 1998, Mignotte *et al.*, 1999) such as SAS images. In this case, the amplitude A is better described by a Weibull law:

$$p_{W_A}(A) = \frac{\delta}{\beta}\left(\frac{A}{\beta}\right)^{\delta-1} \exp\left[-\left(\frac{A}{\beta}\right)^{\delta}\right], \quad A \geq 0 \tag{3.5}$$

with β being the scale parameter and δ representing the shape parameter, strictly positive. These two parameters provide an increased flexibility compared to the Rayleigh law. Note that this law is a simple generalization of the Rayleigh distribution (in the special case $\beta = \sqrt{2}\alpha$ and $\delta = 2$, the Weibull law turns to a simple Rayleigh law).
For a Weibull distribution, the vth-order moment of A is given by:

$$\mu_{A(\nu)} = \beta^{\nu}\Gamma\left(1+\frac{\nu}{\delta}\right) \tag{3.6}$$

Therefore, the proportionality between and still holds, but with a coefficient $k_W(\delta)$ function of δ:

$$k_W(\delta) = \frac{\Gamma(1+1/\delta)}{\sqrt{\Gamma(1+2/\delta)-\Gamma(1+1/\delta)^2}} \tag{3.7}$$

Note that for $\delta = 2$, corresponding to the Rayleigh law, we obtain the same coefficient as in (3.4).
Other more complex non-Rayleigh approaches have been proposed in the literature to statistically describe the bottom reverberation in high-resolution sonar imaging. One of the most famous models is the *K*-distribution. For this model, the number of scatterers in a resolution cell N_d is supposed to be a random variable following a negative binomial distribution. The amplitude A is then described by a *K*-law (called generalized *K*-distribution), a three parameters distribution function given by (Gu & Abraham, 2001):

$$p_{K_A}(A) = \frac{4}{\Gamma(\nu_0)\Gamma(\nu_1)}\sqrt{\frac{\nu_0\nu_1}{\mu}}\left(\sqrt{\frac{\nu_0\nu_1}{\mu}}A\right)^{\nu_0+\nu_1-1} K_{\nu_1-\nu_0}\left(2\sqrt{\frac{\nu_0\nu_1}{\mu}}A\right), \quad A \geq 0 \tag{3.8}$$

where ν_1 is the shape parameter, μ is the scale parameter, and ν_0 is the parameter bound to the number of raw data averaged to figure out the pixel reflectivity. $K_{\nu_1-\nu_0}$ is the modified Bessel function of the second kind and order $\nu_1 - \nu_0$. This distribution describes a rapidly fluctuating Rayleigh component modulated by a slowly varying χ^2 component.
The vth-order moment of A is then given by:

$$\mu_{A(\nu)} = \left(\frac{\mu}{\nu_0\nu_1}\right)^{\frac{\nu}{2}} \frac{\Gamma(\nu_0+\nu/2)\Gamma(\nu_1+\nu/2)}{\Gamma(\nu_0)\Gamma(\nu_1)} \tag{3.9}$$

The relationship between σ_A and μ_A is preserved, with a coefficient depending on the two parameters ν_1 and ν_0:

$$k_K(\nu_0,\nu_1) = \frac{\Gamma(\nu_0+1/2)\Gamma(\nu_1+1/2)}{\sqrt{\nu_0\nu_1\Gamma(\nu_0)^2\Gamma(\nu_1)^2 - \Gamma(\nu_0+1/2)^2\Gamma(\nu_1+1/2)^2}} \tag{3.10}$$

A Rayleigh mixture has been also proposed to describe SAS data, each scattering material within a resolution cell being characterized by one specific Rayleigh distribution (Hanssen *et al.*, 2003).

3.1.3 Application to experimental data

In this section, the performances of the different statistical models are compared using the SAS data presented in section 2 (Fig. 2). For this purpose, two tests are considered: the χ^2 criterion and the Kolmogorov distance (Mignotte *et al.*, 1999). The χ^2 criterion d_{χ^2} is estimated according to the following relation (Saporta, 1990):

$$d_{\chi^2} = \sum_{i=1}^{r} \frac{(k_i - nP_i)^2}{nP_i} \tag{3.11}$$

where k_i is the number of realizations (number of pixels having the value i), P_i is the estimated probability of value i, r is the number of possible values, and n is the number of observations (in our case, the number of pixels).
With k'_i being the number of realizations from 1 to i and p'_i being the value for i of the cumulative distribution function associated with p_i, the Kolmogorov distance is defined as:

$$d_K = \max_{i=1\ldots r} \left| k'_i - np'_i \right| \tag{3.12}$$

The parameters of the Rayleigh and the Weibull laws are evaluated on the SAS image using a maximum-likelihood (ML) estimator. The estimated parameter $\hat{\alpha}_{ML}$ of the Rayleigh law (3.2) is given by the following (Schmitt *et al.*, 1996):

$$\hat{\alpha}_{ML}^2 = \frac{1}{2n}\sum_{i=1}^{n} y_i^2 \tag{3.13}$$

where n is the number of pixels and y_i is the amplitude of pixel i.
Parameters β and δ of the Weibull law (3.5) are estimated by $\hat{\beta}_{ML}$ and $\hat{\delta}_{ML}$, respectively, given by the following (Mignotte *et al.*, 1999):

$$\hat{\delta}_{ML} = \lim_{k\to+\infty} \delta_k \tag{3.14}$$

$$\hat{\beta}_{ML} = \left(\frac{1}{n}\sum_{i=1}^{n} y_i^{\hat{\delta}_{ML}} \right)^{1/\hat{\delta}_{ML}} \tag{3.15}$$

with $\delta_k = F(\delta_{k-1})$, $\delta_0 = 1$ (exponential law), and

$$F(x) = \frac{n\sum_{i=1}^{n} y_i^x}{n\sum_{i=1}^{n}\left(y_i^x \ln y_i\right) - \sum_{i=1}^{n} \ln y_i \sum_{i=1}^{n} y_i^x} \tag{3.16}$$

These estimators are unbiased, consistent, and efficient (Collet *et al.*, 1998).
The estimation of the *K*-law parameters is more problematic. Actually, there is no analytic expression for the derivative of a modified second kind Bessel function. Consequently, no ML estimation can be performed without approximations (Joughin *et al.*, 1993). A moment's method can then be used (3.9), even though it does not offer a closed-form solution.
These estimators are tested on the image presented in section 2 (Fig. 2 left). Fig. 5 presents in solid line the observed distribution (normalized histogram of the image) and compares it with the estimated Rayleigh and Weibull distributions (dashed lines).With a simple visual inspection, one immediately notices that the Weibull distribution and *K*-distribution fit the observed one better than the Rayleigh distribution[1]. This confirms the nonvalidity of the central limit theorem in the case of high-resolution images obtained in SAS imaging. This Rayleigh model will not be used subsequently. It is nevertheless included in this chapter

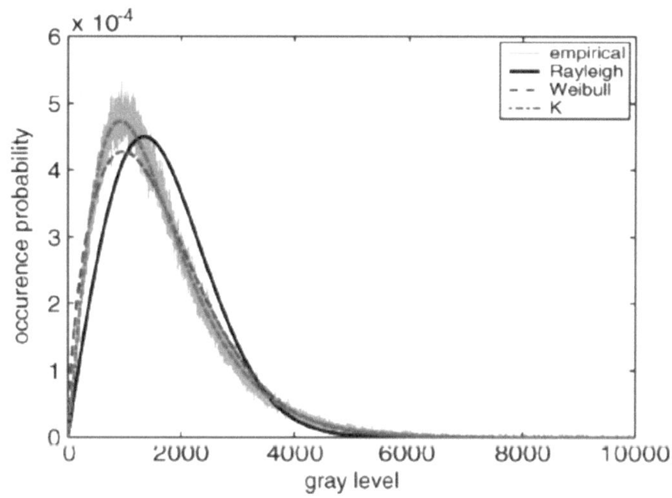

Fig. 5. Rayleigh distribution, Weibull distribution, and K-distribution estimated on Fig. 2

[1] In Fig. 2 a shadow can be seen behind the echoes reflected by the mine. Shadows are present on most sonar images containing underwater mines lying on the seabed. The shadow corresponds to a non illuminated region of the seabed and the sensor receives a weak acoustic wave from this region: the signal related to the shadow area essentially consists of the electronic noise from the processing chain. It can also come from the "differential shadow effect" due to the variation of the shadow zone position during the imaging process. The amplitude *A* of the pixels in this region can thus also be modeled by a Gaussian distribution and the models remain valid.

since the reader dealing with low-resolution sonar images may use the very same detection method proposed in this paper using the Rayleigh distribution. The *K*-distribution seems to provide a better statistical model of the background than the Weibull law. It especially fits better the "head" of the observed distribution. This is confirmed by the quantitative evaluation presented in Table 1. However, the "tail" of the distribution is accurately estimated by both models.

Distribution	Estimated parameters	χ^2 criterion	Kolmogorov distance
Rayleigh	$\hat{\alpha}_{ML} \approx 1482.5$	357.9	2.22 x 10^{-4}
Weibull	$\hat{\beta}_{ML} \approx 1961.7$ $\hat{\delta}_{ML} \approx 1.604$	0.318	1.66 x 10^{-4}
K	$\hat{\mu} \approx 4.300 \times 10^6$ $\hat{\nu}_0 \approx 0.882$ $\hat{\nu}_1 \approx 3.163$	0.044	1.41 x 10^{-4}

Table 1. Comparison of the performances of the distribution on the SAS image of Fig. 2

3.1.4 Choice of the statistical model

Considering the previous remarks, the *K*-distribution seems to be a better model than the Weibull model. However, as we have seen in section 3.1.2, the estimation of *K*-law parameters is more difficult (no ML estimators). The estimators of the *K*-parameters are not optimal and the estimation takes more time.

The difference with the Weibull model is not enough to justify this difficulty. Moreover, Weibull law is largely used in the sonar community and it made its proofs in their applications. That is why we will use the Weibull model in the following, but we keep in mind the existence of other models such as *K*.

3.1.5 Local statistical description

In the previous sections, a global statistical description of the SAS images has been given, ignoring the presence of any echoes. This is fair since the number of target pixels in the image is too small to significantly modify the global statistics. The observed histogram matches indeed very well a Weibull law. In this section, we study local first- and second-order statistical properties. This is achieved by looking at the data through a small sliding window composed of few pixels. In this case, the potential presence of echoes can no longer be ignored. Each echo is modeled as a deterministic element with an amplitude *D* surrounded by a noisy background with a Weibull distribution. We assume that the noise correlation is smaller than the spatial extension of the target echo and that the amplitude fluctuation of the echo is negligible. This is consistent with the experiments where the echoes appear as small sets of connected pixels with an almost constant value. This is justified in Fig. 8(b) with the pixels corresponding to echoes fitting the predicted ellipse in the mean–standard deviation plane.

We note *p* the proportion of deterministic pixels (i.e., pixels belonging to an echo) and $(1 - p)$ the proportion of random values (i.e., pixels belonging to the background) within a small square window (Fig. 6). Considering $\mu'_{D(r)}$, $\mu'_{N(r)}$, and $\mu'_{W(r)}$, the *r*th-order noncentral moments computed on the "echo part" of the window, the "background part" of the window, and the whole window, respectively, the following relation holds:

$$\mu'_{W(r)} = p\mu'_{D(r)} + (1-p)\mu'_{N(r)} \tag{3.17}$$

Considering $\mu_X = \mu'_{X(1)}$ and $\sigma^2_X = \mu'_{X(2)} - \mu'^2_{X(1)}$, the mean and the variance of X, respectively [X can be replaced by D, N, or W, as in (3.17)], we have:

$$\mu_W = p\mu D + (1-p)\mu_N \tag{3.18}$$

$$\sigma_W^2 = p\left(\sigma_D^2 + \mu_D^2\right) + (1-p)\left(\sigma_N^2 + \mu_N^2\right) - \mu_W^2 \tag{3.19}$$

Moreover, echoes are considered as deterministic elements with an amplitude D, leading to:

$$\mu'_{D(r)} = D^r \tag{3.20}$$

and

$$\mu_D = D \text{ and } \sigma_D = 0 \tag{3.21}$$

and, consequently:

$$\mu_W = \mu_N + p(D - \mu_N) \tag{3.22}$$

$$\sigma_W^2 = \sigma_N^2 + \mu_N^2 - \mu_W^2 + p\left(D^2 - \sigma_N^2 - \mu_N^2\right) \tag{3.23}$$

By combining (3.22) and (3.23), we obtain an interesting relationship between σ_W and μ_W:

$$\sigma_W^2 + \mu_W^2 = \left(D + \mu_N - \lambda_{\sigma_N}\right)\mu_W + \left(\lambda_{\sigma_N} - \mu_N\right)D \tag{3.24}$$

with $\lambda_{\sigma_N} = \sigma_N^2 / (D - \mu_N)$. It is important to underline that this relation is independent of p. Also note that in limit cases, this relation remains valid: in the case of $p = 0$ (the window contains only background pixels), $\mu_W = \mu_N$ and $\sigma_W = \sigma_N$; in the case of $p = 1$ (the echo is filling the whole window), $\mu_W = D$ and $\sigma_W = 0$, which is consistent with (3.21). Remember that intermediate values of p correspond to windows being partially composed of echo pixels.

3.2 First and second order parameters: segmentation

The proportional relation of the statistical model describing the sea bed sonar data previously described is used to extract the two first parameters. The local mean and standard deviation are estimated on the SAS image using a square sliding window. These values become the coordinates of the processed pixel in the mean – standard deviation plane.

3.2.1 Mean-standard deviation representation

Inspired by a segmentation tool applied to spectrograms (Hory *et al.*, 2002), this enables the separation of the echoes from the bottom reverberation, both features having different statistical characteristics as stated in section 3.1. In (Ginolhac *et al.*, 2005), the link between first- and second-order statistics is highlighted using this representation.

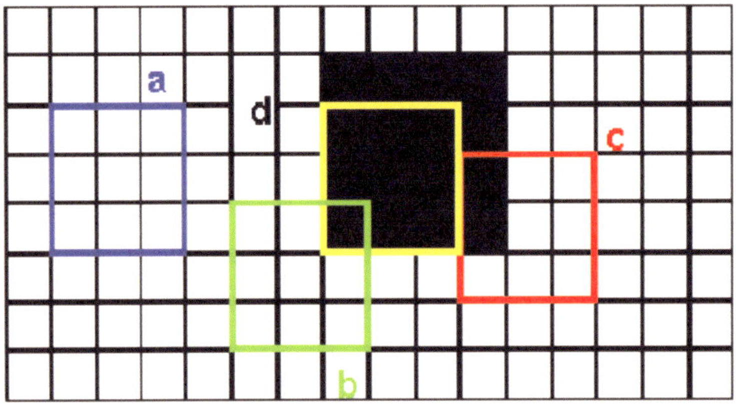

Fig. 6. Modeled echo and various values of the parameter p: (a) $p = 0$, (b) $p = 1/9$, (c) $p = 2/9$, and (d) $p = 1$

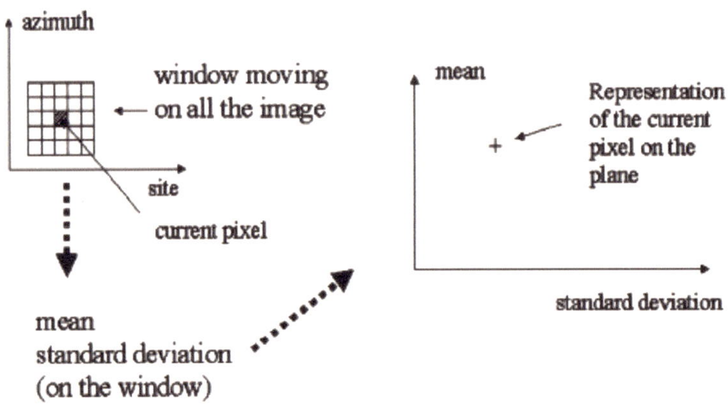

Fig. 7. Building of the mean-standard deviation representation

Whereas in (Ginolhac *et al.*, 2005) this link is simply illustrated as a justification to use first- and second-order statistics, the method presented in this chapter actually performs a segmentation of the mean-standard deviation plane.

The idea is to change the representation space of the data to highlight local statistical properties. The chosen space is the mean-standard deviation plane. For each pixel, the local standard deviation and the local mean are estimated within a square-centered computation window with the following conventional equations:

$$\hat{\mu}_W = \frac{1}{N}\sum_{i=1}^{N} y_i \quad (3.25)$$

$$\hat{\sigma}_W = \sqrt{\frac{1}{N}\sum_{i=1}^{N}\left(y_i - \hat{\mu}_W\right)^2} \quad (3.26)$$

where N is the number of pixels in the computation window ($N = N_x.N_y$ with N_x and N_y being the length and the width of the window, respectively) and y_i is the value of pixel in the window. The pair $(\hat{\sigma}_W, \hat{\mu}_W)$ becomes the coordinates representing the current pixel in the mean-standard deviation plane (Fig. 7). The performances of these estimators are evaluated by computing their moments (Hory *et al.*, 2002). For the mean estimator, the mean $M(\hat{\mu}_W)$ and the variance $V(\hat{\mu}_W)$ are:

$$M(\hat{\mu}_W) = \mu_W \tag{3.27}$$

$$V(\hat{\mu}_W) = \frac{\sigma_W^2}{N} \tag{3.28}$$

The mean estimator is unbiased and consistent with a variance varying as $1/N$. For the standard deviation estimator, the mean $M(\hat{\sigma}_W)$ and the variance $V(\hat{\sigma}_W)$ are given by:

$$M(\hat{\sigma}_W) \approx \frac{1}{2N} \frac{\sqrt{(4N^2 - 3N - 3)\sigma_W^4 - (N-1)m_{W(4)}}}{\sigma_W} \tag{3.29}$$

$$V(\hat{\sigma}_W) = \frac{1}{4N^2} \frac{(N-1)m_{W(4)} - (N-3)\sigma_W^4}{\sigma_W^2} \tag{3.30}$$

with $m_{W(4)}$ being the fourth central moment computed on the window. These equations come from the approximation $V(\sqrt{X}) \approx V(X)/4M(X)$ with X being a random variable (Kendall & Stuart, 1963). This estimator is asymptotically unbiased and is consistent with a variance varying as $1/N$.

The choice of the size of the computation window is a tradeoff. On one hand, the variance of the estimators [see (3.25) and (3.26)] increases for small values of N. N should thus not be too small to enable an accurate estimation. On the other hand, if N is too high, echoes being small elements, the proportion p of the deterministic elements in the computation windows remains low and echoes are lost in the background speckle [see (3.24)]. Consequently, the computation window should be chosen as slightly larger than the spatial extension of the echoes, this size depending on the resolution of the sonar image and the quality of the preprocessing chain.

The mean-standard deviation representation of Fig. 2 image is built with a 5 x 5-cm^2 window and is presented in Fig. 9. A general linear orientation is observed, as well as some pixels distancing the main direction on the right [see Fig. 8(b)]. Three different linear regressions of the data in the mean-standard deviation plane can be computed. They are shown in Fig. 8(a). The first line, with a slope of approximately 1.91, corresponds to the proportionality relation between the mean and the standard deviation estimated when assuming that the bottom reverberation is modeled by a Rayleigh law (3.4). The second line, with a slope of approximately 1.57, corresponds to the proportionality relation estimated with a Weibull model (3.7). At the given computation accuracy, the same line is obtained by a linear regression using a mean square method on the pixels representatives. To describe the global linear orientation of the data in the mean-standard deviation plane, the proportionality coefficient estimated with the Weibull assumption clearly outperforms the estimation based on a Rayleigh law. This confirms the previous results for the case of high-resolution data (Table I). In Fig. 8(b), the curve corresponding to the local relationship between mean-standard deviation on a computation window (3.24) is plotted considering a deterministic element with an amplitude of $D = 3.4 \times 10^{-4}$, approximately corresponding to the typical amplitude of the main echo on the original SAS image. This curve is a part of an ellipse and is a fairly good estimation of the main structure. Results obtained on other data sets could not be presented in this paper for confidentiality reasons. We will see following that each structure can be associated with one echo on the SAS image.

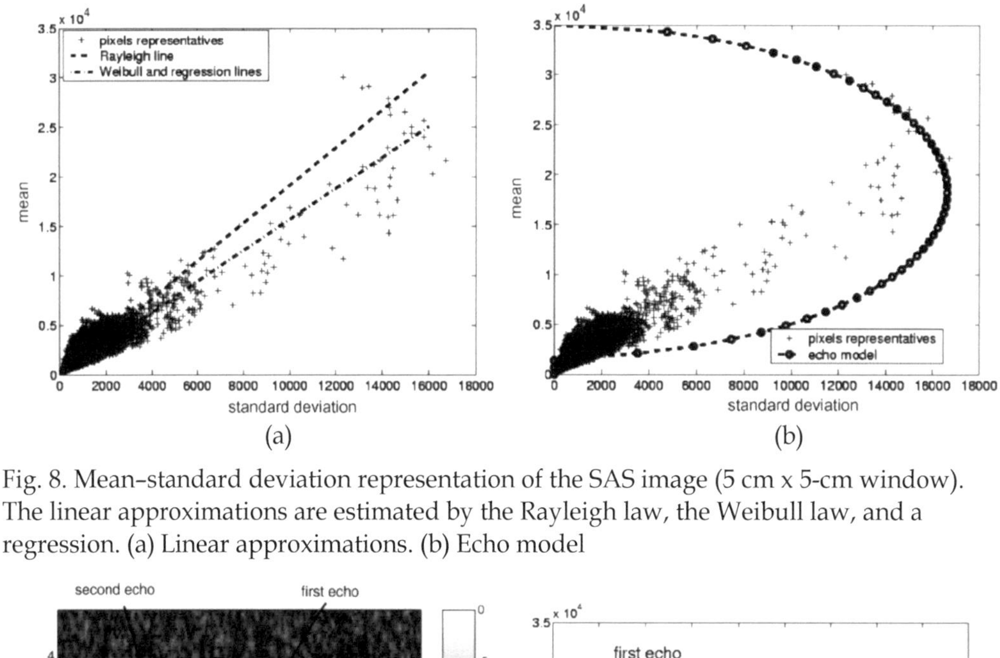

(a) (b)

Fig. 8. Mean-standard deviation representation of the SAS image (5 cm x 5-cm window). The linear approximations are estimated by the Rayleigh law, the Weibull law, and a regression. (a) Linear approximations. (b) Echo model

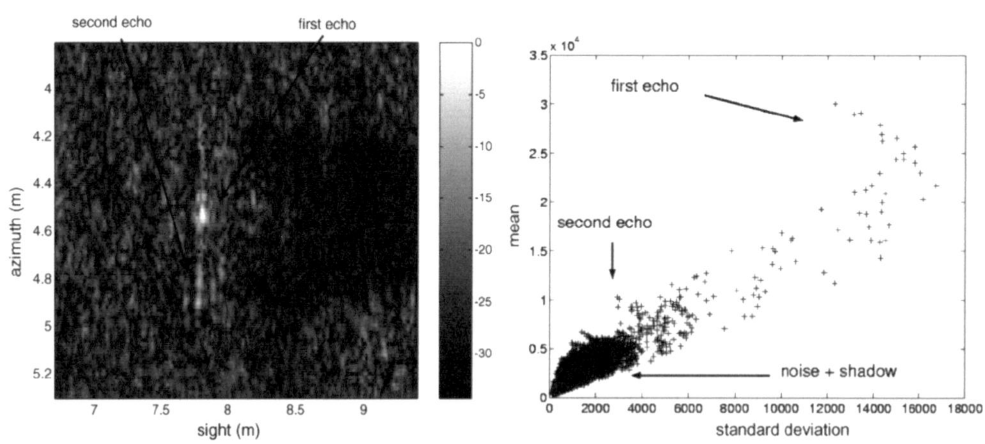

(a) (b)

Fig. 9. Comparison between the SAS image (Fig. 2) and its mean-standard deviation representation. (a) Zoom of the SAS image. (b) Representation

The fact that no pixel is on the *Y*-axis of this representation comes from the size of the computation window sensibly larger than the echoes. Therefore, no window contains only echo pixels and all the windows have a part of background. Moreover, the hypothesis of a constant deterministic echo is not strictly valid, the pixels of one echo having different, but similar, values. However, the value of our model is not called into question to explain the results described previously.

To highlight interesting properties of the mean-standard deviation representation, it is compared with the original image. Fig. 9 presents a zoom of the original image featuring two mine echoes and the corresponding mean-standard deviation representation. For a better understanding, a manual labeling of the sonar image is performed: pixels corresponding to the echoes are selected and corresponding points on the mean-standard deviation representation can be inspected. It turns out that the cluster of points close to the origin of the mean-standard deviation plane corresponds to the bottom reverberation pixels on the SAS image, with low means and low standard deviations. On the contrary, horn-shaped structures (actually parts of ellipses) correspond to the echoes on the sonar image. Two main structures can be seen with different positions and dimensions, each one corresponding to one specific echo. The extremities of these structures correspond to the centers of the echoes which are deterministic elements (high mean and relatively low standard deviation). The intermediary points correspond to the transition between echoes and background (increasing standard deviation and decreasing mean). These properties can be used to classify the different elements on the sonar image by observing the mean-standard deviation plane and the characteristics of the different structures.

3.2.2 Segmentation

Based on the statistical study and the observations previously presented, we propose in this section a segmentation method. The aim is to design an automatic algorithm isolating the echoes from the reverberation background on the sonar images. The proposed method is decomposed into the following steps.

- The Weibull distribution best fitting the observed normalized histogram is estimated with an ML estimator;
- The original amplitude data are mapped in the mean-standard deviation plane;
- In this representation, echoes appear as horn-shaped structures whereas background pixels are closer to the origin (low mean and low standard deviation). Therefore, a double threshold (both in mean and in standard deviation) allows a separation of the echoes pixels from the background pixels. The threshold value in standard deviation is set, either manually or automatically as will be described following;
- Corresponding threshold value for the mean is obtained by multiplying the standard deviation threshold by the proportionality coefficient estimated for the Weibull law [see (3.7)];
- Application of both thresholds in the mean-standard deviation plane isolates corresponding echoes pixels in the original image.

Fig. 10(b) presents an original sonar image. Corresponding mean-standard deviation representation is presented in Fig. 10(a) where the dashed line represents the proportionality coefficient between mean and standard deviation (estimated with the Weibull law), and the solid lines feature the threshold values. Corresponding segmentation of the image is presented in Fig. 10(c): the echoes have been correctly set apart.

To automate the segmentation algorithm proposed, we now propose a method to automatically set the standard deviation threshold value (the threshold value for the mean

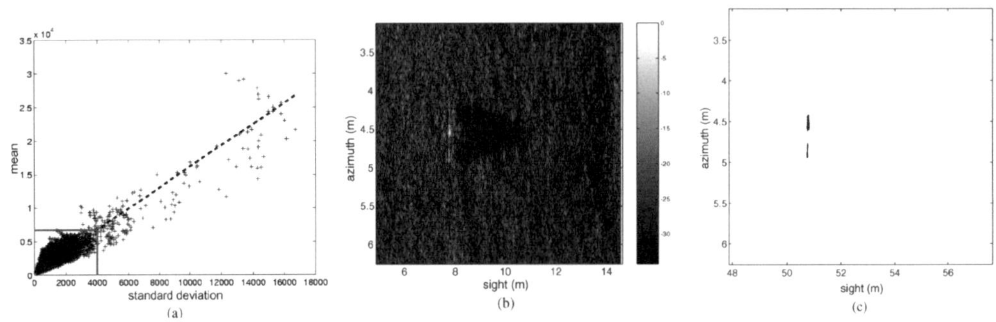

Fig. 10. Segmentation of the SAS image of Fig. 2 (thresholds: standard deviation: 4000; mean: 6751). (a) Thresholds (in thick lines). (b) SAS image. (c) Result of the segmentation

is then set accordingly). This is achieved in stepwise fashion by means of a progressive segmentation: the results obtained with decreasing standard deviation thresholds are computed. For each result, the spatial distribution of the segmented pixels is studied by computing corresponding entropies[2] with respect to the two axes, until a maximum value was reached. For each result, the histograms of the segmented pixels along the *X*- and the *Y*-axis are computed and normalized (so that they sum to 1). See Fig. 12 for one example. Then, the entropy H_{axis} on each axis is computed by the following:

$$H_{axis} = -\sum_{i \in I} p_{axis}(i) \log_2 p_{axis}(i) \tag{3.31}$$

with $p_{axis}(i)$ being the number of segmented pixels (after normalization) in the column (respectively, the line) number i, $I = \{I = 1 \ldots N_{axis} \text{ with } p_{axis} \neq 0\}$, and N_{axis} being the number of columns (respectively, lines) of the original image. These entropies characterize the spreading of the segmented pixels in the SAS image: a uniform distribution of the segmented pixels over the image leads to high entropies, whereas much localized regions lead to small entropies.

As a consequence, a decrease of the threshold value (more pixels are segmented) leads to an increase of the entropy (segmented pixels tend to distribute over the whole image). However, this increase is not regular (see Fig. 13): for instance, two slope break points clearly appear in the entropy evolution along the azimuth axis and one appears for the sight axis (they are pointed out by arrows in Fig. 13. They correspond to the standard deviation threshold of about 6250 and 4000, respectively). For a better understanding of these irregularities, the segmentations corresponding to different threshold values are presented in Fig. 11: when the threshold progressively decreases, the first echo begins to be segmented; then, the second echo is segmented as well which explains the rapid increase of the entropy (break point 1). Finally, the random background reverberation is reached, with segmented pixels spread all over the image. This explains the sharp increase of entropy (break point 2). Note that with the two segmented echoes being parallel to the azimuth axis, the first slope breaking is only visible on the azimuth axis (the sight axis only "sees" one echo).

---

[2] Entropy-based segmentation algorithms have already been proposed in the literature. For example, Pun used an entropy criterion, evaluated on the gray level histogram (Pun, 1980, 1981).

As a conclusion, the optimal segmentation, detecting both echoes with a maximal size but with no background element, is obtained with a threshold corresponding to the highest slope breaking (with a lower threshold, structures from the background are segmented generating false alarms in the system). This optimal threshold value is automatically detected from the derivative profile of the entropy. For this purpose, the maximum[3] of the two entropies defined previously is computed [see Fig. 14(a)]. The maximum of the derivative points out the highest slope breaking. However, to detect the real beginning of this slope breaking, the threshold corresponding to the half of this maximum is selected [see Fig. 14 (b)]. The result obtained on the SAS image from this threshold is presented in Fig. 12: the two echoes are correctly segmented.

Note that the computed threshold value is used in the following for the fusion process.

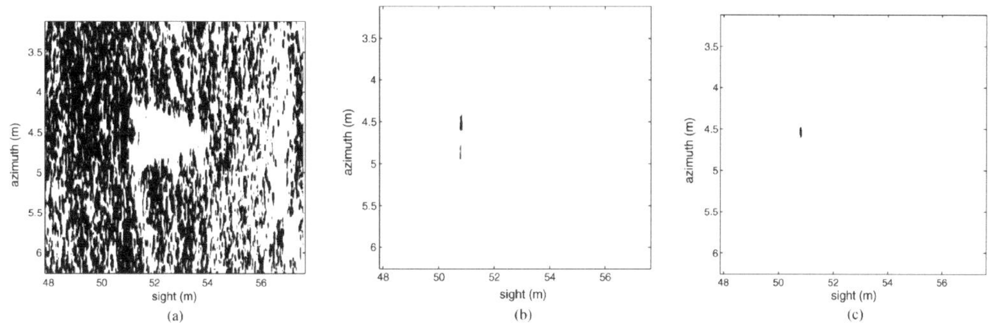

Fig. 11. Segmentation results for different thresholds in standard deviation. (a) Threshold 1000. (b) Threshold 5000. (c) Threshold 8000

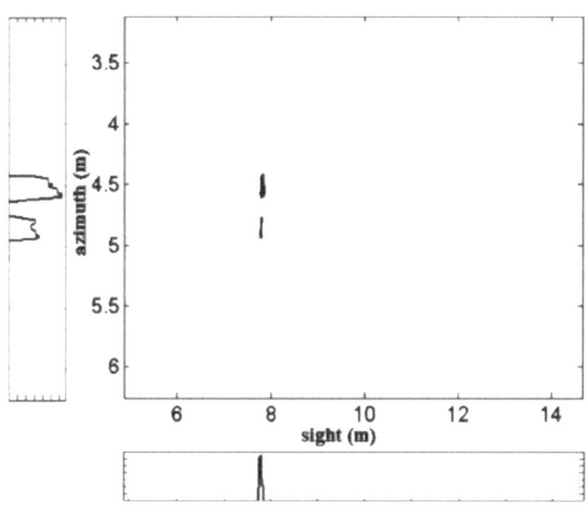

Fig. 12. Segmented SAS image and repartition of the segmented pixels according to the two axes. Computed entropies: X-axis: 3.46; Y -axis: 4.58

[3] Similar results are obtained with other combination operators (simple sum, quadratic sum, etc.).

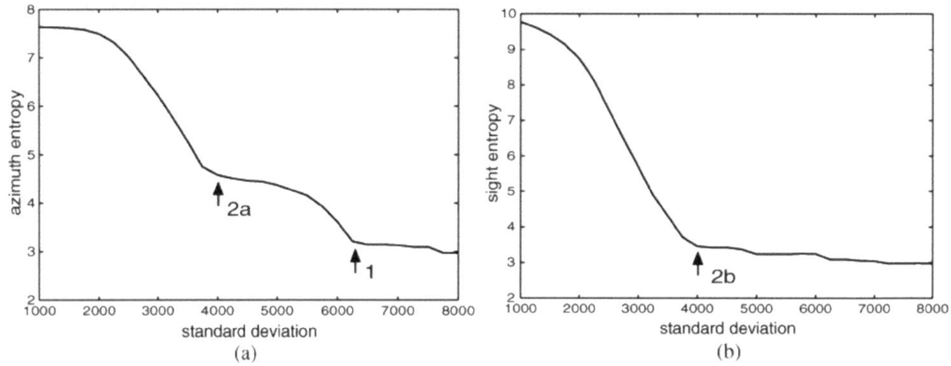

Fig. 13. Entropy variation on the two axes in function of the threshold in standard deviation. (a) Azimuth axis. (b) Sight axis

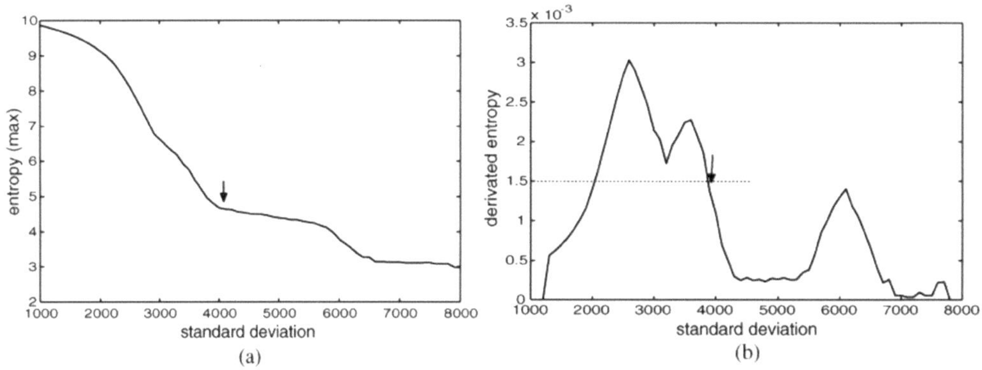

Fig. 14. Maximum of the entropies, its derivative, and setting of the threshold (see the arrows). (a) Entropy (max). (b) Derivate entropy

3.3 Higher order statistics

Pertinent information regarding SAS data can also be extracted from higher-order statistics (HOSs). In particular, the relevance of the third-order (skewness) and the fourth order (kurtosis) statistical moments for the detection of statistically abnormal pixels in a noisy background is discussed in (Maussang *et al.*, EURASIP, 2007). In this previous work, an algorithm aiming at detecting echoes in SAS images using HOS is described. It basically consists in locally estimating the HOS on a square sliding window.

3.3.1 HOS estimators

The two most classically used HOSs are the skewness (derived from the 3rd-order moment) and the kurtosis (derived from the 4th-order moment) (Kendall & Stuart, 1963). One should underline that beyond these two standard statistics, other statistics with an order greater than 4 can be mathematically defined. However, these statistics are extremely difficult to estimate in a reliable and robust way and are thus practically never used. Noting $\mu_X(r)$ as the rth order central moment of a random variable X, the definition of the skewness is given by:

$$S_X = \frac{\mu_{X(3)}}{\mu_{X(2)}^{3/2}} \tag{3.32}$$

A definition of the kurtosis is given by:

$$K_X = \frac{\mu_{X('4)}}{\mu_{X(2)}^{2}} - 3 \tag{3.33}$$

The skewness measures the symmetry of a random distribution, while the kurtosis measures whether the data distribution is peaked or flat relative to a normal distribution. These statistics are theoretically zero for the normal distribution.

To estimate the skewness and the kurtosis on a sample X of finite size N, k-statistics $k_{X(r)}$ can be used. k_r is defined as the unique symmetric unbiased estimator of the cumulant $\kappa_{X(r)}$ on X (Kendall & Stuart, 1963). An unbiased estimator of the skewness is then given by:

$$\hat{S}_X = \frac{\mu_{X(3)}}{\mu_{X(2)}^{3/2}} \tag{3.34}$$

Defining the rth sample central moment of X by the following expression:

$$m_{X(r)} = \frac{1}{N}\sum_{i=1}^{N}\left(x_i - \bar{x}\right)^r \tag{3.35}$$

where $\bar{x} = (1/N)\sum_{i=1}^{N} x_i$ and x_i are the N samples of X, we can derive another definition of this estimator. Actually, considering the relationships between $k_{X(r)}$ and $m_{X(r)}$, we have:

$$\hat{S}_X = \frac{\sqrt{N(N-1)}}{N-2}\frac{m_{X(3)}}{m_{X(2)}^{3/2}} \tag{3.36}$$

In the same way, we derive the following estimator for the kurtosis:

$$\hat{K}_X = \frac{k_{X(4)}}{k_{X(2)}^{2}} = \frac{(N+1)(N-1)}{(N-2)(N-3)}\frac{m_{X(4)}}{m_{X(2)}^{2}} - \frac{3(N-1)^2}{(N-2)(N-3)} \tag{3.37}$$

Asymptotic statistical properties are studied for high values of N. Firstly, we can mention that these estimators are biased in the first order and that they are correlated (the bias being dependent on higher-order moments). However, exact results can be derived in the Gaussian case. In this case, M and V being the mean and the variance respectively, we have:

$$M\left(\hat{S}_X\right) = 0$$

$$M\left(\hat{K}_X\right) = 0$$

$$V\left(\hat{S}_X\right) = \frac{6N(N-1)}{(N-2)(N+1)(N+3)} \approx \frac{6}{N} \tag{3.38}$$

$$V\left(\hat{K}_X\right)=\frac{24N(N-1)^2}{(N-3)(N-2)(N+3)(N+5)}\approx\frac{24}{N}$$

In the general case, there is no analytical expression for unbiased estimators independently from the probability density function of the random value. However, one should note that in the case of a normal distribution, the estimators are unbiased. Nevertheless, variances of these estimators are relatively high and it is well known that a reliable estimation requires a large set of samples.

3.3.2 Application on sonar images

We have seen in section 3.1 that a good statistical model of the background noise in the case of high resolution sonar images is given by the Weibull law. With such a non-Gaussian distribution, background values of the skewness and the kurtosis are not null anymore. On real SAS data, δ [see (3.5)] is function of the resolution of the image, but it is generally approximated by 1.65 (Maussang *et al.,* 2004). This corresponds to skewness and kurtosis values close to 1 (Fig. 15).

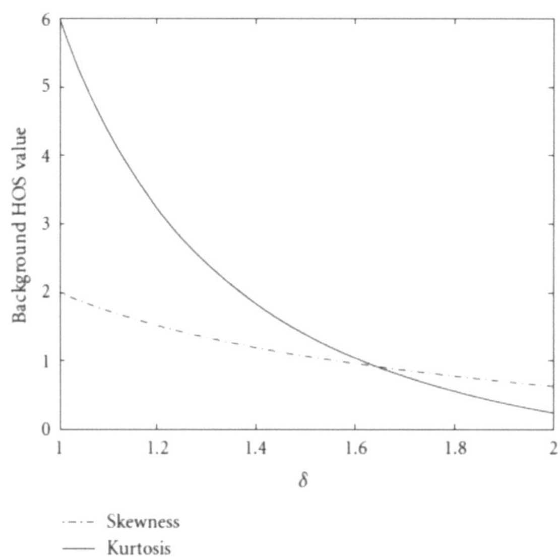

Fig. 15. Weibull background HOS values in function of the parameter δ of the Weibull law

Considering the echoes generated by the mines as deterministic elements, the SNR is sufficiently high to have higher values of the HOS if the calculus window contains an echo (Maussang *et al.*, EURASIP, 2007).

Fig. 16 presents the kurtosis results obtained on SAS image of Fig. 3 where all the objects of interests are framed by high values of the kurtosis, the size of the frame being linked to the size of the computation window. A theoretical model of these frames is used to perform a matched filtering and thus refocus the detection precisely at the center of the objects of interest. The last step consists in rebuilding the objects using a morphological dilation (Maussang *et al.,* EURASIP, 2007). The corresponding detection result is presented in Fig. 16(c): all the objects of interest are marked by high values, thus providing a good detection.

However, some false alarms remain, and the detection is not as accurate as with the first algorithm (see section 3.2). This will be taken into account for the fusion process (section 4).

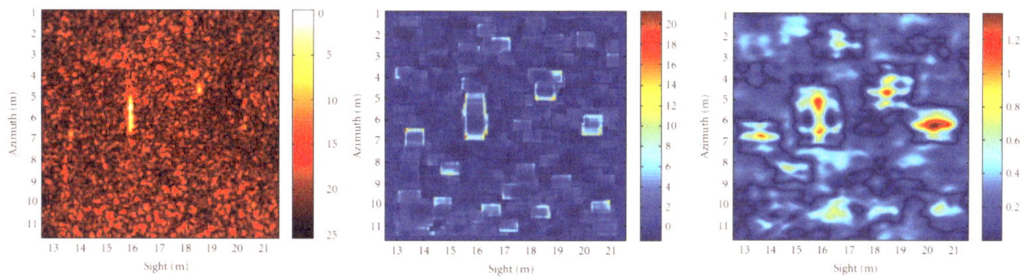

(a)SAS image (b) Kurtosis (c) Detection

Fig. 16. Detection on the SAS data of Fig. 3 (kurtosis 11 × 11, matched filtering 15 × 15, SD = 3).

4. Underwater mines detection using belief function theory

In the previous section, we have presented two algorithms aiming at detecting echoes in SAS images. In order to further improve the detection performances, we present a fusion scheme taking advantage of the different extracted parameters. The combination of parameters in a fusion process can be addressed using probabilities. This popular framework has a solid mathematical background (Duda & Hart, 1973). Numerous papers have been written on this theory using modeling tools (parametric laws with well-studied properties) and model learning. However, these methods are affected by some shortcomings. Firstly, they do not clearly differentiate doubt from conflict between sources of information. Single hypothesis being considered, the doubt between two hypotheses is not explicitly handled and the corresponding hypotheses are usually considered as equiprobable. Conflict is handled in the same way. Moreover, probabilities-based fusion methods usually need a learning step using a large amount of data, which is not necessarily available for an accurate estimation.

Another solution consists in working within the belief function theory (Shafer, 1976). The main advantage of this theory is the possibility to deal with subsets of hypotheses, called propositions, and not only with single hypothesis. It allows to easily model uncertainty, inaccuracy, and ignorance. It can also handle and estimate the conflict between different parameters. Regarding the problem of detection, this theory enables the combination of parameters with different scales and physical dimensions. Finally, the inclusion of doubt in the process is extremely valuable for the expert who can incorporate this information for the final decision. As a conclusion, the belief function theory is selected to address the considered application. The proposed fusion scheme is described in the next subsection.

4.1 Fusion scheme and definition of the mass functions

For the detection of echoes in SAS images, the frame of discernment Ω defined for each pixel is composed of the two following hypotheses:

i. "object" (O) if the pixel belongs to an echo reflected by an object;

ii. "nonobject" (NO) if it belongs to the noisy background or a shadow cast on the seabed.

The set of propositions 2^{Ω} is thus composed of four elements: the two single hypothesis, also called singletons, O and NO, the set Ω = {O,NO}, noted O U NO (U means logical OR) and called "doubt," and the empty set called "conflict." In this application, the world is obvious closed (Ω contains all the possible hypotheses).

The proposed fusion process uses the local statistical parameters extracted from the SAS image, as presented in section 3. These parameters are fused as illustrated on Fig. 17: the relationship between the first two statistical orders is taken into account by using the thresholds in standard deviation and mean estimated by the automatic segmentation, the third and fourth statistical moments are used after focusing and rebuilding operations.

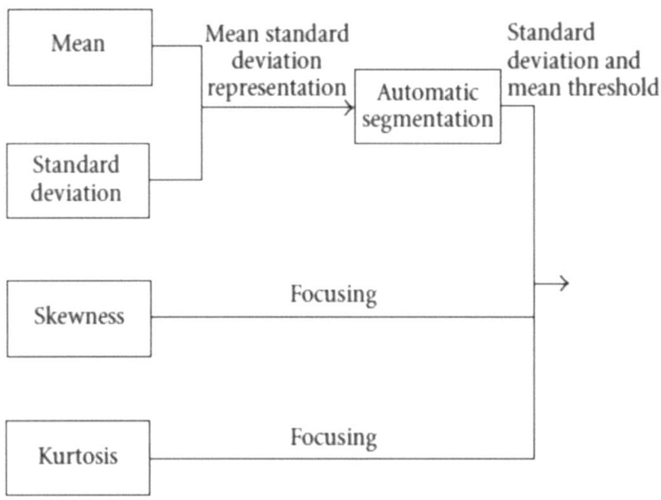

Fig. 17. Main structure of the proposed detection system

The mass of belief is the main tool of the belief function theory as the probability for the probability theory. The definition of the mass functions enables to model the knowledge provided by a source on the frame Ω. In this application, every parameter is used as a source of information. For one given source i, a mass distribution m_i^t on 2^Ω is associated to each value t of the parameter. This type of functions verifies the following property: $\sum_{A \subset 2^\Omega} m(A) = 1$. We propose to define each mass function by trapezes or semi trapezes. In the considered application, only the three propositions (O), (NO), and (O U NO) are concerned. Four thresholds must thus be defined namely, t_i^1, t_i^2, t_i^3, and t_i^4 (see Fig. 18). They are set using knowledge on local and global statistics of sonar images. They also take into account the minimization of the conflicts while preserving the detection performances (no *nondetection*).

The first mass function concerns the two first statistical orders simultaneously because they are linked by the proportional relationship. In order to build the trapezes, we consider the pair (mean; standard deviation) as used for the automatic segmentation. Fig. 19 illustrates the design of the corresponding mass function, based on the mean standard deviation representation. We first describe this function in the general case, the setting of the parameters being described afterward.

Pixels with a local standard deviation below t_1^1 are assigned a mass equal to one for the proposition "nonobject" and a mass equal to zero for the others. Pixels with a local standard deviation between t_1^1 and t_1^2 are assigned a decreasing mass (from one to zero) for the proposition "nonobject," an increasing mass (from zero to one) for the proposition "doubt,"

meaning "object OR nonobject" (O U NO). These variations are linear in function of the standard deviation.
The construction of the mass functions goes in a similar way for t_1^3 and t_1^4. This mass function is function of the standard deviation, but, considering the proportional relation holding between the mean and the standard deviation, an equivalent mass function can easily be designed for the mean. Then the mass function corresponding to the mean being redundant with the standard deviation is not computed.
We propose to set the different parameters of these mass functions using the following expressions:

$$t_1^1 = M_W(\hat{\sigma}_B);$$
$$t_1^2 = M_W(\hat{\sigma}_B) + \sqrt{V_W(\hat{\sigma}_B)};$$
$$t_1^3 = \sigma_s - \frac{1}{2}\sqrt{V_W(\hat{\sigma}_B)}; \quad (4.1)$$
$$t_1^4 = \sigma_s + \frac{1}{2}\sqrt{V_W(\hat{\sigma}_B)}$$

where $\hat{\sigma}_B$ stands for the background standard deviation estimated, using the Weibull model previously computed, on a region of the image without any echo. σ_s is the threshold in standard deviation fixed by the algorithm described in section 3.2. M_W and V_W are the mean and variance of the standard estimators (Kendall & Stuart, 1963) applied on σ_s considering the size of the computation window used for mean standard deviation building. This allows taking into account the uncertainty in the statistical parameters estimation by the fuzziness of the mass distributions.

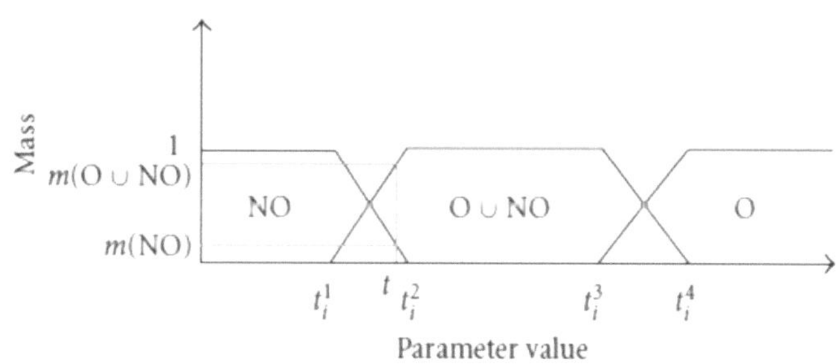

Fig. 18. Definition of the mass functions

The two other mass functions concern the HOSs: the skewness and the kurtosis, respectively. As mentioned in section 3.3, the corresponding detector provides less accurate results, which prevent a precise definition of the areas of interest. Furthermore, some artifacts generate false alarms. As a consequence, the information provided by these parameters will only be considered to assess the certainty of belonging to the background. A null mass is thus systematically assigned to the proposition "object," whatever are the values of the HOS. The mass is distributed over the two remaining propositions: "nonobject" and "doubt." This is illustrated on Fig. 20: only two parameters remain t_2^1 and t_2^2.

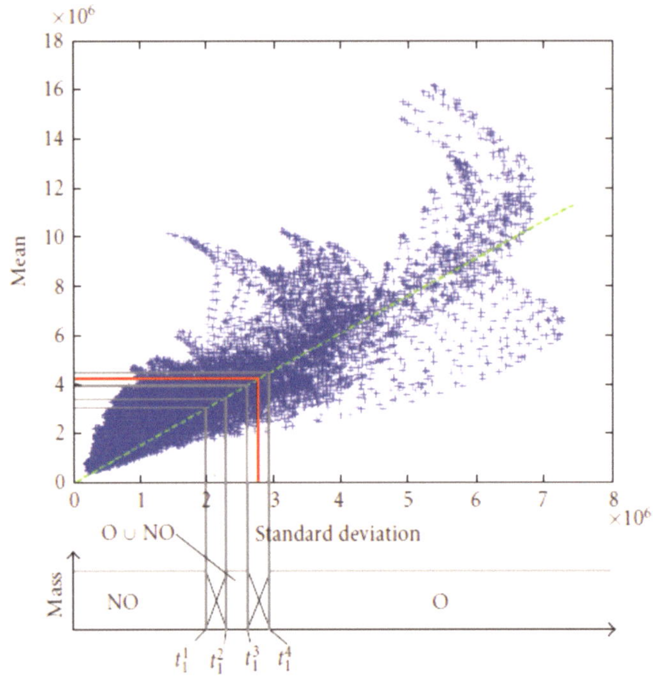

Fig. 19. Definition of the mass functions for the first two-order statistical parameters: $t_1^2 - t_1^1 = t_1^4 - t_1^3 = \sqrt{V(\hat{\sigma}_W)}$ (the thresholds obtained from the automatic segmentation are in red. The mean standard deviation graph given on this figure has been calculated on the image presented on Fig. 4).

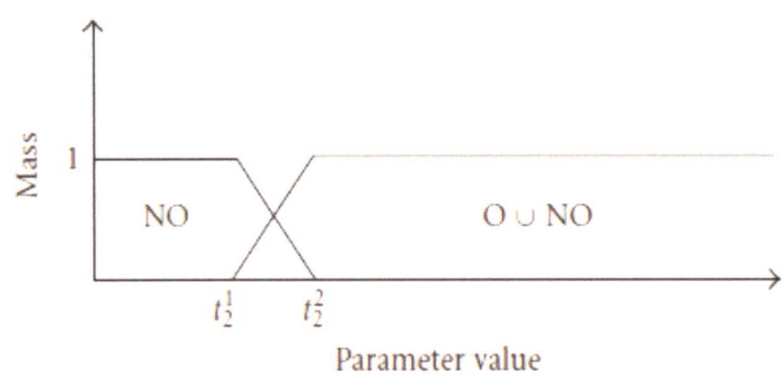

Fig. 20. Definition of the mass functions for the higher-order statistics (the graphic is valid for definition of t_3^1 and t_3^2)

Parameters t_2^1 and t_2^2 (skewness) are set by considering the normalized cumulative histogram, noted $H(t)$, of the HOS values over the whole SAS image. This is illustrated in Fig. 21. Considering that pixels with low HOS values necessarily belong to the noisy background and that pixels with high values (that might belong to an echo of interest) are extremely rare, the following expressions are used:

$$t_2^1 = H^{-1}(0.75)$$

$$t_2^2 = H^{-1}(0.90) \tag{4.2}$$

These equations are valid for t_3^1 and t_3^2 (kurtosis). This assumes that at least 75% of the image belongs to the background, which is easily fulfilled. Similarly, the 10% pixels with the highest HOS values are considered as potential objects (doubt has a mass equal to one).

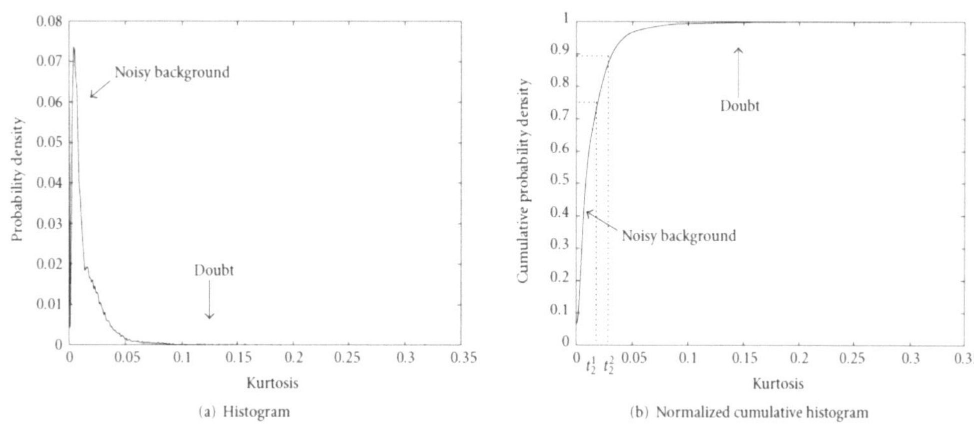

Fig. 21. Example of histogram and cumulative histogram of a kurtosis image (after focusing and rebuilding, as presented on Fig. 16(c))

Based on the local statistical moments of the data, three mass functions have been defined. The data fusion aims at improving the detection performances and eases the final decision by the expert. It is performed using the following conjunctive rule:

$$m_{1,2,3} = m_1 \oplus m_2 \oplus m_3 \tag{4.3}$$

where $\oplus$ is the conjunctive sum: $(m_1 \oplus m_2)(A) = \sum_{B \cap C = A} m_1(B) m_2(C)$ with A a proposition.

Note that, for the sake of simplicity, the superscript t of the mass m_i^t, corresponding to the parameter value, is removed from the notations.

A conflict between the different sources can appear during this combination phase. This information is preserved as it is valuable to assess the adequacy of the fused parameters. If one of the fused parameters provides irrelevant information, the conflict is high. Further investigations are then required to determine the cause of this situation (bad estimation of the parameter, limits of the data, etc.).

4.2 Decision

The results of the fusion step can be used in different ways, producing different end user products:

i. "binary" representations can be generated, providing segmented images and giving a clear division of the image into regions likely to contain objects or not;
ii. "enhanced" representations of the original SAS image can also be constructed from the results of the fusion.

These representations should somehow underline the regions of interest while smoothing the noise, but leave the decision to the human expert. The "binary" representations only use the results of the fusion process in order to classify each pixel according to the belief, the plausibility, or the pignistic value. A simple solution consists in thresholding the belief or plausibility for the proposition "object," for instance, all the pixels with a belief above 0.5 are assigned to the class "echo." A binary image is obtained, separating the echoes from the background. However, this method requires the setting of the threshold by the user. In order to overcome this shortcoming, another strategy consists in associating each pixel to the hypothesis ("object" or "nonobject," resp.) with the highest belief. This unsupervised method also provides a binary image. The same methods can be used with the plausibility. However, since the space of discernment only contains two elements, plausibility and belief actually provide the same results. The corresponding results are presented on different datasets on Fig. 22(d) and Fig. 23(d).

Beyond the binary result, a more precise classification can be constructed by assigning every pixel to the class with the highest mass of evidence (including the conflict). The resulting image is divided into four classes: "object," "nonobject," "doubt," and "conflict." Corresponding results are presented on Fig. 22(c) and Fig. 23(c). This nonbinary representation leaves more flexibility to the expert for the final interpretation. A similar strategy has been used in the frame of medical imaging in (Bloch, 1996).

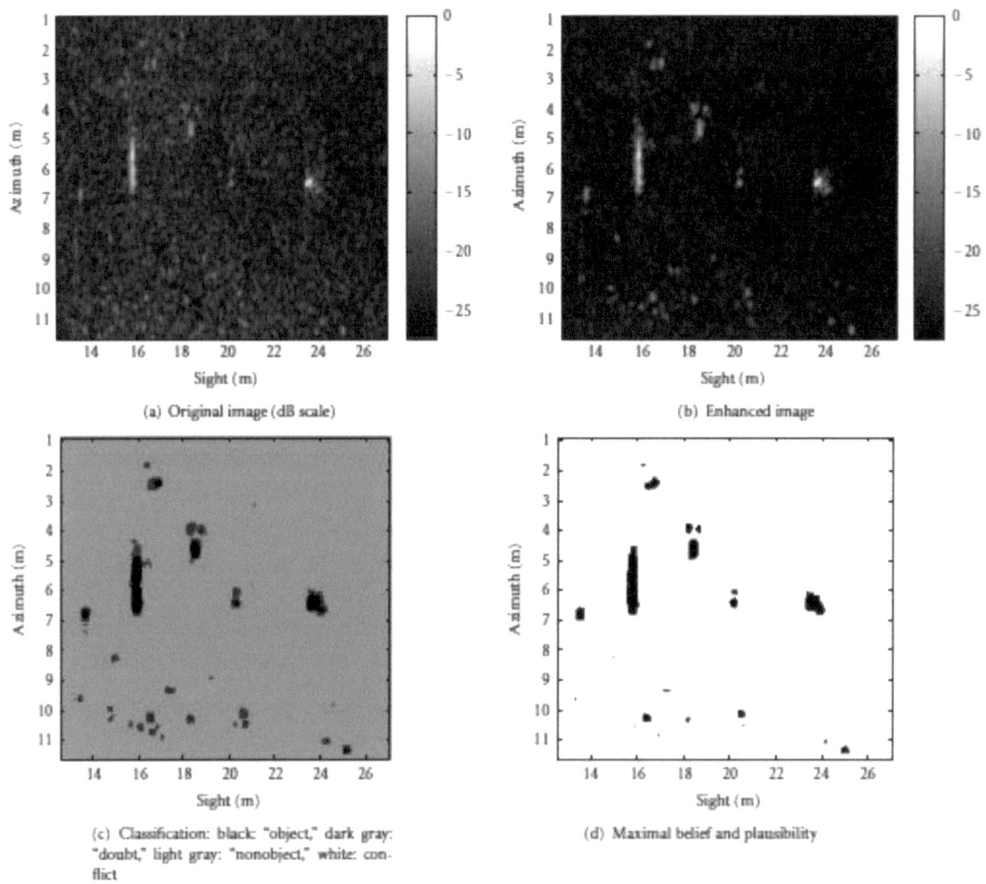

(a) Original image (dB scale)

(b) Enhanced image

(c) Classification: black: "object," dark gray: "doubt," light gray: "nonobject," white: conflict

(d) Maximal belief and plausibility

Fig. 22. Presentation of the fusion results—image of Fig. 3

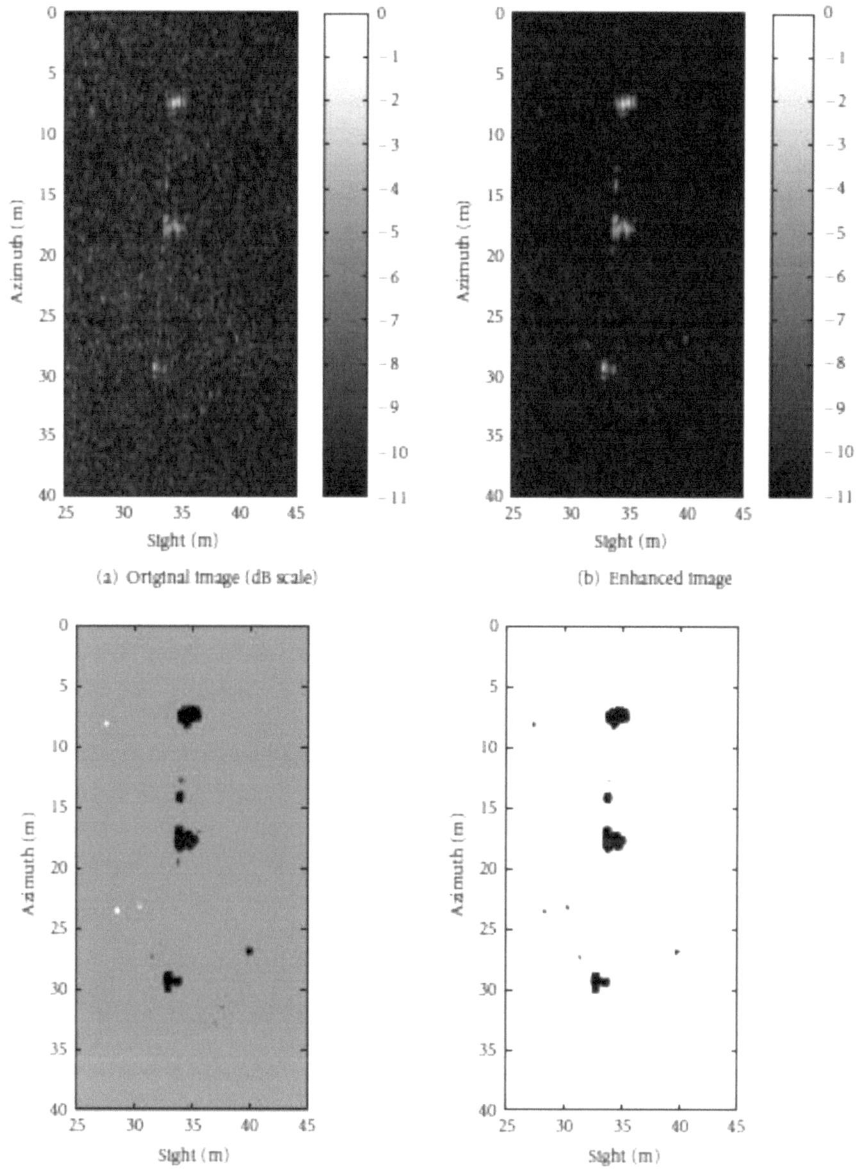

(a) Original image (dB scale)

(b) Enhanced image

(c) Classification: black: "object," dark gray: "doubt," light gray: "nonobject," white: conflict

(d) Maximal belief and plausibility

Fig. 23. Presentation of the fusion results—image of Fig. 4

These representations are well suited for a "robotics oriented" detection: regions of interest are defined, and an automatic system, such as an autonomous underwater vehicle (AUV), can be sent to identify the objects. However, such representations lose a lot of potentially valuable information (environment, relief, intensity, etc.). Such information may be useful for a human expert to actually identify the objects and solve some ambiguities. Therefore, other representations can be considered. For instance, we propose to combine the results of the fusion with the original image in order to enhance information. This is achieved by weighting the pixels of the original image by a factor linearly derived from the belief (or the plausibility) of the class "object." The intensity of pixels likely NOT being echoes is decreased (low belief), thus enhancing the contrast with the pixels most likely being echoes. For instance, Fig. 22(b) and Fig. 23(b) feature the resulting images with the weighting factor linearly ranging from 0.3 for a null plausibility to 1 for a plausibility of 1. On these images, the background tends to disappear, but all potential objects of interest are preserved. Finally, another solution consists in performing an adaptive filtering of the sonar image in function of the belief. This is described in (Maussang *et al.*, 2005).

4.3 Performance estimation

The decision coming from the results of the fusion process is valid only if the algorithm generating these results is sufficiently efficient. That is why, assessing the performances of the proposed algorithm is a crucial problem. In this section, we propose and discuss different approaches. A manually labelled ground truth can be taken into account or not, the evaluation can work on a direct analysis of the mass functions or on the classification results.
Note that in this section $m_i(A)$ denotes the mass value associated to the proposition A for the pixel i after fusion.

4.3.1 Intrinsic qualities of the mass functions

The evaluation of the detection performances can first be addressed by directly considering the quality of the resulting mass distribution. The first criterion is the nonspecificity (Klir & Wierman, 1999). This value estimates the ambiguity remaining in the mass distribution: it is low if the largest part of the mass of evidence is on a singleton or a single hypothesis (certain response); it is high if the mass is on a proposition of higher cardinal (doubt on several hypotheses). The nonspecificity is defined on a mass m by the following expression:

$$N(m) = \sum_{A \in 2^{\Omega}} m(A).\log_2 |A| \tag{4.4}$$

where $|A|$ is the cardinal of the subset A. The nonspecificity can take values in the following interval:

$$0 \leq N(m) \leq \log_2 |\Omega| \tag{4.5}$$

The bottom limit (zero) is reached in the case of $a_i \in \Omega$ with $m(\{a_i\}) = 1$ (total certainty). It reaches the upper limit with $m(\Omega) = 1$ (total ignorance). The lower is the nonspecificity, the better and more accurate is the detection. A nonspecific mass function gives few false responses (limited risk), but brings limited information (all the hypotheses can be true). On the contrary, a specific response is accurate, but has a higher risk of error.
For the addressed application, the space of discernment is composed of two hypotheses. The nonspecificity is only computed for the mass associated with the proposition "doubt." Moreover, this value is bound with each pixel of an image. We choose then to define the density of nonspecificityestimating the quality of the fusion result on the whole image. It is defined by the following expression:

$$d_{N(m)} = \frac{1}{n}\sum_{i=1}^{n} m_i(O \cup NO) \tag{4.6}$$

with n the size of the image (in pixels) and m_i(O U NO) the mass of "doubt" for the pixel i after the fusion. The values of this density are between 0 and 1. The lower is this density, the more certain is the response of the fusion.

On the other hand, the higher is the specificity, the higher is the risk of conflict. Consequently, the conflict between sources must be analyzed. As previously, we define a density of conflict with the following expression:

$$d_{C(m)} = \frac{1}{n}\sum_{i=1}^{n} m_i(\emptyset) \tag{4.7}$$

This density is between 0 and 1. Obviously, the lower is this density, the more coherent are the sources of information and the more reliable is the result.

4.3.2 Assessing the quality of the mass functions when a ground truth is available

In order to validate the results of the fusion process, additional information can be used. For instance a ground truth can be designed by the expert. Fig. 24(b) features such a segmented image where the expert roughly isolated the pixels likely to correspond to actual echoes (class "objects" (O), in black) from the background (class "nonobjects" (NO), in white).

If B denotes the environment truth ($B \in \Omega$, e.g., see Fig. 24: B = O if the pixel is in black, B = NO if the pixel is in white), we define the rate of nonspecificity knowing the environment truth B:

$$N(m/B) = \sum_{A/B\subset\Omega} m(A).\log_2|A| \tag{4.8}$$

$N(m/B)$ corresponds to the sum of the elements including B, weighted by their cardinal. For instance, for one given pixel, if B = O, the rate of nonspecificity is estimated using the masses of A = O and A = O U NO, respectively.

This expression is applied to SAS images and a rate of nonspecificity density associated with the hypothesis B can be defined by:

$$d_{N(m/B)} = \frac{1}{n}\sum_{i=1}^{n} m_i(O \cup NO).\delta_i \tag{4.9}$$

with $\delta_i(B)$ = 1 if the pixel i has label B ("object" or "nonobject") in the environment truth, $\delta_i(B)$ = 0, otherwise. In this way, only the pixels with the correct assignment B are taken into account in the density estimation. This density consequently allows to characterize the nonspecificity previously estimated (see (44)): it can either come from doubt on object detection (the most dangerous situation) or on the background.

It is obvious that addition of the density B = O and the density B = NO is equal to the density of nonspecificity of (4.6):

$$d_{N(m)} = d_{N(m/O)} + d_{N(m/NO)} \tag{4.10}$$

In a similar way, we define a rate of error knowing the environment truth by the following expression:

$$Er(m/B) = \sum_{A \cap B = \emptyset} m(A) \log_2 (|A| + 1) \tag{4.11}$$

Considering our application, the rate of density of error, associated with B, is defined by:

$$d_{Er(m/B)} = \frac{1}{n} \sum_{i=1}^{n} m_i(\overline{B}) \delta_i(B) \tag{4.12}$$

with $B \in \{O, NO\}$ and $\overline{B}$ the complementary set of B. As a matter of fact, for a given B, only the mass associated to B is taken into account: B and O U NO have at least one common element with B. The total error density can also be calculated by adding the two errors corresponding to B = O and B =NO:

$$d_{Er(m)} = d_{Er(m/O)} + d_{Er(m/NO)} \tag{4.13}$$

where $d_{Er(m)}$ is an estimation of the detection quality, considering potential mistakes on the pixel nature ("object" or "nonobject"). It should be as low as possible.

In this part, it is assumed that the designed ground truth actually corresponds to the truth. However, in real cases, this might be different as the expert might hesitate on the actual nature of some pixels (fuzzy boundaries of the objects of interest, false alarms, etc.). This results in errors that appear in the error parameters.

A last criterion measuring the performance comes from the assumption that a decision is taken for each pixel, considering the corresponding mass functions. The images of belief and plausibility associated with the hypothesis "object" are segmented by applying a threshold. 50 different threshold values, between 0 and 1, are applied. For each threshold, the detection and false-alarm probabilities are computed on the resulting binary image. This is achieved by comparing the segmentation with the environment truth. The plot of these 50 points in the false-alarm rate versus detection probability plane features the ROC curve that is classically used to assess the performances of detection systems in sonar imagery.

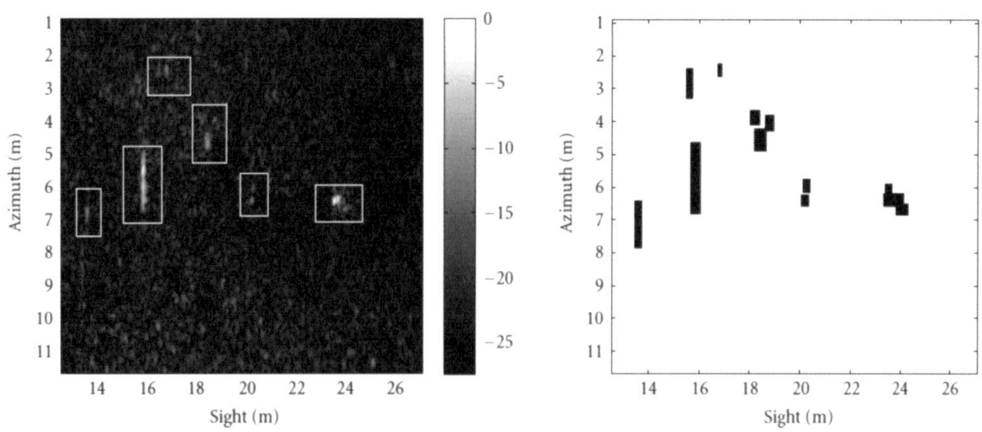

(a) Sonar image (b) Ground truth designed by the expert

Fig. 24. Example of image used for the environment truth

4.4 Results on sonar images

The fusion process presented in this paper is applied on the sonar images presented in section 2.

The first image (Fig. 3) features several buried or partially buried objects. In this image, the echoes are hardly visible apart from a partially buried cylindrical mine on the left (at 16m in sight). For each pixel and for each parameter, we firstly estimate the mass associated to each proposition ("object," "doubt," and "nonobject," resp.) by using the mass functions previously defined (Fig. 25, Fig. 26, and Fig. 27, resp.). These images are combined using the orthogonal rule in order to obtain the mass images associated to each proposition (Fig. 28). This results in an image of belief (corresponding to the mass of the class "object") and an image of plausibility (corresponding to the sum of the masses of the classes "object" and "doubt") associated to the proposition "object" (Fig. 29). One should underline that all the objects in the image are efficiently detected: belief and plausibility are close to 1 in the regions likely to contain echoes. The plausibility highlights some spurious regions at the bottom of the image. These regions have a small area and could be easily removed, for instance, by a morphological filter. The first- and second-order parameters are complementary to the third- and fourth-order ones. Actually, the doubt on Fig. 25 (1st and 2nd order) is decreased by the mass "nonobject" brought by the higher-order statistical parameters (Fig. 26 and Fig. 27). On the other hand, the doubt coming from HOS is limited by the mass "object" and "nonobject" provided by the first orders. The first-order parameters provide precise information, but with some false alarms (Fig. 25), whereas higher orders provide a few false alarms (consider the "doubt" image), but imprecise information (Fig. 26 and Fig. 27). It illustrates the usual duality between certainty and accuracy, and how a fusion process can take advantage of multiple complementarities sources.

Some conflict appears in the result of the fusion (Fig. 28(d)). However, it remains low (the sum of the masses of the focal elements is strictly inferior but close to 1), and isolated. This result shows the good concordance of the parameters.

4.4.1 Evaluation of the performances on the sonar image

A first evaluation of the fusion process consists in analyzing the contribution of each parameter to the final result. This is achieved by combining the parameters two by two. As previously observed, the addition of one HOS parameter decreases the mass "doubt" (compare Fig. 30(b) with Fig. 25(b)). The fusion of three parameters further decreases this mass (Fig. 29(b)). The more parameters are added to the fusion process, the more accurate is the response. Note that the addition of one parameter to the fusion process "selects" more accurately the masses: the "object" mass that differs from the values 0 or 1 are fewer.

A quantitative evaluation can also be completed by estimating conflict and nonspecificity densities, independently from the environment truth, or a combination of these values (rate of density of nonspecificities and error). The results are listed in Table 2. The results confirm the previous qualitative remarks as follows:

i. nonspecificity decreases when new parameters are added. Note that this density is high for the two HOS parameters and their fusion;
ii. conflict can increase with the addition of one parameter, but this is not obvious in this application.

That proves the good reliability of the chosen parameters.

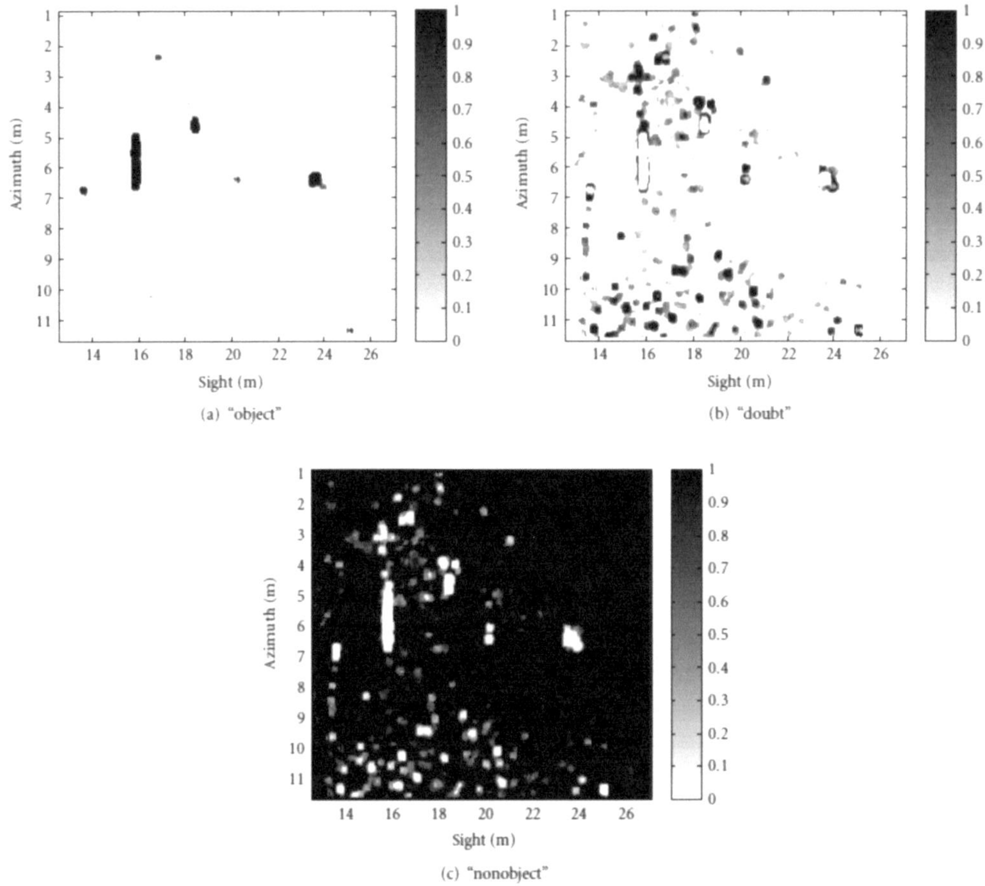

Fig. 25. Mass images obtained for each proposition after the fusion of the mean standard deviation parameter (segmentation) in Fig. 3

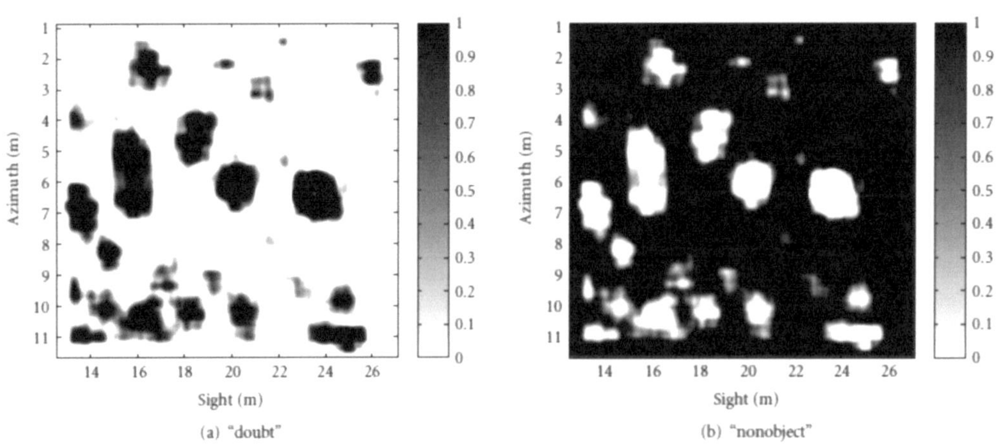

Fig. 26. Mass images obtained for each proposition with the skewness parameter in Fig. 3

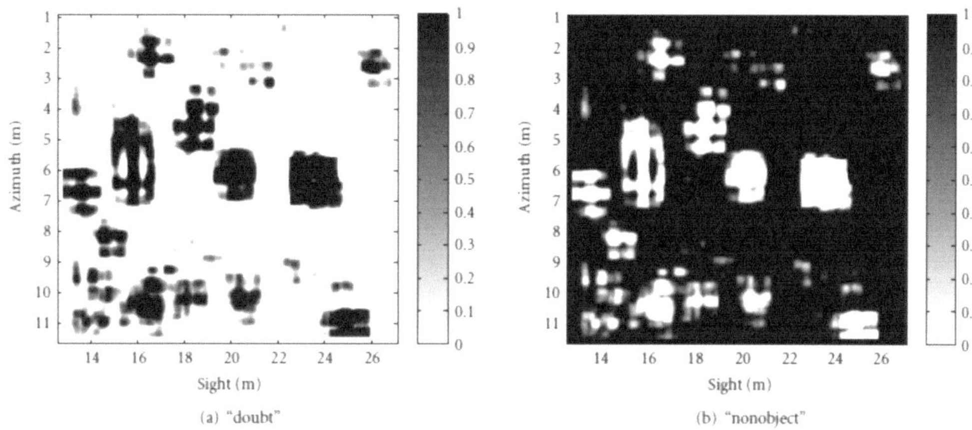

Fig. 27. Mass images obtained for each proposition with the kurtosis parameter in Fig. 3

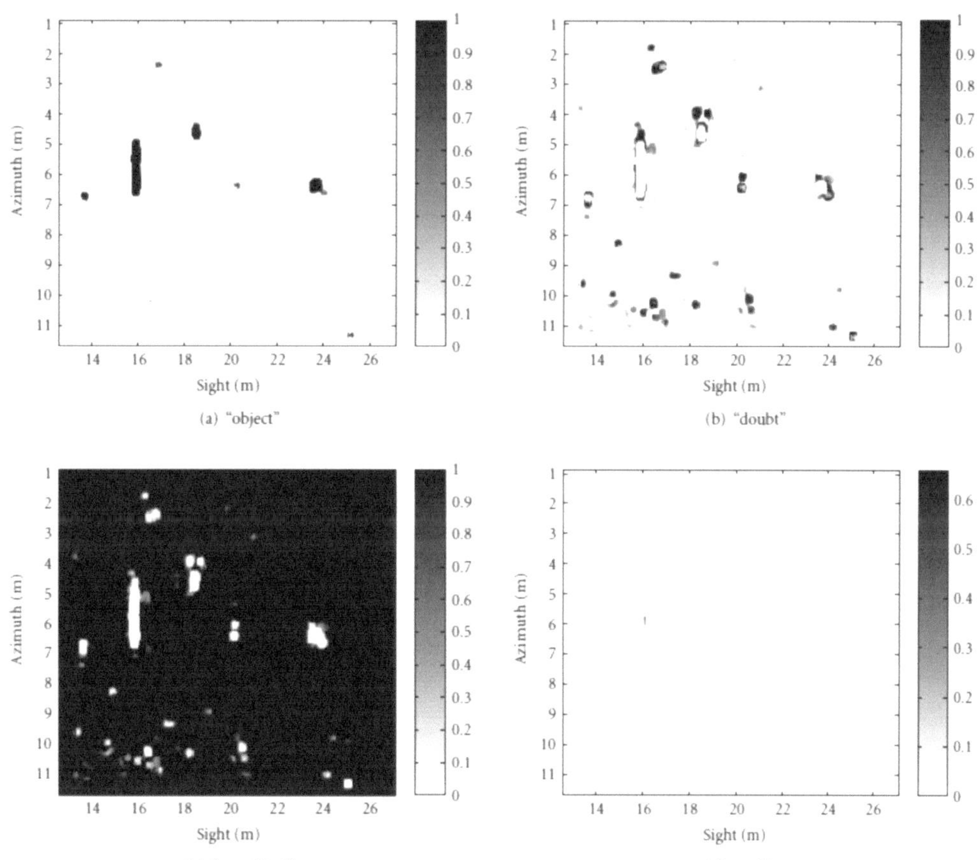

Fig. 28. Mass images obtained after fusion of the three parameters in Fig. 3

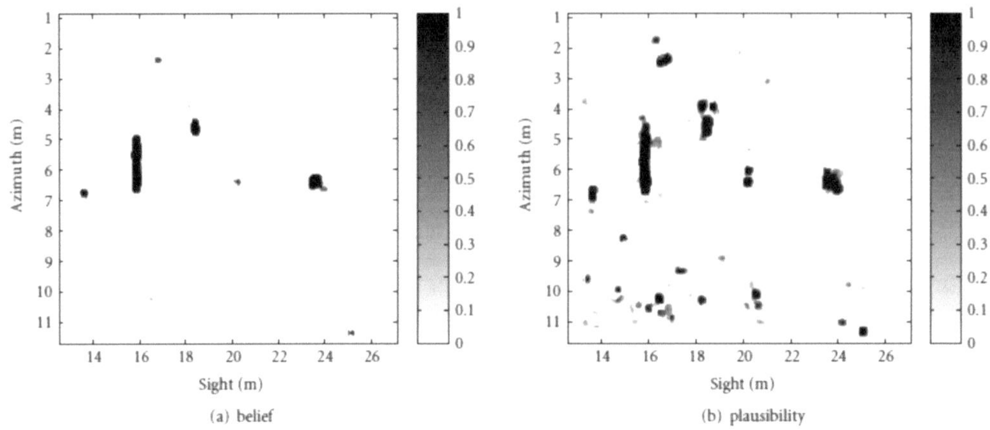

Fig. 29. Belief and plausibility object images obtained after fusion of the three parameters in Fig. 3

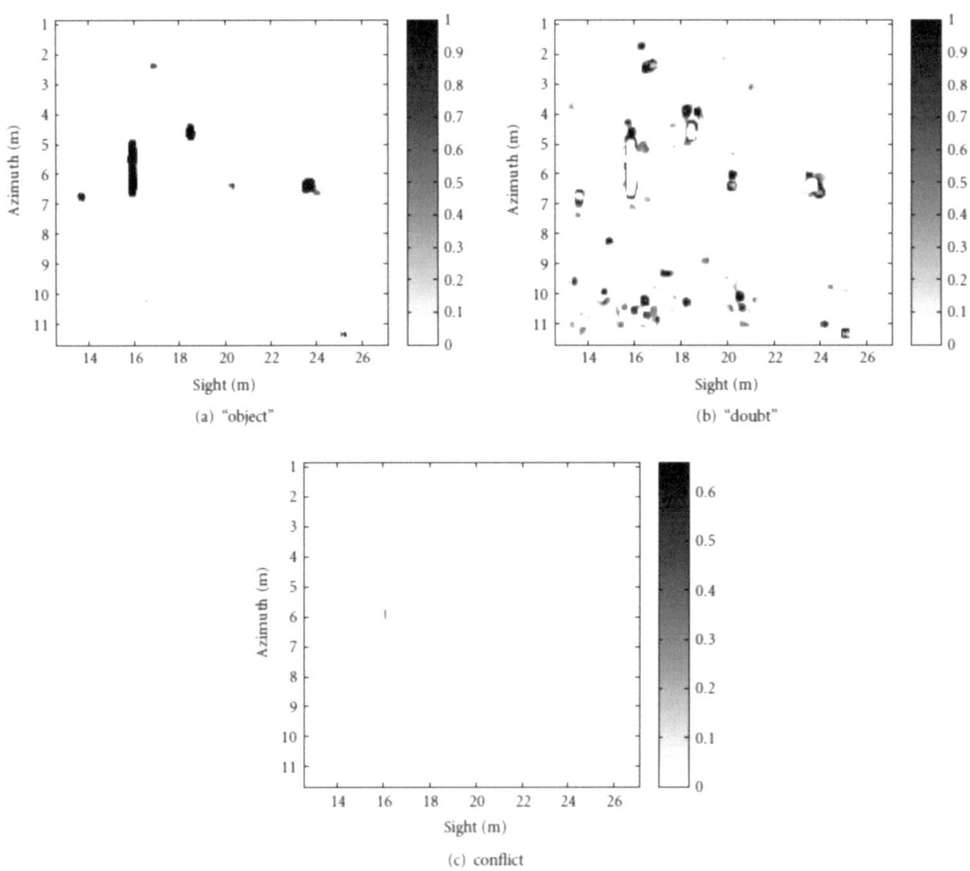

Fig. 30. Mass images obtained for each proposition after the fusion of the mean standard deviation (segmentation) and kurtosis parameters in Fig. 3

These values also estimate the amount of information brought by each parameter: if adding one parameter does not significantly decrease the density of nonspecificity, the corresponding parameter can be considered as bringing very little information. Moreover, if the density of conflict increases, this parameter is contradictory with the others and the reliability of this parameter (or one of the other) should be questioned.

The environment truth is a source of information that can be used to assess the performances of the system. The addition of one HOS parameter slightly decreases the error, which remains low for the HOS. As a matter of fact, the fuzzy definition of the mass functions keeps the error bounded (if the mass "doubt" is 1, the error is null). On the contrary, the relatively high value of error on the areas selected as "object" can be explained by the large size of the regions selected by the expert. This rough selection actually includes a part of the region selected as "background" by the fusion process; but this should not be considered as a bad detection: the echoes are well detected, but are only smaller than the masks of the original reference image. This will be confirmed by the ROC curves (the maximum detection probability is smaller than one).

The nonspecificity is greater for the "nonobject" pixels on the reference image than for the "object" pixels. This is a promising conclusion for the fusion process: the result is more accurate if a potentially dangerous object is present.

Finally, ROC curves of the fusion results are built and compared with the curves obtained with each parameter alone (segmentation with the 1st and 2nd order, the skewness, or the kurtosis). They are also compared with the ROC curves obtained with the standard detector consisting in directly thresholding the original data.

The first comment on the results presented in Fig. 31 concerns the lack of points between low values of false alarms (until 0.03) and the point of probability equal to 1. This is a consequence of the pixels declared as "echo" by the expert, but classified as "nonobject" by the system. In order to include these pixels as "object" by the system, all the pixels of the image must be selected (this is, achieved with a threshold of zero). These pixels are not significant at all and come only from the rough design of the regions containing echoes. This results in the maximum false-alarm and detection probabilities being far from the point (1, 1) (see the arrow on Fig. 31 (b)). In the same way, minimum detection and false-alarm probabilities exist for belief and plausibility obtained with a threshold of 1.

Densities ($\times 10^{-3}$)	1- 2	3	4	1-2 + 3	1-2 + 4	3 + 4	1-2 + 3 + 4
conflict	**0**	**0**	**0**	**0.0299**	**0.0885**	**0**	**0.105**
nonspecificity	**51.0**	**166.1**	**166.1**	**23.8**	**20.9**	**121.3**	**19.0**
/O	7.9	18.8	17.5	6.7	6.1	17.4	6.1
/NO	43.0	147.3	148.6	17.1	14.8	103.9	12.9
error	**5.0**	**2.2**	**3.5**	**6.2**	**6.8**	**3.5**	**6.8**
/O	4.5	2.2	3.5	5.7	6.3	3.5	6.3
/NO	0.557	0	0	0.520	0.535	0	0.518

Table. 2. Performances of the fusion in Fig. 3,
(1-2: mean standard deviation (segmentation), 3: skewness, 4: kurtosis)

The second comment is that the false-alarm rates and detection probabilities are lower for belief than plausibility. This is linked to the certainty/accuracy duality previously mentioned. Moreover, note that the plausibility and the belief curves are both above all the other curves: this assesses the improvement of the detection performances obtained thanks to the fusion process.

4.4.2 Results on other data

In this section, the proposed fusion process is tested on two more SAS images. Image of Fig. 4 (Fig. 32) represents a region of 40m × 20m of the seabed with a pixel size of about 4 cm in both directions (see section 2). It contains three cylindrical mines: one mine is lying on the sea floor (top of image), another one is partially buried (approximately in the middle of the image), and the last one is completely buried under the sea floor (lower part of image).
Fig. 32 represents the belief and plausibility after fusion, and Fig. 33 presents the corresponding ROC curves. Moreover, quantitative criteria estimated for this image are presented in Table 3 and can be compared with the results of the first image. The fusion process has been performed with mass functions defined previously, in function of the corresponding standard deviation thresholds and higher order statistics histogram.
The same comments and conclusions hold for this new image. The detection performances are improved (in particular, see the belief image). However, the fusion with the skewness parameter does not significantly affect the result in image of Fig. 4: the nonspecificity, error, and conflict densities are similar whether two or three parameters are aggregated.

Densities (x 10^{-3})	1- 2	3	4	1-2 + 3	1-2 + 4	3 + 4	1-2 + 3 + 4
conflict	**0**	**0**	**0**	**0.406**	**0.527**	**0**	**0.528**
nonspecificity	**8.1**	**159.0**	**162.8**	**3.9**	**3.5**	**122.0**	**3.4**
/O	1.8	13.5	12.0	1.8	1.5	12.0	1.5
/NO	6.3	145.4	150.8	2.2	1.9	110.0	1.9
error	**6.5**	**0.0446**	**1.6**	**6.0**	**6.2**	**1.6**	**6.2**
/O	4.4	0.0446	1.6	4.4	4.6	1.6	4.6
/NO	2.0	0	0	1.6	1.6	0	1.6

Table 3. Performances of the fusion in Fig. 4,
(1-2: mean standard deviation (segmentation), 3: skewness, 4: kurtosis)

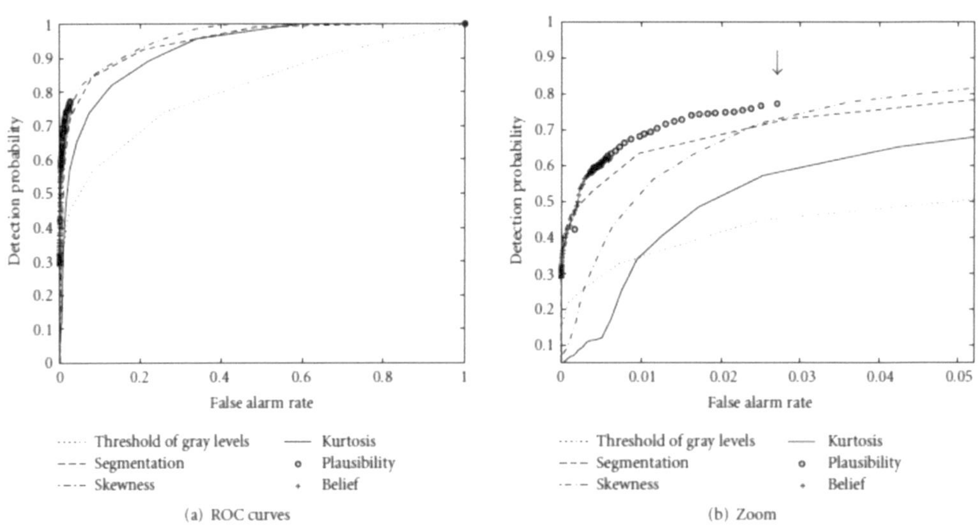

Fig. 31. ROC curves of each of the three parameters compared with the results of the fusion process (belief and plausibility) in Fig. 3

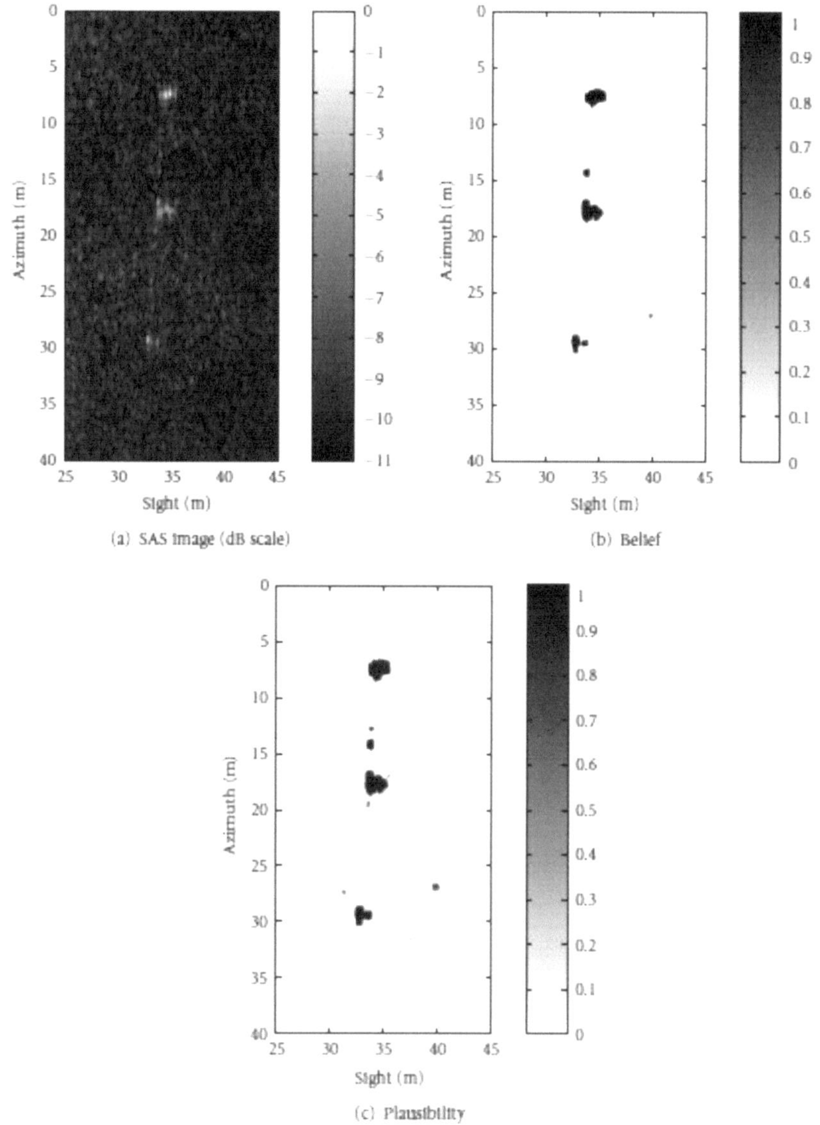

Fig. 32. Belief and plausibility images obtained after fusion of the three parameters in Fig. 4

5. Conclusion and perspectives

This chapter presented the interest of the use of high resolution images formed thanks to SAS system, and proposed a fusion architecture aiming at taking advantage of the complementary properties of sources, based on statistical properties, in order to improve the detection performances.

Being able to handle conflicts between sources and doubt between different hypotheses, the belief theory is well suited to represent and characterize the information provided by the different sources. It also provides a fusion rule. The fused data can be used either to take a decision or to enhance the data adaptively, leaving the final decision to an expert.

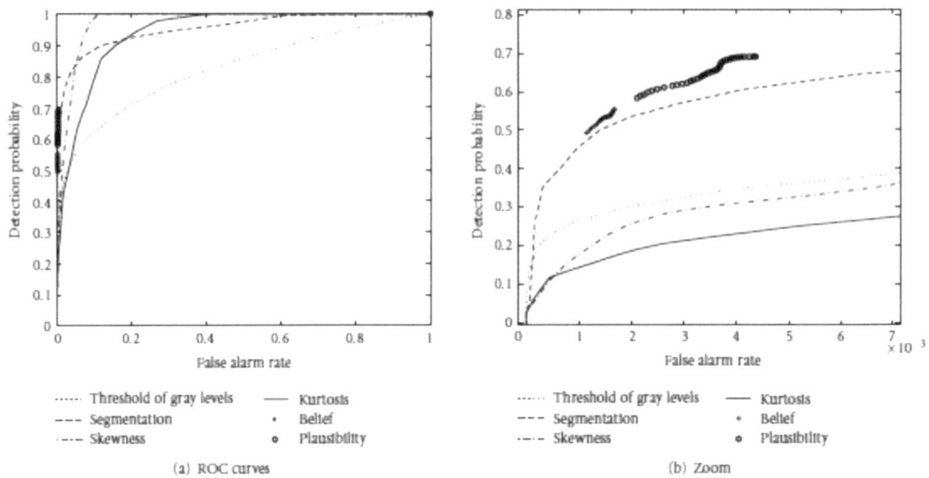

Fig. 33. ROC curves of each of the three parameters compared with the results of the fusion process (belief and plausibility) in Fig. 4

The design of the mass functions is fairly simple and flexible. A general knowledge about the acquisition system and the induced statistical properties on the SAS image enables the setting of the few parameters (trapeze-shaped functions). Confronted to different datasets, these settings were not modified, thus assessing the robustness of the whole procedure.

The evaluation of the proposed architecture is based on new parameters, some of them classically taking a manually labelled ground truth into account, some others being independent from this ground truth and aiming at directly assessing the quality of the available information.

These last criteria determine intrinsic properties of the mass functions, such as nonspecificity and conflicts densities. The first set of criteria concerns the properties conditioned by the ground truth: rates of nonspecificity and error densities, probabilities of detection and false alarm.

The fusion architecture has been tested on two real SAS images and convincing results have been obtained: the fusion actually improves the detection performances of the different sources taken separately.

The proposed process may be improved by incorporating new parameters (statistical, morphological, criteria characterizing the spatial distribution of the features, etc.) coming either from a deeper knowledge of the data or from new sonar images (multiple acquisitions). The interest of such a fusion structure lies in its flexibility: the addition of new parameters is easy to work out and does not need any change of structure or parameterization. Moreover, it is possible to estimate the quantity of information brought by each of the new parameter. This allows to reach the next levels in the detection and classification process, as described in the introduction, by deciding if the regions previously segmented actually contain a sought object and by identifying this object (mine, kind of mine, etc.).

6. Acknowledgements

The authors wish to thank Groupe d'Etudes Sous-Marines de l'Atlantique (DGA/DET/GESMA, France) and TNO, Security and Safety (The Netherlands) for providing SAS data in this work supported by GESMA.

7. References

Abbot, J.G. & Thurstone, F.L. (1979). Acoustical speckle: Theory and experimental analysis. *Ultrason. Imag.*, no. 1, 1979, pp. 303 - 324

Bloch, I. (1996). Some aspects of Dempster-Shafer evidence theory for classification of multi-modality medical images taking partial volume effect into account, *Pattern Recognition Letters*, vol. 17, no. 8, pp. 905–919, 1996

Collet, C.; Thourel, P. ; Mignotte, M. ; Pérez, P. & Bouthemy, P. (1998). Segmentation markovienne hiérarchique multimodèle d'images sonar haute résolution, *Traitement du Signal*, vol. 15, no. 3, pp. 231–250, Oct. 1998

Cutrona L. (1975). Comparison of sonar system performance achievable using synthetic-aperture techniques with the performance achievable by more conventional means. *J.Acoust. Soc. Am.*, vol 58 (2), pp 336 - 348, 1975

Cutrona L. (1977). Additional characteristics of synthetic aperture sonar systems and a further comparison with nonsynthetic-aperture sonar systems. *J.Acoust. Soc. Am.*, vol 61 (5), pp 1213 - 1217, 1977

Duda, R. & Hart, P. (1973). *Pattern Classification and Scene Analysis*, JohnWilley & Sons, New York, NY, USA, 1973

Gilmour, G. (1978). *Synthetic aperture side-looking sonar system.* US patent n° 4088978, May 1978.

Ginolhac, G.; Chanussot, J. & Hory, C. (2005). Morphological and statistical approaches to improve detection in the presence of reverberation, *IEEE J. Ocean. Eng.*, vol. 30, no. 4, pp. 881–899, Oct. 2005

Goodman, J. W. (1976). Some fundamental properties of speckle, *Journal of Optical Society of America*, vol. 66, no. 11, pp. 1145–1150, 1976

Gough P. & Hayes M. (2004). Synthetic aperture sonar : the past, the present and the future, *Proc. Institute of Acoustics, Sonar Signal Processing*, vol 26 (5), 2004.

Gu M. & Abraham, D.A. (2001). Using McDaniel's model to represent non-Rayleigh reverberation, *IEEE J. Ocean. Eng.*, vol. 26, no. 3, pp.348–357, Jul. 2001

Hanssen, A.; Kongsli, J.; Hansen, R.E. & Chapman, S. (2003). Statistics of synthetic aperture sonar images, *Proc. MTS/IEEE OCEANS Conf.*, San Diego, CA, Sep. 2003, pp. 2635–2640

Hayes, M.P. & Gough, P.T. (1999). Using synthetic aperture sonar for mine detection, *Proceedings of Austral.-Amer. Joint Conf. Technol. Mines Mine Countermeas. Syst.*, pp. 1.1 - 1.10, Sydney, Australia, July 1999

Hétet, A. (2000). Evaluation of specific aspects of synthetic aperture sonar, by conducting at sea experiments with a rail, in the frame of mine hunting systems. *Fifth European Conference on Underwater Acoustics, ECUA 2000*, Lyon, France, 2000

Hétet, A. (2003). *Contribution à la détection de mines enfouies dans le sédiment marin par synthèse d'ouverture basse fréquence.* Ph.D. thesis, University Paris 6, July 2003

Hétet, A.; Amate, M.; Zerr, B.; Legris, M.; Bellec, R.; Sabel, J.C. & Groen, J. (2004). SAS processing results for the detection of buried objects with a ship-mounted sonar. *Seventh European Conference on Underwater Acoustics, ECUA 2004*, Delft, The Netherlands, July 2004

Hory, C.; Martin, N. & Chehikian, A. (2002). Spectrogram segmentation by means of statistical features for non-stationary signal interpretation, *IEEE Trans. Signal Process.*, vol. 50, no. 12, pp. 2915–2925, Dec. 2002

Joughin, I.R.; Percival, D.B. & Winebrenner, D.P. (1993). Maximum likelihood estimator of K distribution parameters for SAR data, *IEEE Trans. Geosci. Remote Sens.*, vol. 31, no. 5, pp. 989–999, Sep. 1993

Kendall, M. G. & Stuart, A. (1963). *The Advanced Theory of Statistics*, 2nd ed. London, U.K.: Griffin, 1963, vol. 1

Klir, G. J. & Wierman, M. J. (1999). Uncertainty-Based Information, Physica, Heidelberg, Germany, 1999

Maussang, F.; Chanussot, J. & Hétet, A. (2004). On the use of higher order statistics in SAS imagery," in *Proceedings of IEEE International Conference on Acoustics, Speech and Signal Processing (ICASSP '04)*, vol. 5, pp. 269–272, Montreal, Quebec, Canada, May 2004

Maussang, F.; Chanussot, J., Visan, S.C. & Amate, M. (2005). Adaptive anisotropic diffusion for speckle filtering in SAS imagery, in *Proceedings of the Oceans Conference (OCEANS '05)*, vol. 1, pp. 305–309, Brest, France, June 2005

Maussang, F.; Chanussot, J.; Hétet, A. & Amate, M. (2007). Mean-standard deviation representation of sonar images for echo detection: application to SAS images. *IEEE Journal of Oceanic Engineering*, vol. 32, no. 4, October 2007, pp. 956 - 970

Maussang, F.; Chanussot, J.; Hétet, A. & Amate, M. (2007). Higher-order statistics for the detection of small objects in a noisy background application on sonar imaging. *EURASIP Journal on Advances in Signal Processing*, Vol. 2007, Article ID 47039, 2007, 17 pages, doi:10.1155/2007/47039

Maussang, F.; Rombaut, M. ; Chanussot, J.; Hétet, A. & Amate, M. (2008). Fusion of local statistical parameters for buried underwater mine detection in sonar imaging. *EURASIP Journal on Advances in Signal Processing*, Vol. 2008, Article ID 876092, 2008, 19 pages, doi:10.1155/2008/876092

Mignotte, M.; Collet, C.; Pérez, P. & Bouthemy, P. (1997). Unsupervised Markovian segmentation of soar images, *Proceedings of the 22nd IEEE International Conference on Acoustics, Speech, and Signal Processing (ICASSP'97)*, Vol. 4, pp. 2781 - 2784, Munich, Germany, April 1997

Mignotte, M.; Collet, C.; Pérez, P. & Bouthemy, P. (1999). Three-class Markovian segmentation of high-resolution sonar images, *Comput. Vis. Image Understanding*, vol. 76, no. 3, pp. 191–204, Dec. 1999

Milisavljević, N.; Bloch, I. & Acheroy, M. (2008). Multi-sensor Data Fusion Based on Belief Functions and Possibility Theory: Close Range Antipersonnel Mine Detection and Remote Sensing Mined Area Reduction, In: *Humanitarian Demining. Innovative Solutions and the Challenges of Technology*, Habib M.K. (Ed.), pp. 95 - 120, I-Tech Education and Publishing, ISBN 978-3-902613-11-0,Vienna, Austria

Pun, T. (1980). A new method for grey-level picture thresholding using the entropy of the histogram, *Signal Process.*, vol. 2, no. 3, pp. 223–237, Jul. 1980

Pun, T. (1981). Entropic thresholding, a new approach, *Comput. Graphics Image Process.*, vol. 16, pp. 210–239, 1981

Saporta, G. (1990). *Probabilités, analyse des données et statistique.* Paris, France: Technip, 1990

Schmitt, F.; Mignotte, M.; Collet, C. & Thourel, P. (1996). Estimation of noise parameters on sonar images, *Proc. SPIE Conf. Signal Image Process.*, Denver, CO, Aug. 1996, vol. 2823, pp. 1–12

Shafer, G. (1976). *A Mathematical Theory of Evidence*, Princeton University Press, Princeton, NJ, USA, 1976

Spiess F. & Anderson V. (1983). *Wide swath precision echo sounder.* US patent n° 4400803, august 1983.

U.S. Department of the Navy (2000), *The Navy unmanned undersea vehicle (UUV) master plan*, available: http://www.npt.nuwc.navy.mil/UUV

Walsh G. (1969). *Acoustic Mapping Apparatus.* US patent n° 3484737, 1969.

Sonar Sensor Integration

7

Multi-Sonar Integration and the Advent of Senor Intelligence

Edward Thurman, James Riordan and Daniel Toal
Mobile & Marine Robotics Research Centre
University of Limerick,
Ireland

1. Introduction

The subsea environment represents the last major frontier of discovery on Earth. It is envisioned that exploration of the seabed, in both our deep-ocean and inshore waters, will present a multitude of potential economic opportunities. Recent interest in the ever-expanding exploration for valuable economic resources, the growing importance of environmental strategies and the mounting pressure to stake territorial claims, has been the main motivation behind the increasing importance of detailed seabed mapping, and rapid advancements in sensor technology and marine survey techniques (McPhail, 2002; Nitsche et al., 2004; Desa et al., 2006; Niu et al., 2007).

Over the past decade, there has been an increasing emphasis on the integration of multiple sonar sensors during marine survey operations (Wright et al., 1996; Laban, 1998; Pouliquen et al., 1999; Yoerger et al., 2000; Duxfield et al., 2004; Kirkwood et al., 2004). The synergies offered by fusing and concurrently operating multiple acoustic mapping devices in a single survey suite underpin the desire for such an operational configuration; facilitating detailed surveying of the ocean environment, while enabling the information encoded in one instrument's dataset to be used to correct artefacts in the other.

Innovative advancements in the intelligence of sensors have permitted time-critical decisions to be made based on the assessment of real-time environmental information. In-mission data evaluation and decision making allows for the optimisation of surveys, improving mission efficiency and productiveness.

While low-frequency (<200kHz) sonar has a long range imaging capability, the generated datasets are inherently of low resolution, reducing the ability to discriminate between small-scale features. Conversely, high-frequency (>200kHz) imaging sonar generates high-resolution datasets, providing greater detail and improving data analysis. High-frequency sonar systems are therefore the desired sensor systems used during seabed survey missions. However, seawater severely restricts acoustic wave propagation, reducing the range (field of view) of high-resolution sonar in particular. Consequently, high-resolution survey sensors must be deployed in close-proximity to the seabed. UUVs are ideal platforms for providing the near-seabed capability required, and often demanded, by marine survey operations (McPhail, 2002). Furthermore, recent technological advancements have allowed UUVs to provide high-resolution survey capabilities for the largely unexplored deep-water environments, previously considered uneconomical or technically infeasible (Whitcomb, 2000).

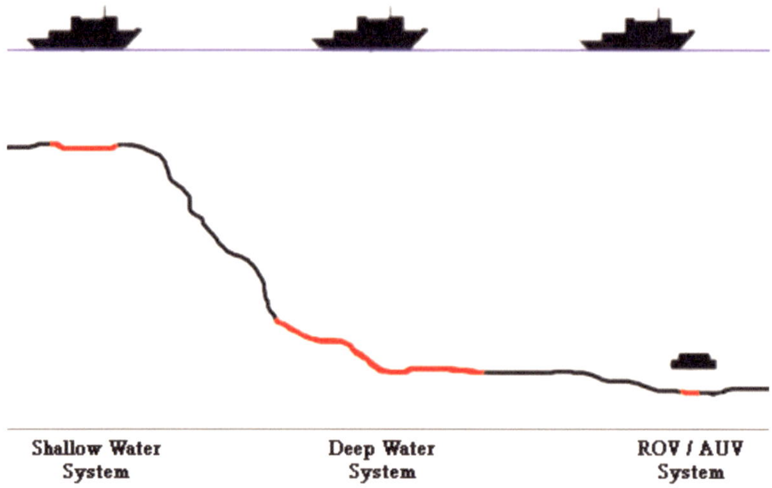

Fig. 1. Comparison of sonar systems operating at different depths. Notice the increasing footprint as the distance increases. However, as the distance increases, the operating frequency of the sonar must decrease, as seawater severely restricts acoustic wave propagation, resulting in lower resolution datasets.

	Water Depth	Operating Frequencies	Resolution	Swath Coverage	Remarks
Shallow Water Systems	< 100m	> 200kHz	Medium - High	Low - Medium	Continental shelf, inshore-water seabed surveying
Deep Water Systems	> 200m	< 200kHz	Low	High	Wide-area, deep-ocean seabed surveying
ROV/AUV Systems	5m – 4000m	200kHz – 500kHz	High	Low	Detailed, high-resolution seabed surveying

Table 1. Comparison of typical operating specifications for sonar systems operating at different depths.

However, the operation of multiple co-located, high-frequency acoustic sensors results in the contamination of the individual datasets by cross-sensor acoustic interference. The development of sensor control routines and 'intelligent' sensors helps to avoid this sensor crosstalk.

This chapter details the modern sonar technologies used during survey operations of today and the integration of these sensors in modern marine survey suites. The problems associated with integration of multiple sonar sensors are explained, and the sensor control routines employed to avoid such problems are discussed. Finally, the future direction of payload senor control and the development of intelligent sensor routines are presented.

2. Sonar technologies

Due to the high attenuation of electromagnetic waves underwater, video and radar are unsuitable for wide-area mapping of the subsea environment. Instead, acoustic waves are the only practical way to chart wide areas of the seafloor. Sonar technology is an essential part of a modern marine survey system and has been successfully employed to record the composition, physical attributes, and habitat and community patterns of our ocean seabeds. Today, there are numerous acoustic devices available for charting the seafloor including multibeam echosounders, sidescan sonar, interferomteic sonar and synthetic aperture sonar. These systems differ in their acoustic mapping techniques and capabilities, and provide diverse interpretations of the seabed. The different acoustic techniques, applications and survey capabilities of modern sonar technologies are briefly detailed below:

2.1 Multibeam echosounders

Multibeam echosounders are capable of collecting highly accurate seafloor depth information. Over the last number of decades these systems have been successfully used for gathering high-resolution seafloor bathymetric data in shallow- and deep-water regions (Hammerstad et al., 1991; Laban, 1998; Kloser, 2000; Parnum et al., 2004). The multibeam sonar system emits an acoustic pulse wide in the across-track field and narrow in the along-track field, producing a "cross fan" beam pattern to obtain detailed coverage of the bottom. The receive beam pattern is wide in the along-track field and narrow in the across-track field. The resulting product of the transmit and receive beams is a narrow beam that ensonifies an area of the seafloor, providing range-angle couplets of sample points (over 500 individual points in some systems) along the swath. Multibeam sonar systems are also capable of supplying acoustic backscatter imagery, by recording the intensity of the backscattered signal as it is swept along the seabed. However, the image is of lower resolution and poorer quality than the sidescan sonar backscatter image (Smith & Rumohr, 2005). Multibeam systems are also expensive and require high processing power.

2.2 Sidescan sonar

Sidescan sonar is an acoustic imaging device used to produce wide-area, high-resolution backscatter images of the seabed, under optimal conditions it can generate an almost photo-realistic, two-dimensional picture of the seabed. This acoustic instrument is used for charting seabed features and revealing special sediment structures of both biogenic and anthropogenic origin (McRea, 1999; Brown et al., 2004; Smith & Rumohr, 2005). Sidescan does not usually produce bathymetric data. However, it does provide information on sediment texture, topology, bedforms and the low grazing angle of the sidescan sonar beam over the seabed makes it ideal for object detection (Kenny, Cato et al. 2003). One disadvantage of sidescan sonar is that it does not provide reliable information on the position of seabed features.

2.3 Interferometric sonar

Interferometric systems are capable of providing high-resolution, wide-swath bathymetry in shallow water with swaths of 10 - 15 times the instrument altitude (Gostnell et al., 2006). Interferometry is the technique of superimposing (interfering) two or more waves, to detect differences between them. Measurement of the difference in acoustic path allows the accurate assessment of the angular direction. Interferometric technology could prove highly beneficial to seabed mapping programmes. However, it is still considered a developing technology within the marine industry. While there have been numerous papers written on the theoretical functionality of these systems and a variety of manufacturer studies conducted, there have been few independent analyses of their in situ performance (Gostnell, 2005).

2.4 Synthetic aperture sonar

The Synthetic Aperture Sonar (SAS) is a high-resolution acoustic imaging technique that combines the returns from several consecutive pings to artificially produce a longer sonar array. With the use of sophisticated processing, the data is used to produce a very narrow effective beam. The most important attribute of an SAS system is its along-track resolution being independent of both range and frequency. SAS is a direct analogue of synthetic aperture radar (SAR) processing, which is well established in both airborne and spaceborne applications (Curlander & McDonough, 1992) providing vast area coverage, imagery and bathymetry at high spatial resolution. For a generation, engineers have attempted to replicate SAR concepts with sidescan seafloor imaging sonars. However, SAS has long been considered a purely theoretical concept (Lurton, 2002) and its implementation was thought to be untenable due to lack of coherence in the ocean medium, precise platform navigation requirements and high computation rates. With advances in innovative motion compensation and autofocusing techniques, signal processing hardware, precise navigation sensors, and stable submerged autonomous platforms, SAS is now beginning to be used in commercial survey and military surveillance systems (Sternlicht & Pesaturo, 2004).

3. Sensor integration

The integration of multiple sonar sensors into a marine survey suite allows for the simultaneous collection, and fusing, of individual datasets of the same seafloor region. Accordingly, the provision, and combined analysis, of complementary and comparative datasets affords a more accurate representation of the seafloor, the removal of possible dataset ambiguities and improved data analysis and interpretation (Wright et al., 1996; Evans et al., 1999; Hughes Clarke et al., 1999; Dasarathy, 2000; Fanlin et al., 2003; Duxfield et al., 2004; Nitsche et al., 2004; Shono et al., 2004).

Data fusion is the process of taking information from multiple, independent datasets and combining it to extract information not available in single datasets; the combined analysis of contoured bathymetry maps, generated from multibeam echosounders, and the sidescan sonar acoustic reflectivity images permit the geologic interpretation of multibeam bathymetry data to be enhanced by providing an acoustic characterisation of the seafloor from which geologic composition can be inferred, while the bathymetric information improves the representation of the seafloor relief in sidescan imagery by providing the geometric configuration of the seabed (de Moustier et al., 1990; Pouliquen et al., 1999).

An integrated interpretation of acoustic datasets is presented by Nitsche et al. (Nitsche et al., 2004). According to the authors, the integrated examination of sidescan, sub-bottom and high-resolution bathymetry data enable the clear distinction of different seabed facies, and hence, an understanding of the related processes, vastly improving data interpretation and classification. Shono et al. (Shono et al., 2004) explore the synergies offered by an integrated hydro acoustic survey scheme, in which the survey region is mapped using a multibeam echosounder and a sidescan sonar. The bathymetry data and sidescan imagery present complementary datasets of the seabed region, enhancing the individual, and combined, dataset analysis; affording a greater understanding of the seafloor region. The author also concludes that the integrated approach provides for a more economical and efficient survey. Wright et al, (Wright et al., 1996) present methods for performing multi-sensor data fusion. Through their investigations, the authors demonstrate that data fusion aids the classification and identification of seabed features, minimises dataset ambiguities and improves upon positional accuracy of the features present.

The integration of multiple sensors onto a single platform, such as a UUV, also minimises the relative positional error between features evident in the various datasets, as the target region is ensonified by the sensors under the same environmental conditions and geo-referenced by the same navigational data. Simultaneous multi-sonar operation also eliminates the need to conduct separate surveys for each instrument, as well as the collection of supporting data required to fully understand the operating environment during each individual survey, thereby significantly reducing the survey duration and consequently the survey costs (Thurman et al., 2007).

Reports of successful AUV survey missions suggest that bathymetric mapping, sidescan imaging, magnetometer survey and sub-bottom profiling are the principle mission of the new survey-class AUVs (Whitcomb, 2000). To execute these missions, modern AUVs are typically equipped with a range of survey sensors, integrated into the single marine survey suite. The synergies offered by integrating and concurrently operating multiple acoustic mapping devices on a UUV underpin the desire for such an operational configuration, facilitating high-resolution surveys of the deep-ocean environment, while enabling the information encoded in one instrument's dataset to be used to correct artefacts in the other.

4. Acoustic interference avoidance

The reception circuitries of sonar transducers are typically frequency band-limited to prevent acoustic interference from parallel operating instruments of different frequencies. However, the high-resolution versions of most imaging sonar operate within the same frequency band, with typical working frequencies for high-frequency multibeam echosounders being 200kHz-400kHz, and high-frequency sidescan sonar ranging from 200kHz to 500kHz. While some instruments, such as multibeam echosounders, can be depth gated to filter spurious returns, sidescan sonar records can be severely distorted by sensor crosstalk, as they rely on the full temporal trace of the returned backscatter to construct an intensity image. Consequently, the simultaneous operation of multiple high-frequency sonar is prohibited by the inherent complication of cross-sensor acoustic interference (de Moustier et al., 1990; Ishoy, 2000; Kirkwood, 2007).

The integration and concurrent operation of multiple sonar sensors in a marine survey suite creates issues of cross-sensor acoustic interference. The contamination caused by sensor crosstalk severely degrades the resulting datasets of the parallel operating sensors. Traditionally, compromises were sought to avoid this sensor crosstalk and more recently, in particular in the operation of UUV platforms, survey sensor control routines have been developed.

Surveys requiring multiple high-resolution datasets typically require a compromise of mobilising separate survey vessels for each sensor (Parrott et al., 1999; McMullen et al., 2007). Conducting a survey of the same seafloor region for each of the interfering sensors is uneconomical and inefficient. Evans et al. (Evans et al., 1999) investigated the advantages of single or dual vessel solutions for hydrographic surveys requiring multiple datasets of a region. The team concluded that although the dual vessel solution allowed the gathering of the multibeam data at higher survey speeds, the single vessel solution to conducting multibeam and sidescan sonar surveys proved more economical and improved the hydrographic analysis and understanding of the data. However, during the single vessel survey, the sidescan sonar deployed was of lower frequency (100kHz) to the sidescan sonar deployed during the dual vessel survey (300kHz), reducing the sidescan imagery resolution and data integrity.

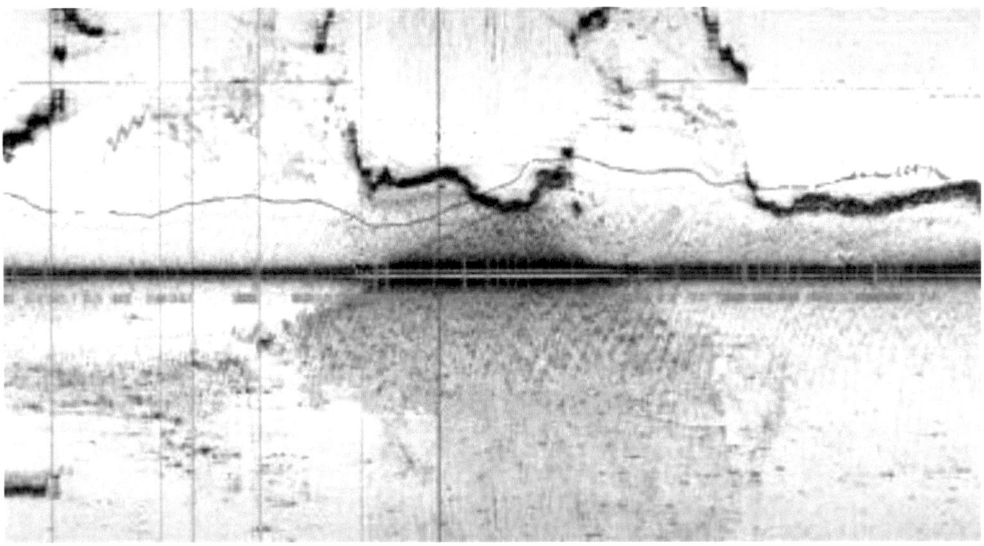

Fig. 2. Crosstalk can be seen on this sidescan sonar image where the backscatter is very low. The interference was caused from a simultaneously operating sonar.

Others have also attempted to avoid cross-sensor acoustic contamination by separating the operating frequency of the payload sonar sufficiently far that they are undetectable from one another (Pouliquen et al., 1999; Lurton & Le Gac, 2004). As a result, the sonar systems employed are a combination of high-frequency and low-frequency sonar. The low frequency systems significantly degrade the quality of the generated datasets, with the result that small-scale features may not be evident, thereby compromising the data interpretation process. An advanced solution must be utilised that will enable the simultaneous operation of high-frequency acoustic sensors to provide detailed datasets that are demanded by today's needs and standards.

Temporally separating the transmission-reception cycles of similar frequency sonar has been attempted in (de Moustier et al., 1990), which reports the concurrent acquisition of multibeam and sidescan sonar data using co-frequency 12kHz systems by interleaving their pings. The described algorithm takes into account the timing requirements of both systems and schedules the multibeam transmit cycles around the fixed sidescan timing events using a sound synchronisation unit. The sound synchronisation unit interleaves the transmission-reception cycle of each sensor, thus avoiding acoustic interference. However, because of the fixed transmission rates, the system is best suited for long-range, deep-water applications and does not provide optimal ping repetition rates.

Other triggering modules have been developed by a number of marine technology companies, such as GeoAcoustics' Timer2 module, to allow asynchronous triggering of multiple sonar systems while avoiding the effects of sensor crosstalk. Operator specified timing schedules are used to trigger the individual systems at fixed intervals. The interleaving of the otherwise interfering pulses avoids dataset contamination, enabling the simultaneous use of multiple high-frequency sonar sensors.

5. Remote payload control

The deployment of UUVs has, by their very nature, necessitated the development for remote payload sensor control routines. Typically, command and control of payload sensors are pre-programmed and/or operator based. C&C Technology's HUGIN 3000, a third generation AUV manufactured by Kongsberg Simrad, interfaces to the payload sensors through the HUGIN Payload Processor (Hagen & Kristensen, 2002). Survey specifications are programmed before deployment and take control of all sensor operations onboard. Another well-proven AUV, the Atlas Maridan's SeaOtter, enables the synchronised operation of multiple sensors by specifying the repetition rate, delay and duty cycle for each sensor during survey planning. The values are sent over the vehicle network to the Local Trigger Manager (LTM), which generates the signals required for each instrument during deployment (Ishoy, 2000). The Monterey Bay Aquarium Research Institute (MBARI) has developed the DORADO AUV, capable of conducting simultaneous multibeam bathymetry, sidescan sonar and sub-bottom surveys of an area of interest. The AUV is integrated with Reson's 7100 multibeam echosounder (200kHz), Edgetech's 110/410kHz chrip sidescan sonar and an Edgetech 2 - 16kHz chrip sub-bottom profiler. Simultaneous operation of the multiple sensors is managed by the Reson propriety timing algorithm. The multibeam echosounder acts as the master system, and along with the other integrated systems, is pinged using a fixed 1 pulse per second (PPS) clock, made available by the navigation system (Kirkwood, 2007).

Survey results have shown that the described systems have successfully completed surveys of an area of interest in which simultaneously operating sonar are deployed (George et al., 2002; Wernli, 2002; Kirkwood et al., 2004; Desa et al., 2006). However, the integrated systems do not allow for optimal surveys, leading to deficient datasets; the acoustic sensors used are not all of high-frequency and the payload control is pre-programmed and non-adaptable.

The intelligence of sensors is becoming increasingly sophisticated. Innovative developments in sensor technology have enabled the real-time data acquisition, processing and decision making based on the collected and processed data, of sensor systems during survey operations. Researchers at the Mobile and Marine Robotics Research Centre (MMRRC), University of Limerick, have developed an approach to the real-time adaptive control of multiple high-frequency sonar survey systems for UUVs (Thurman et al., 2008). This approach is based around a centralised sensor payload controller which manages the integrated sensors during survey missions, facilitating the operation of co-located, high-frequency sonar. The Multibeam is the master system and supplies the raw data to be processed in real-time to provide a priori bathymetry data to auxiliary acoustic sensors. The automated system is based on the interleaving of the sonar transmission-reception cycles to avoid issues of cross-sensor acoustic interference, permitting the integration of multiple acoustic sensors operating in parallel. By dynamically adapting the ping rates of the payload sensors, the system optimises the execution of the seabed mapping survey and improves the quality of the resulting data, thereby significantly increasing survey productivity.

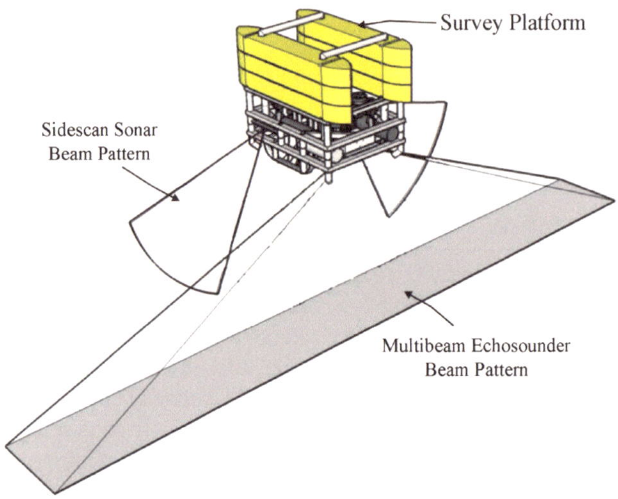

Fig. 3. System arrangement; the multibeam echosounder is mounted at the fore of the platform, with the sidescan sonar mounted in the rear, permitting the multibeam echosounder to provide a priori data to the controller.

6. Integrated acoustic controller system

Previously, multibeam bathymetric data was collected and stored during survey operations, with processing performed post-survey. However, recent advances in computational technology have enabled real-time processing of multibeam data. In-survey processing of the multibeam data allows time-critical survey control decisions to be made based on the assessment of real-time environmental information. The Integrated Acoustic Controller System utilises the modern computational resources and real-time processing techniques to enable synchronised multi-sonar operation through the prediction and temporal separation of each of the UUV's payload sonar's transmission-reception window.

Unlike traditional sensor triggering routines, which operate on fixed timing schedules, the system dynamically adapts the time separation between successive pings. The sensor triggering timing schedule is calculated as a function of each sensor's imaging geometry, the range between each sensor and the ensonified seafloor, the survey vessel velocity, and the desired resolution of the collected dataset. With the imaging geometry and mounting configuration of each instrument known, the required set of parameters is completed by analysis of the navigation and bathymetric data streams collected during the survey. The terrain-adaptive timing schedule enables optimal use of each sensor's available transmission-reception cycle windows; providing the capability to interleave the pings of multiple acoustic sensors, thus avoiding acoustic contamination while still adhering to high-resolution survey requirements.

The system exploits the fact that, due to the slow forward speed of the UUV platform, typically 2 - 4 knots, there occurs a high ping-to-ping coherence between successive multibeam swaths. This permits the duration of the next multibeam transmission-reception window to be predicted with a high degree of accuracy. The multibeam transducer is mounted to the fore of the survey platform such that the geometry of the region of the seafloor to be interrogated by the sidescan sonar will already have been mapped, providing the a priori information needed to predict its transmission-reception window.

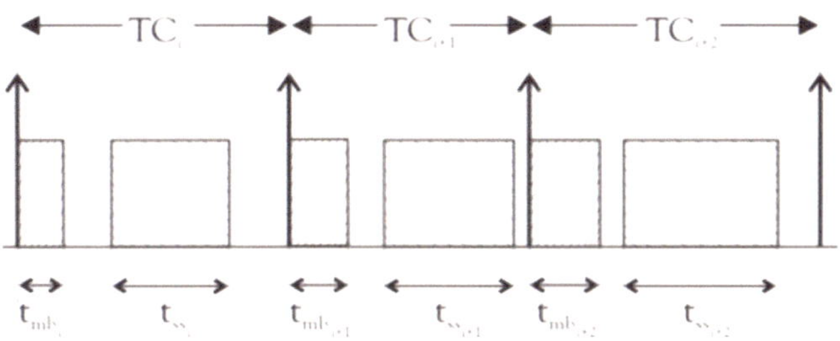

Fig. 4. Timing diagram of the Integrated Acoustic Controller System; within each Triggering Cycle, TC, the transmission-reception cycles of the multibeam echosounder, t_{mb}, and the sidescan sonar, t_{ss}, are scheduled. Separation of the transmit-receive windows enable the concurrent operation of the high-frequency sonar.

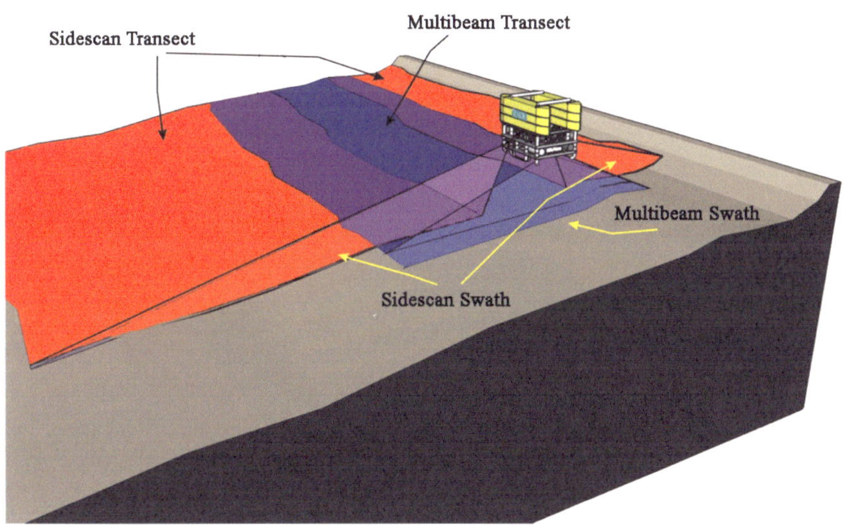

Fig. 5. Sonar transects during dual sonar operation.

The system is comprised of the multibeam sonar and data acquisition module, the sidescan sonar and data acquisition module, the position and orientation sensor and data acquisition module and the multi-sonar synchronisation module. The multibeam sonar system is the master system and provides to the survey controller the raw data that determines the multi-sensor triggering routine. The multibeam sonar data acquisition module reads in the raw seafloor data, filters for outliers and extracts each individual beam's time and angle couplet. In parallel, the navigational sensors provide concurrently generated high-frequency time-stamped Motion Reference Unit data, which is queued in the memory buffer. Both streams are fused by selecting the navigational message relating to the time-stamp encoded in the multibeam data. A transformation matrix is constructed and converts the body-fixed multibeam data to earth-fixed seafloor depth samples. A select number of geo-referenced depth points are then used to generate and populate a Digital Terrain Model (DTM) of the surveyed region (in calculating the adaptive timing schedule it is not typically required to build a fully populated DTM, thereby reducing the processor's computational workload). By analysing the region of the DTM within the seafloor footprint of the payload sonar's reception beam the optimal ping rate of each individual sonar is calculated for each swath.

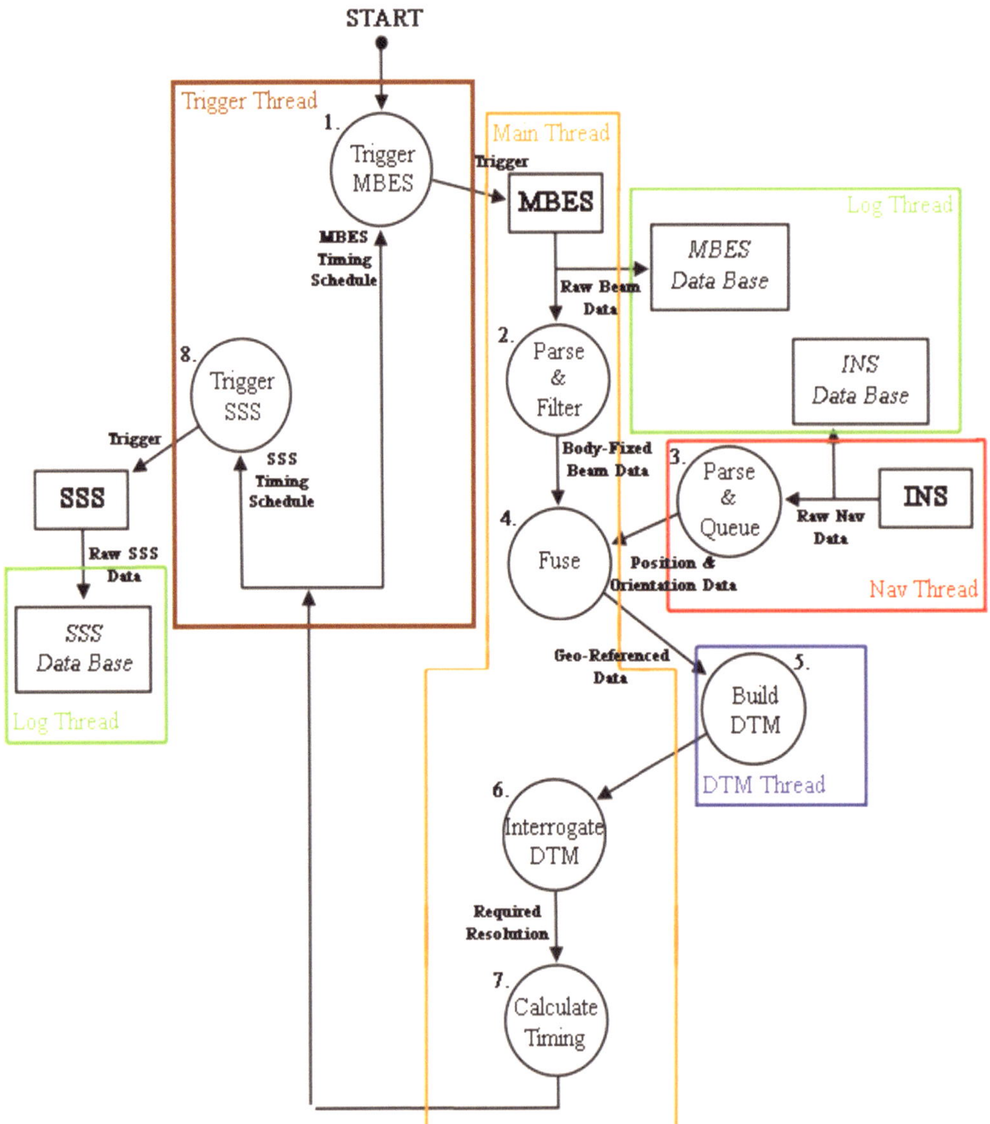

Fig. 6. Software architecture; the system is decomposed into a multi-threaded framework to enable independent modules to execute in parallel.

The system benefits are manifold and are of significant interest to the marine and off-shore communities:

- The system adapts to the varying geometry of the seafloor, optimising the use of the individual sensors.

- Survey productivity is increased due to the considerable reduction in survey duration and cost; the area of interest is surveyed, along with the supporting data being collected, only once.
- The simultaneous acquisition of multiple datasets improves the data interpretation process by allowing the combined analysis and interpretation of independent datasets of the same region.

The relative positional error between features evident in the datasets is also minimised, as the target region is ensonifed by the instruments integrated on the same platform under the same environmental conditions and geo-referenced by the same navigational data, promoting the straightforward co-registration of the acoustic signature of features across multiple datasets.

7. Conclusion

Increased interest in the detailed exploration of our ocean seabeds has spurred the development and technological advancements in sonar technology. Sonar is an essential part of a modern marine survey system and has been successfully employed to record the composition, physical attributes, and habitat and community patterns of the seafloor. The integration of multiple sonar sensors into a marine survey suite allows for the simultaneous collection of individual datasets of the same seafloor region. A move towards multi-sensor integration is becoming more and more apparent in the marine industry, allowing for the enhancement of decision making and data analysis by exploiting the synergy in the information acquired from multiple sources.

However, the integration and concurrent operation of multiple sonar sensors in a marine survey suite creates issues of cross-sensor acoustic interference. The contamination caused by sensor crosstalk severely degrades the resulting datasets, and hence, the data examination and understanding. Traditionally, compromises were sought to avoid this sensor crosstalk by mobilising separate surveys for each of the interfering sensors or by separating the operating frequency of the sonar sufficiently far that they are undetectable from one another, and more recently, in particular in the operation of UUV platforms, survey sensor control routines have been developed. Nevertheless, solutions to the problem of sensor crosstalk remain inadequate and inefficient.

The intelligence of sensors is advancing rapidly. Innovative developments in sensor technology have enabled the data acquisition, processing and decision making to occur in real-time during survey operations. An approach to the real-time adaptive control of multiple high-frequency sonar systems was presented in this chapter. This approach is based around a centralised sensor payload controller which manages the integrated sensors during survey missions, facilitating the operation of co-located, high-frequency sonar. The multibeam is the master system and supplies the raw data to be processed in real-time to provide a priori bathymetry data to auxiliary acoustic sensors. The automated system is based on the interleaving of the sonar transmission-reception cycles in a non-interfering fashion.

By allowing real-time decision making to be made based on real-time mission data, the system optimises the execution of the seabed mapping survey and improves the quality of the resulting data, thereby significantly increasing survey productivity, and consequently, the data analysis and interpretation.

8. References

Brown, C. J.; Hewer, A. J.; Meadows, W. J.; Limpenny, D. S.; Cooper, K. M. & Rees, H. L. (2004). Mapping seabed biotopes at Hastings Shingle Bank, eastern English Channel. Part 1. Assessment using sidescan sonar. *Marine Biological Association of the United Kingdom*, Vol. 84, No. (2004), pp. 481-488

Curlander, J. C. & McDonough, R. N. (1992). *Synthetic Aperture Radar: Systems and Signal Processing*, Wiley, 978-0-471-85770-9, New York

Dasarathy, B. V. (2000). Industrial applications of multi-sensor multi-source information fusion, *proceedings of IEEE Industrial Technology 2000*, pp. 5-11 vol.1, 2000, Goa, India

de Moustier, C.; Lonsdale, P. F. & Shor, A. N. (1990). Simultaneous operation of the Sea Beam multibeam echo-sounder and the SeaMARC II bathymetric sidescan sonar system. *Oceanic Engineering, IEEE Journal of*, Vol. 15, No. 2, (1990), pp. 84-94

Desa, E.; Madhan, R. & Maurya, P. (2006). Potential of autonomous underwater vehicle as new generation ocean data platforms. *Current Science*, Vol. 90, No. 9, (2006), pp. 1202-1209

Duxfield, A.; Hughes Clarke, J. E.; Martin, B. A.; Legault, J.; Comeau, M. & Monahan, D. (2004). Combining multiple sensors on a single platform to meet multi-purpose nearshore mapping requirements, *proceedings of Canadian Hydrographic Conference*, pp. 2004, Ottawa, Canada

Evans, R. E.; Morton, R. W. & Simmons, W. S. (1999). A Dual or Single Vessel Solution to Conducting Multibeam and Sidescan Surveys for NOAA in the Gulf of Mexico: A Lessons Learned Approach, *proceedings of U.S. Hydrographic Conference*, pp. 1999, Mobile, Alabama

Fanlin, Y.; Jingnan, L. & Jianhu, Z. (2003). Multi-beam Sonar and Side-scan Sonar Image Co-registration and Fusing. *Marine Science Bulletin (English Edition)*, Vol. 5, No. 1, (2003), pp. 16-23

George, R. A.; Gee, L.; Hill, A. W.; Thomson, J. A. & Jeanjean, P. (2002). High-Resolution AUV Surveys of the Eastern Sigsbee Escarpment, *proceedings of Offshore Technology Conference*, pp. 2002, Houston, Texus

Gostnell, C. (2005). Efficacy of an Interferometric Sonar for Hydrographic Surveying: Do interferometers warrant an in-depth examination? *The Hydrographic Journal*, Vol. 118, No. (2005), pp. 17-24

Gostnell, C.; Yoos, J. & Brodet, S. (2006). NOAA Test and Evaluation of Interferometric Sonar Technology, *proceedings of Canadian Hydrographic Conference 2006*, pp. 2006, Halifax, Canada

Hagen, P. E. & Kristensen, J. (2002). The HUGIN AUV "Plug and play" payload system, *proceedings of MTS/IEEE Oceans '02*, pp. 156-161, 2002, Biloxi, Mississippi

Hammerstad, E.; Pøhner, F.; Parthiot, F. & Bennett, J. (1991). Field testing of a new deep water multibeam echo sounder, *proceedings of MTS/IEEE OCEANS '91*, pp. 743-749, 1991, Honolulu, Hawaii

Hughes Clarke, J. E.; Mayer, L.; Shaw, J.; Parrott, R.; Lamplugh, M. & Bradford, J. (1999). Data handling methods and target detection results for multibeam and sidescan data collected as part of the search for Swiss Air Flight 111, *proceedings of Shallow Survey - 99*, pp. 1999, Australian Defence Science and Technology Organization, Sydney, Australia

Ishoy, A. (2000). How to make survey instruments "AUV-friendly", *proceedings of MTS/IEEE OCEANS 2000*, pp. 1647-1652 vol.3, 2000, Providence, Rhode Island

Kirkwood, W. J. (2007). Development of the DORADO Mapping Vehicle for Multibeam, Subbottom, and Sidescan Science Missions. *Field Robotics, Journal of*, Vol. 24, No. 6, (2007), pp. 487-495

Kirkwood, W. J.; Caress, D. W.; Thomas, H.; Sibenac, M.; McEwen, R.; Shane, F.; Henthorn, R. & McGill, P. (2004). Mapping payload development for MBARI's Dorado-class AUVs, *proceedings of MTS/IEEE OCEANS '04*, pp. 1580-1585, 2004, Kobe, Japan

Kloser, R. J. (2000). Optimal seabed habitat mapping using multibeam acoustics with associated physical and visual sampling devices - at sea trials, *proceedings of Acoustics 2000*, pp. 15-17, 2000, Joondalup, Australia

Laban, C. (1998). Seabed Mapping, *Hydro International*, Vol. 2, No. 1, pp. 6-9

Lurton, X. (2002). *An introduction to Underwater Acoustics, Principles and Applications*, Springer-Praxis, 3-540-42967-0, London

Lurton, X. & Le Gac, J.-C. (2004). The CALIMERO project: Scientific objectives and first at-sea results, *proceedings of SeaTechWeek 2004*, pp. 21-22, 2004, Brest, France

McMullen, K. Y.; Poppe, L. J.; Signell, R. S.; Denny, J. F.; Crocker, J. M.; Beaver, A. L. & Schattgen, P. T. (2007). Surficial geology in central Narragansett Bay, Rhode Island; interpretations of sidescan sonar and multibeam bathymetry, *U.S. Geological Survey Open-File Report 2006-1199*,

McPhail, S. (2002). Autonomous Underwater Vehicles: Are they the Ideal Sensor Platform for Ocean Margin Science?, In: *Ocean Margin Systems*, G. Wefer, D. Billett, D. Hebbelnet al (Ed.), pp. 79-97, Springer-Verlag, Berlin

McRea, J. E. (1999). Mapping marine habitats with high resolution sidescan sonar. *Oceanologica Acta*, Vol. 22, No. 6, (1999), pp. 679-686

Nitsche, F. O.; Bell, R.; Carbotte, S. M.; Ryan, W. B. F. & Flood, R. (2004). Process-related classification of acoustic data from the Hudson River Estuary. *Marine Geology*, Vol. 209, No. 1-4, (2004), pp. 131-145

Niu, H.; Adams, S.; Husain, T.; Bose, N. & Lee, K. (2007). Applications of Autonomous Underwater Vehicles in Offshore Petroleum Industry Environmental Effects Monitoring, *proceedings of 8th Canadian International Petroleum Conference*, pp. 2007, Calgary, Canada

Parnum, I. M.; Siwabessy, P. J. W. & Gavrilov, A. N. (2004). Identification of Seafloor Habitats in Coastal Shelf Waters Using a Multibeam Echosounder, *proceedings of Acoustics 2004*, pp. 181-186, 2004, Gold Coast, Australia

Parrott, R.; Clarke, J. H.; Fader, G.; Shaw, J. & Kamerrer, E. (1999). Integration of multibeam bathymetry and sidescan sonar data for geological surveys, *proceedings of MTS/IEEE OCEANS '99*, pp. 1129-1133, 1999, Seattle, Washington

Pouliquen, E.; Zerr, B.; Pace, N. G. & Spina, F. (1999). Seabed segmentation using a combination of high frequency sensors, *proceedings of MTS/IEEE OCEANS '99*, pp. 888-893, 1999, Seattle, Washington

Shono, K.; Komatsu, T.; Sato, Y.; Koshinuma, J. & Tada, S. (2004). Integrated hydro-acoustic survey scheme for mapping of sea bottom ecology, *proceedings of MTS/IEEE OCEANS '04*, pp. 423-427, 2004, Kobe, Japan

Smith, C. J. & Rumohr, H. (2005). Imaging Techniques, In: *Methods for the Study of Marine Benthos*, A. Eleftheriou and A. D. McIntyre (Ed.), pp. 87 - 111, Blackwell,

Sternlicht, D. & Pesaturo, J. F. (2004). Synthetic Aperture Sonar: Frontiers in Underwater Imaging, *Sea Technology*, Vol. 45, No. 11, pp.

Thurman, E.; Riordan, J. & Toal, D. (2007). Automated Optimisation of Simultaneous Multibeam and Sidescan Sonar Seabed Mapping, *proceedings of IEEE OCEANS 2007*, pp. 1-6, 2007, Aberdeen, Scotland

Thurman, E.; Riordan, J. & Toal, D. (2008). Approach to the Real-Time Adaptive Control of Multiple High-Frequency Sonar Survey Systems for Unmanned Underwater Vehicles, *proceedings of IFAC Navigation, Guidence and Control of Underwater Vehicles (NGCUV) 08*, pp. 2008, Kilaloe, Ireland

Wernli, R. L. (2002). AUVs - A Technology Whose Time Has Come, *proceedings of Underwater Technology 2002*, pp. 309-314, 2002, Tokyo, Japan

Whitcomb, L. L. (2000). Underwater robotics: out of the research laboratory and into the field, *proceedings of IEEE International Conference on Robotics and Automation (ICRA) 2000*, pp. 85 - 90, 2000, San Francisco, California

Wright, J.; Scott, K.; Tien-Hsin, C.; Lau, B.; Lathrop, J. & McCormick, J. (1996). Multi-sensor data fusion for seafloor mapping and ordnance location, *proceedings of Autonomous Underwater Vehicle Technology '96, Internation Symposium on*, pp. 167-175, 1996, Monterey, California

Yoerger, D. R.; Bradley, A. M.; Singh, H.; Walden, B. B.; Cormier, M.-H. & Ryan, W. B. F. (2000). Multisensor mapping of the deep seafloor with the Autonomous Benthic Explorer, *proceedings of Underwater Technology 2000, International Symposium on*, pp. 248-253, 2000, Tokyo, Japan

8

On the Benefits of Using Both Dual Frequency Side Scan Sonar and Optical Signatures for the Discrimination of Coral Reef Benthic Communities

Tim J Malthus and Evanthia Karpouzli
School of GeoSciences, University of Edinburgh, Drummond St, Edinburgh United Kingdom

1. Introduction

The importance of coral reef ecosystems is well established (McManus and Noordeloos, 1998). The threats to these highly diverse and endangered communities are well known and a large number of reports document the dramatic effects of climate change and particularly global seawater warming, coastal development, pollution, and impacts from tourism, overfishing, and coral mining on them (Grigg & Dollar, 1990; Holden & LeDrew, 1998; Lough, 2000; Buddemeier, 2002; Knowlton, 2001; Sheppard, 2003). To protect these ecosystems the extent of their degradation must be documented through large scale mapping programmes, and inventories of existing coral reef areas are particularly important (Riegl & Purkis, 2005; Mora et al., 2006). Such programmes are essential so that the health of these ecosystems can be assessed and local and global changes over time can be detected (Holden & LeDrew, 1998).

Seagrass beds are also recognized as playing a pivotal role in coastal ecosystems. They are crucial to the maintenance of estuarine biodiversity, the sustainability of many commercial fisheries, for stabilizing and enriching sediments and providing an important food resource and spawning areas for many marine organisms (Powis & Robinson, 1980; Bell & Pollard, 1989, Dekker et al., 2006). Unprecedented declines in seagrass beds have occurred in temperate and tropical meadows throughout the world; their global decline highlights the need for monitoring programmes to manage their conservation and sustainable use (Short & Wyllie-Echeverria, 1996; Ward et al., 1997).

Coral reefs, seagrass, and macroalgal habitats are commonly found in association with, and in close proximity to each other, and are linked by many pathways such as sediment deposition mechanisms, the primary productivity cycle, and the migration of many fish species (Mumby, 1997). Due to their nutritional biology and photosynthetic requirements, coral reefs generally exist in clear tropical waters and this makes them highly suited for optical remote sensing (Mumby, 1997; Green et al., 2000). Although less confined to them, macroalgal and seagrass habitats are also found in such environments. Under stress, both coral and seagrass ecosystems may retreat and become replaced by macroalgal or less productive and biologically diverse sedimentary or bare rocky habitats. Such impoverishment adversely affects biodiversity and productivity and ultimately local, tourist-based, economies.

The potential of marine remote sensing as an alternative mapping tool to conventional methods like *in situ* diving surveys is now well understood and recognized. Tropical coastal environments are well suited to optical remote sensing because sunlight is minimally attenuated compared to other marine regions penetrating up to depths of 25 m or greater (Mumby, 1997; Green et al., 2000; Isoun et al., 2003). The technique is recognised as being the most cost-effective and feasible means of mapping and monitoring of tropical coastal ecosystems over large areas (Bouvet et al., 2003; Maeder et al., 2002; Green et al., 2000; Luczkovich et al., 1993).

The sensors used to monitor reef ecosystems generally can be divided into either passive or active systems. Passive remote sensors measure reflected sunlight in a given bandwidth of the electromagnetic spectrum and constitute traditional optical systems, while active remote sensing systems generate their own source of energy, and measure the reflected energy. Examples of active remote sensing systems include imaging sonars (e.g. side scan sonar) and Synthetic Aperture Radar (SAR).

The mapping of both temperate and tropical marine benthic habitats using medium and high spatial resolution optical satellite systems shows that generally only a few classes can be discriminated on the basis of spectral signatures alone (Holden & Ledrew, 1999; 2001; Hochberg & Atkinson, 2000), owing to the limited spectral information available in conventional optical instruments and the similarities in reflectance of many species and habitats (Hochberg & Atkinson, 2000; 2003; Holden, 2001; Hochberg et al., 2003; Karpouzli et al., 2004). Whilst higher spectral resolution data may increase the power of habitat discrimination, limited availability of such data in future spaceborne systems restricts its application to coarse coverage only. In the last few years, high spatial resolution data from commercial satellites such as IKONOS and QuickBird has shown to be well suited for mapping coral reef systems (Maeder et al., 2002; Andrefouet 2003; Capolsini et al., 2003). In particular, the incorporation of additional information on small scale variability in higher spatial resolution remotely sensed data has been shown to improve on the accuracies of spectral centred classifications (Dustan et al., 2001; Jakomulska & Stawiecka, 2002; Palandro, 2003).

A major challenge to optical remote sensing in both temperate and tropics regions is cloud cover which reduces the number of images available over a period of time over an area of interest (Jupp et al., 1981). The attenuation of light by water also significantly limits the technique in deeper and more turbid waters (Holden, 2001; 2002). These limitations have been drivers to develop and use active remote sensing systems for imaging the seabed such as acoustic systems. However, in comparison to satellite or airborne optical sensors, acoustic systems have rarely been used to map and monitor tropical marine habitats (Prada, 2002, White et al., 2003; Riegl & Purkis, 2005) and their potential is still in need of evaluation (Bouvet, 2003). Acoustic systems such as imaging sonars may offer further advantages over optical systems such as the provision of structural information of different habitat types, and geomorphological zonation. This additional information may improve the discrimination of spectrally similar but structurally different bottom types.

Despite the increasing evidence of the benefits to be gained there is presently a lack of studies on the synergistic use of alternative remote sensing approaches for mapping shallow water marine near shore habitats (Malthus & Mumby, 2003). The most obvious advantage of using acoustic and optical methods in combination is the different depth ranges that each of the systems operate; optical systems perform best in shallow waters (generally up to a maximum of 25 m in the clearest waters), while the deployment of sonar systems (single and multibeam) can be used to depths of hundreds of metres, but it is very much limited in shallow waters.

Few studies have attempted to integrate side scan sonar data with optical data to exploit the complementarity of the two systems and which have been in temperate waters (Pasqualini et al., 1998; Piazzi et al., 2000). These studies used visual photo-interpretation methods and occasional automated methods to classify the optical imagery, and to establish the upper boundary limits, whilst the sonograms were used for detecting lower depth limits. To date, no studies have tested the potentially improved accuracy of habitat classification when the optical and acoustic signatures are used in combination.

Although the potential of incorporating additional information on small scale variability in higher spatial resolution data to improve spectrally centered classifications has been recognized by a limited number of researchers (Jakomulska & Stawiecka, 2002), few studies have incorporated textural and spectral parameters for classifying benthic habitats simultaneously where these parameters have originated from high spatial resolution multiband acoustic and optical datasets. This study represents a first attempt to test the discrimination of coral reef habitats based on textural and spectral parameters derived from side scan sonar and IKONOS datasets. The overall aim is to statistically evaluate optical and acoustic remote sensing in discriminating reef benthic communities and their associated habitats, both in isolation and in combination.

2. Methods

2.1 Study area

The study site, selected for its conservation importance and for the availability of ancillary data, was focused on the littoral habitats of San Andres island (12° 34′ N; 81° 43′ W, land area 24 km^2), Colombia, situated within the San Andres, Old Providence and Santa Catalina Archipelago in the western Caribbean Sea. Approximately 180 km east of the Nicaraguan coast and 800 km northeast of the Colombian coast, the Archipelago comprises a series of oceanic islands, atolls and coral shoals (Figure 1). The submerged habitats of the Archipelago were designated a UNESCO Biosphere Reserve in 2000. The main extent of the sublittoral platform of San Andres is to the east and northeast and is bordered by a barrier reef, where depths range between 1 and 15 m before dropping rapidly to >1000 m (Geister & Diaz, 1997). The typical submerged habitats found around San Andres are seagrass (mainly Thalassia and Syringodium genera) and algal beds in different proportions, soft and hard coral habitats, as well as sandy and rocky substrates. These communities have seen high levels of mortality during the last two decades, with studies reporting overall reductions in live coral by more than 50% and corresponding increases in algal cover and biomass of such species as Dictyotaceae and *Halimeda* (Diaz et al., 1995; Zea et al., 1998). These changes coincide with significant increases in the human population of San Andres which has risen from 5,675 inhabitants in 1952 to around 80,000 by 1992 making it the most densely populated island in the whole of the Caribbean (Vollmer, 1997).

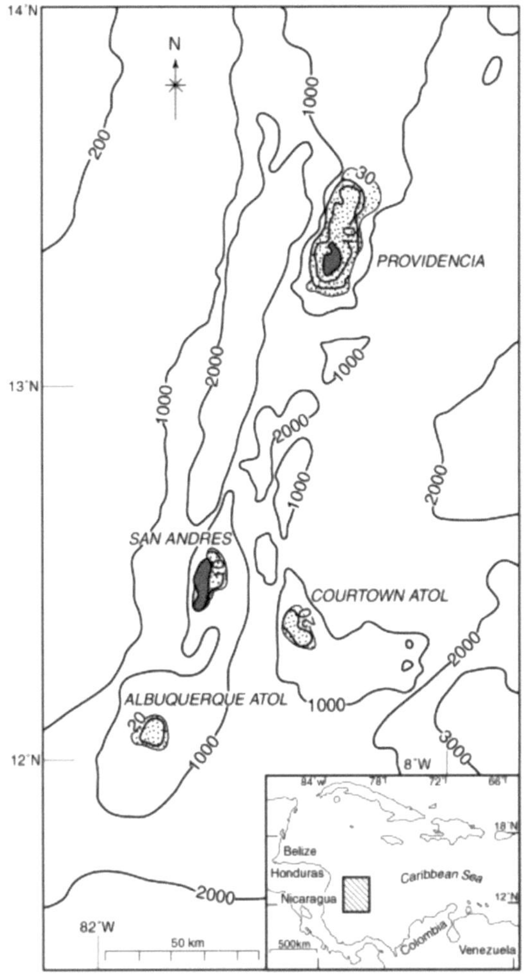

Fig. 1. Map of the western Caribbean Sea showing the location of the Archipelago of San Andres and Providencia.

2.2 Optical imagery

An IKONOS multispectral satellite image (11-bit radiometric resolution, 4-m spatial resolution) was acquired on the 9th September 2000 to coincide with a side scan sonar survey, and ground-truthing biological surveys. The weather conditions at the time of acquisition were fair, with limited patchy cloud overlying the terrestrial part of the image. An independent geometric correction (linear quadratic transformation, nearest neighbour resampling) was performed to improve on the geometric correction, based on 21 DGPS-determined ground control points, and which yielded an RMS error of 0.87 m. To atmospherically correct the image, the empirical line method was employed, based on a number of pseudo-invariant land targets of varying brightness from which *in situ* reflectances were determined at the time of image acquisition, as outlined in detail in Karpouzli and Malthus (2003).

Following land masking a water column correction was applied based on the semi-analytical approach of Maritorena et al., (1994) and Maritorena (1996) using independently obtained *in situ* water column spectral attenuation (k) and depth (z) estimates. An earlier study on the spatial variation of the attenuation of downwelling Photosynthetically Active Radiation, k_d(PAR), in the littoral zone of San Andres showed that attenuation is highly variable (Karpouzli et al., 2003). For this reason, pixel specific values of k for each of the visible IKONOS bands were estimated on the basis of field measurements and simple models (Karpouzli, 2003; Karpouzli et al., 2003). The application of Maritorena's model resulted in an image with enhanced bottom reflectance where the influence of varying bathymetry was greatly reduced, particularly where accurate estimates of depth and attenuation existed (Karpouzli, 2003).

2.3 The acoustic dataset

Dual frequency (100 and 500 kHz) Side Scan Sonar (SSS) backscatter and bathymetric data were acquired in the coastal waters of San Andres using a towed GeoAcoustics, fully digital, side scan sonar system (model: 159D, SS941, linked to an EOSCAN real time acquisition and image processing system) with the survey designed to overlap with parts of the IKONOS image and to encompass the full range of the marine habitats found in the area (coral, seagrass, algal and sediment habitats). The acoustic data were collected using a 7 m survey vessel equipped with a Trimble Global Positioning System. Differential correction of the navigation signal was conducted in real-time using the Omnisar satellite network resulting in a horizontal accuracy of about 0.5 m. Anamorphic and slant-range corrections were applied to the raw data in real time, thus eliminating lateral and longitudinal distortions in the sonograms, as were geometric corrections to reference the towed fish in respect to the position of an onboard dGPS system. Survey lines were spaced 10-30 m apart (depending on depth) to achieve a 100% backscatter cover of the selected survey areas.

Depth was determined concurrently using a FURUNO haul-mounted echo-sounder at intervals of one second, equating to a data point approximately every metre. After tidal corrections, these data were merged with soundings obtained from digitised hydrographic charts of the area and interpolated using a radial basis function in Surfer (version 7.02). The resulting DEM matched the spatial resolution of the IKONOS image and was used to undertake the water column correction of the satellite image.

A further geometric correction using an affine model was applied and the paired 100 kHz and 500 kHz subsets were corrected using image to image registration with the optical IKONOS dataset.

Texture layers - The sonograms acquired were used for extracting a number of acoustic parameters on the basis of which the discrimination of a number of habitat classes was tested. These included the mean signal intensity of the 2 original sonar bands (at 100 and 500 kHz frequencies: I_{100} and I_{500}) and two statistical models of texture which created four extra data layers for the sonograms, two for each frequency available. Firstly, a circular variance filter was passed over each frequency backscatter layer where the value of the central effective unit in the data (the 'texel, after Linnett et al., 1993) was the variance of the moving window over the original backscatter data. The result represents a measure of texture with a localised area (Var_{100} and Var_{500}). The size and the shape of the moving window was optimized for the coral dominated classes; the high spatial resolution of the sonograms enabled the visual discrimination of individual coral mounds and the size of the kernel was designed to coincide with the 3 m mean radius of coral mound aggregations found in the study site. In this context, variance is a 2nd-order algorithm, and the two textural layers produced were effectively measuring the coarseness or roughness of the original data bands in the two frequencies of operation. Homogeneous areas (e.g. sand beds) had a low variance while heterogeneous areas (e.g. rubble, coral patches) were characterised by high variance values.

The second textural parameter was represented by the standard deviation of the texels within signature areas (SD_{100} and SD_{500}). This parameter describes effectively the variation in texture coarseness and represents a measure of global nature as opposed with the previous parameter which has a local nature. The size of the signature areas varied between 60 m^2 and 400 m^2 (20 x 20 m) depending on the habitat type, and was adjusted to ensure that it was not straddling class boundaries and encompassing adjacent classes.

2.3 Biological surveys

Biological surveys were carried out for the purpose of groundtruthing the satellite and side scan sonar images. The survey methodology was optimized for the 0.35 m side scan sonar as opposed to the 4 m resolution IKONOS data, employing a combination of transects and 1 x 1 m and 0.25 x 0.25 m quadrats. The principal attribute recorded was percentage cover of the top layer of the dominant vegetation/lifeforms since this is the attribute recorded by optical remote sensors. In the case of seagrass and calcareous green algae, density was also recorded. Due to the penetrative nature of certain sonar frequencies (100 kHz or less, Blondel and Murton, 1997) which may penetrate the top few centimeters of sediments, detailed *in situ* notes of overlapping lifeforms and substrate to a depth of 5 cm below were also made, even when they were not exposed. Substrate types were recorded in the categories: bedrock, rubble, sand (coralline, terrigenous), mud, dead coral, boulders, and thin layer of sand on bedrock. Lifeforms were recorded most often to species level for hard corals, macroalgae and seagrass species and to a higher level taxon for soft corals, sponges, sea anemones, sea urchins and sea cucumbers.

The biological surveys were conducted in two phases. The first phase (April-July 1999) collected rapid spot-check data to identify the broadscale habitats found around San Andres. Three replicate 1 x 1 m quadrats were employed at each site. General notes were also made of the type of habitats in the area within approximately a 5 m radius. Depth and position (DGPS) of each site were also recorded. A total of 57 spot-check sites were surveyed. During the second phase (September-October 2000) more detailed targeted surveys were conducted at 17 sites around the island, covering all available habitats, and guided by the results of the rapid spot check surveys. Each site measured 500 square metres (50 x 10 m), to match the average range of the sonar tracks (50 m). A 50 m transect, laid within a homogeneous patch of each habitat type at each site, was sampled at 10 m intervals, totalling 10 quadrats per site. These quadrats were not considered replicates since the distance of 10 m between them was enough in a number of cases to cause a change in the composition of the observed communities, and totalled to 170 samples. The depth at the individual sites, as with the spot-check surveys, ranged between 1 and 16 m. The start and the end of the central transect lines were positioned using DGPS, and the position of each quadrat was also determined. Video footage and still images of the whole transect and each quadrat were produced using a Hi-8 Sony Video recorder. Although the footage was not used to directly estimate benthic cover, it was used as a permanent record of the site and to support the surveyor's results.

Classification scheme - With minor modifications, the hierarchical habitat classification system of Mumby and Harborne (1999) was used for this study (Table 1). This system was found suitable because it is based on field data from Caribbean habitats, and due to its hierarchical structure it could accommodate both the variable availability of data from the two phases of the groundtruthing surveys, as well as the different spatial scales of the satellite and sonar images. Some amendments were made in thresholds of different classes to account for regional differences. Only the first and second tiers (coarse and medium descriptive resolution) of the ecological component of the classification system were used to assign each spot-check or survey quadrat to a benthic class. The third tier (detailed descriptive resolution) of the scheme was considered too detailed for using with the remotely sensed data. In total, 4 coarse habitats were identified (coral, algal dominated, bare substratum, and seagrass dominated classes), and 20 bottom types in the medium descriptive resolution.

2.4 Discriminant function analysis for habitat classification

Discriminant function analysis (DFA) was used to test the discrimination of the sub-littoral habitat types found around San Andres, based on the IKONOS optical data and acoustic data in isolation and in combination. Water column corrected spectral and SSS acoustic backscatter and textural signatures for 125 of the detailed, ground-truthing sites were extracted. Only the areas for which acoustic data were also available were selected, so that the synergy of the two datasets could be tested. The DFA analysis was performed at both the coarse and medium descriptive habitat classification levels. To build a model of discrimination, individual bands or sonogram layers were chosen as independent variables within the DFA by a forward stepwise selection process. Confusion matrices were produced to assess the accuracy of the classifications at each level and to identify misclassifications, and overall accuracy rates and user's accuracy for the individual classes were calculated. This was done by comparing the *a posteriori* classification of the habitat members with their *a priori* membership. Although this is a biased measure of discrimination, since the same datasets were used to derive the functions and evaluate the accuracy of the predictions (as opposed to using an independent dataset), this was necessary given the restricted number of data points available. These values were, however, useful for comparing accuracies between the results of the DFA based on the acoustic, optical, and combined datasets, as well as between the classifications at different descriptive resolutions. The study is also primarily concerned with estimating relative rather than absolute separability so this approach is a justifiable one.

3. Results and discussion

3.1 Optical results

At both classification levels the stepwise DFA selected only the Blue IKONOS band, as best discriminating the habitat types and discarded the green and red bands as statistically redundant for separating the classes. Whilst this is expected for the red band as red light is attenuated within the top 1-3 m of the water column due to absorption by water itself, it was more surprising for the green band. The classification results are discussed bearing in mind that they were achieved on the basis of the blue band brightness alone.

Coarse level Label and characteristics	Medium descriptive level Label and characteristics
1. Coral classes	1.1 Branching corals - Majority of corals are branching (eg Acropora spp.)
> 1% hard coral cover	1.2 Sheet corals - Majority of corals are branching (eg Agaricia spp.)
	1.3 Blade fire corals with green calcified algae - Majority of corals are blade fire corals
	1.4 Massive and encrusting corals - Majority of corals are massive and encrusting
	1.5 Dead coral - Dead coral ≥ live coral
	1.6 Gorgonians - Gorgonians ≥ hard coral
2. Algal dominated	2.1 Green algae - Majority of algae are green
≥ 50% algal cover &	2.2 Fleshy brown and sparse gorgonians - Majority of algae are fleshy brown
< 1% hard coral cover	2.3 Lobophora - Monospecific Lobophora beds
	2.4 Red fleshy and cructose algae - Majority of algae are red. Encrusting sponges present
3. Bare substatum	3.1 Bedrock and rubble with dense gorgonians - ≥ 30% gorgonians and ca 30% algal cover
> 50% bare substratum	3.2 Bedrock and rubble with sparse gorgonians - < 30% gorgonians and little algal cover
< 1% hard coral	3.3 Sand and rubble with sparse algae - Both sand and rubble present and occasionally boulders; No gorgonians
	3.4 Sand with sparse algae - No rubble present
	3.5 Mud - Mud is the predominant substrate
	3.6 Bedrock - Bedrock is the predominant substrate; Sand and algae met be present but sparse
4. Seagrass dominated	4.1 Sparse seagrass - 10-30 % seagrass
> 10% seagrass cover &	4.2 Medium density seagrass - 31-69 % seagrass
< 50% algae	4.3 Dense seagrass - > 70 % seagrass
	4.4 Seagrass with distinct coral patches

Table 1. The hierarchy of classes contained within the ecological component of the modified classification scheme by Mumby and Harborne (1999) with quantitative diagnostic descriptors.

The Discriminant function analysis (DFA) results from the extracted IKONOS signature data yielded an overall accuracy of 29% at the medium resolution level (10 classes) and 40% for the coarse level of descriptive resolution (4 classes). The greater accuracy at the coarser level is in agreement with findings of classification accuracies of similar habitats from optical imagery (Mumby and Edwards, 2002). Individual class (user's) accuracies ranged from 12-100% at the medium level, and 0-58% at the coarse level (Tables 2 and 3). At the medium descriptive level, the highest user's accuracy was achieved for dense seagrass (100%) followed by sand and algae (50%), massive corals (45%) and sparse seagrass (43%). Least discrimination was achieved for the dead coral class (12%), sheet corals (16%) green algae

(17%), and medium density seagrass (18%). Most confusion at this descriptive level occurred between the different coral classes, the algal, coral and seagrass classes, the medium and dense seagrass classes, and the sand and sheet coral classes.

Actual class	Predicted class										Row Total
	1.2	1.4	1.5	2.1	3.1	3.3	3.4	4.1	4.2	4.3	
1.2 Sheet corals (mainly *Agaricia*) > 1%	4										4
1.4 Massive and encrusting corals		13	4		1	3					21
1.5 Dead coral (Dead > live coral cover)	3	4	2			1	1	1			12
2.1 Green algae (≥ 50% algal cover, majority green algae)		2		1	1	1			1		6
3.1 Bedrock and rubble with dense gorgonians (> 50% bare)			1		3	1					5
3.3 Sand & rubble with some algae (> 50% bare)	2		3		3	2	1	1			12
3.4 Sand with some algae (> 50% bare)	15		1	2	4		2	2	1		27
4.1 Sparse seagrass and algae (<50%)		3	6		1			3			13
4.2 Medium density seagrass and algae (<50%)		3		3					2		8
4.3 Dense seagrass and algae (<50%)		4							7	3	14
Column Total	24	29	17	6	13	8	4	7	11	3	122
User classification accuracy (%)	**16**	**45**	**12**	**17**	**23**	**25**	**50**	**43**	**18**	**100**	

Table 2. Classification error matrix for spectral signatures extracted from IKONOS imagery at the medium descriptive resolution (10 classes). Cases in row categories are classified into column predicted classes. Overall classification accuracy: 29%.

Although the coarser descriptive resolution achieved a greater overall accuracy of 40%, this is still poor for scientific or management applications. The best discrimination was achieved for the bare substratum class (58%, Table 3) which is in agreement with other IKONOS case studies that report sand to be always well classified (Andrefouet et al., 2003). Two-way misclassification existed between the bare substratum and algal dominated classes. The next best discriminated class was seagrass (53%) with the main confusion of this class being with algae. This is not surprising considering that the algal class here was dominated by green algae and which was spectrally similar to the seagrass class. These similarities contributed to the entire misclassification of the algal class. Similarly, the coral class was wholly misclassified as seagrass or bare substratum.

There are difficulties in comparing classification accuracies between different reef studies. Green et al., (1996) pointed out this difficulty due to inconsistencies in the classification schemes used, the different number and type of classes, differences in *in situ* data collection, image processing methods, and the means by which accuracy is assessed. Furthermore, the particulars of the different sites, e.g. in depth and differences in the dominant species, will greatly influence the accuracy rates. However, our results are in general agreement with those obtained elsewhere using real or simulated IKONOS data where sand-dominated habitats are the best discriminated, while coral classes are generally poorly classified, especially when algal-dominated areas or dense seagrass beds that are spectrally similar to deep corals are included in the analysis (Hochberg and Atkinson, 2003; Andrefouet et al., 2003).

Actual class	Predicted class				
	1. Coral classes	2. Algal dominated	3. Bare substratum	4. Seagrass dominated	Row total
1. Coral classes	0	7	15	17	39
2. Algal dominated		0	3	4	7
3. Bare substratum	3	15	25	1	44
4. Seagrass dominated	1	9	0	25	35
Column Total	4	31	43	47	125
User classification accuracy (%)	0	0	**58**	**53**	

Table 3. Classification error matrix for spectral signatures extracted from IKONOS imagery at the coarse descriptive resolution (4 classes). Cases in row categories are classified into column predicted classes. Overall classification accuracy: 40%.

The key factors which contribute to poor classification are the similarity in spectral signatures between many habitat classes and the limited umber of IKONOS wavebands. Seagrasses, algae and reef habitats are dominated by photosynthetic organisms resulting in similar spectral signatures. Differences between classes are often subtle and require high spectral resolution and often spectral derivative analysis for segregation (Clark et al., 2000; Hochberg and Atkinson, 2000; Hochberg et al., 2003; Karpouzli et al., 2004). IKONOS spectral bands are too broad and poorly placed to detect subtle differences needed to discriminate between such classes.

Mumby and Edwards (2002) and many of the case studies in Andrefouet et al., (2003) reported higher overall classification accuracies, and higher coral class accuracies using IKONOS imagery. However, our results are not directly comparable with these studies which employed supervised image classifications rather than classifications from extracted spectral signatures using DFA. After supervised classification has been applied to an image it is possible to improve the map accuracy for the classified image by contextual editing where contextual rules can be used to reclassify misclassified classes to the correct categories to optimise results (Andrefouet et al., 2003). However, another study that did not use contextual editing to improve the accuracy of the classification of IKONOS data reported similarly disappointing user's accuracies for coral classes of 9.7% and 9.5% for a nine-class medium resolution scheme and a four-class coarse resolution scheme, respectively (Capolsini et al., 2003).

3.2 Acoustic results

Example sonograms of the two backscatter intensity bands (100 kHz and 500 kHz) from some of the habitat classes surveyed are shown in Figure 2. Areas of high backscatter appear bright, while low backscatter appears dark. Rocks and coral mounds can be seen as distinct features yielding a strong and highly textured return (Figure 2a, d). Due to their complex morphology and bathymetric relief, areas of acoustic shadows (appearing black in the SSS data) can be observed adjacent to them. The result is that for coral targets both 100 and 500 kHz images appear highly textured with rough and irregular surfaces; signatures from these areas would be expected to be characterized by high variance. Sand is less reflective and more homogeneous in character compared to the highly textured coral patch (Figure 2a). The 500 kHz sonogram shows a better definition of the coral mounds and higher reflectivity over sand. This is due to the fact that higher frequency sound experiences less attenuation in penetrating sediment than lower frequency sound, resulting in overall less penetration and a higher backscatter (Mitchell, 1993).

Syringodium generates bright but fuzzy backscatter indicative of a strong echo return while the fuzziness indicates that the return is not from a distinct hard object (Figure 2b). Seagrass typically returns a strong echo in comparison to the sediment surrounding the beds, especially in the 100 kHz sonogram attributed to the existence of lacunae in the seagrass leaves (Sabol et al., 2002; Siljestrom et al., 2002). Mixed algal cover exhibits highly textured and strong backscatter of an intermediate coarseness (Figure 2c). *Halimeda* has also been reported to have a strong backscatter in other studies (Blondel and Murton, 1997), possibly due to its highly calcified leaves and stems.

Medium-fine sand produces a much weaker acoustic return than either coral or rubble showing a fine texture especially in the 100 kHz band (Figure 2d). In this example, the gorgonian reef patch shows higher reflectivity in the direction of sonification and the long shadows of sparse massive coral mounds suggest a change in the angle of the seabed, and therefore the higher relief of the reef patch. This was confirmed by dive survey. Similar findings for rocky, gravel and sand substrates are reported from Barnhardt et al., (1998).

The Discriminant function analysis (DFA) results for the acoustic data yielded higher accuracies compared with the optical data. At the medium level of resolution (10 classes) an overall accuracy of 34% was achieved, which was 5% higher than achieved for the optical data at the same level (Table 4). Similar to the DFA results of the optical data, classification accuracy increased at the coarse level of descriptive resolution reaching 50%, 10% more than achieved with the optical data at that level (Table 5). This finding is in agreement with findings of classification accuracies of similar habitats from acoustical single beam data (White et al., 2003).

Individual class user's accuracies ranged from 22-50% at the medium descriptive level, and 5-78% at the coarse level of descriptive resolution (Tables 4 and 5). However, the results of the DFA at the medium level suggest that it is not possible to discriminate all classes on the basis of their acoustic properties alone. Among them the hard coral classes were best discriminated with user accuracies ranging from 40 - 50%, followed by dense gorgonian habitats (40%), the sand and rubble classes (40%), and the sand classes (40%). The seagrass classes had the lowest accuracies (22 - 29%) with misclassifications occurring between them, as well as between them and sand or rubble habitats with some algal cover. Most confusion occurred between the sparse seagrass classes and the sand and rubble with sparse algae classes, and the medium dense, and sparse seagrass classes (Table 4). The green algae class (2.1) was also poorly discriminated with a classification rate of only 25%. Confusion existed between that class and sand and algae (3.4), the sand with rubble class (3.3), the two seagrass classes, and the massive coral class (1.4). With the exception of the latter, the rest of the classes showed a similar coarseness in texture in the sonograms, especially the seagrass and algal habitats which may partly explain their poor discrimination. The typical green algal habitats found around San Andres were on sandy substrate and most often mixed with seagrass species which may explain their apparently similar acoustic responses. Coarse sand

and seagrass also had similar textures with the main difference being in the intensity of the backscatter and their different response in the two frequencies.

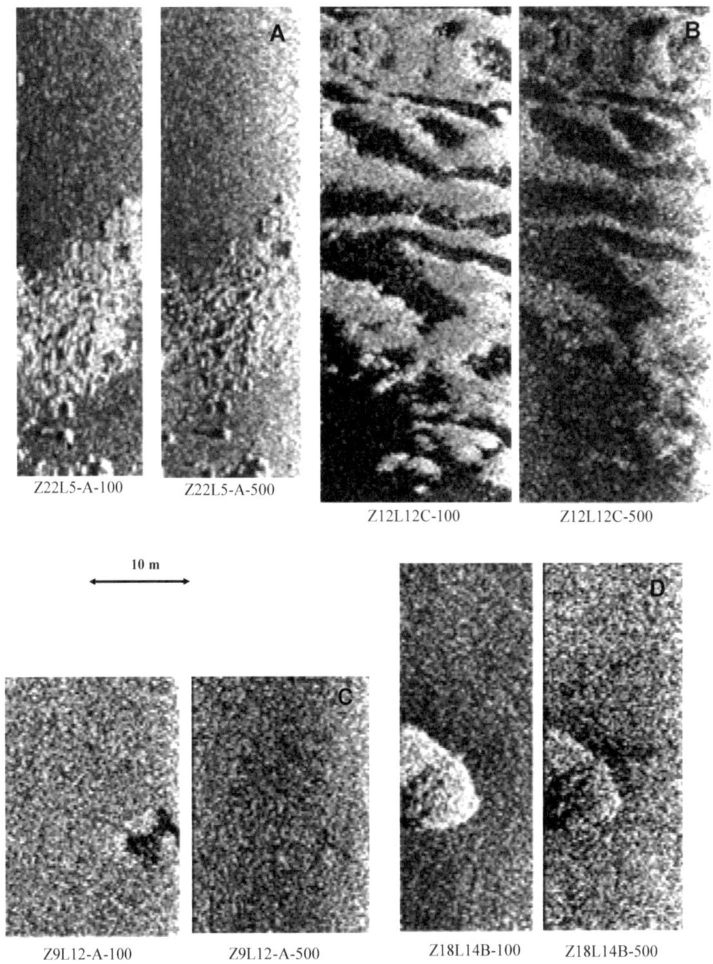

Fig. 2. 100 kHz and 500 kHz side scan sonar images from a number of habitat classes surveyed. Frequency is indicated by "100" or "500" at the end of the area ID. Areas of high backscatter are bright, low backscatter is dark. A) Medium density massive and encrusting coral amongst sand, B) *Syringodium* seagrass among fine sand, C) Medium algal density (principally *Dictyota* and *Halimeda*) mixed with *Syringodium*, D) Medium fine sand with oval patch of dense gorgonian coral on bed rock.

Reducing the descriptive resolution of the DFA increased the accuracies of all classes except for the algal class which was reduced to 5% from 25% (Table 5). This class was largely misclassified as seagrass although many seagrass cases were misclassified as algal classes. However, seagrass classification was improved over the medium level, demonstrating the potential for seagrass discrimination using acoustic data at a coarser level where subclasses of different densities are not considered. The bare substratum class also showed an increase of accuracy at the coarser level of resolution, to 52%.

Actual class	**Predicted class**										Row Total
	1.2	1.4	1.5	2.1	3.1	3.3	3.4	4.1	4.2	4.3	
1.2 Sheet corals (mainly Agaricia) > 1%	3	1									4
1.4 Massive and encrusting corals	2	6	2	3	2	2			4		21
1.5 Dead coral (Dead > live coral cover)	1	2	6	1				1	1		12
2.1 Green algae (≥ 50% algal cover, majority green algae)				4		2					6
3.1 Bedrock & rubble with dense gorgonians (>50% bare)	1	1			2				1		5
3.3 Sand & rubble with some algae (>50% bare)						8		3	1		12
3.4 Sand with some algae (>50% bare)		2	4	4	1	4	2	2	5	3	27
4.1 Sparse seagrass and algae (<50%)		2		2		4		4		1	13
4.2 Medium density seagrass and algae (<50%)									5	3	8
4.3 Dense seagrass and algae (<50%)		1		2			3	4	2	2	14
Column Total	7	15	12	16	5	20	5	14	19	9	122
User classification accuracy (%)	**43**	**40**	**50**	**25**	**40**	**40**	**40**	**29**	**26**	**22**	

Table 4. Classification error matrix for acoustic signatures extracted from the acoustic data at the medium descriptive resolution (10 classes). Cases in row categories are classified into column predicted classes. Overall classification accuracy: 34%.

The class best discriminated on the basis of its textural parameters was coral with a user's accuracy of 78% at the coarse descriptive resolution (Table 5). Many of the processes that drive coral reef dynamics such as recruitment processes or hurricane damage result in patchy distributions which, together with variable three dimensional structures, contribute to this class showing the greatest variance measures (Mumby & Edwards 2002).

Few other studies have reported accuracy rates for mapping coral reef habitats using acoustic remote sensing methods (White et al., 2003; Riegl & Purkis, 2005, Lucieer, 2007). Whilst the results this study are not strictly comparable with those obtained using AGDS by White et al., (2003) and Riegl and Purkis (2005) their single beam acoustic signatures measured parameters of "roughness" and "hardness" of the habitats under investigation. White et al., (2003) reported similarly poor (28%) overall accuracies for a 10 class level of resolution and a higher overall accuracy (60%) at a coarser level of four classes. At the coarse level coral was the best discriminated class with a user's accuracy of 68%, comparable to the results of this study. Riegl and Purkis (2005) presented similar classification accuracies of 56% when attempting to classify 4 classes (coral, rock, algae and sand) on the basis of two signal frequencies, 50 and 200 kHz.

Actual class	Predicted class				
	1. Coral classes	2. Algal dominated	3. Bare substratum	4. Seagrass dominated	Row total
1. Coral classes	25	1	9	3	39
2. Algal dominated		1	2	4	7
3. Bare substratum	7	5	18	14	44
4. Seagrass dominated		12	5	18	35
Column Total	32	19	35	39	125
User classification accuracy (%)	**78**	**5**	**52**	**46**	

Table 5. Classification error matrix for acoustic signatures extracted from the acoustic data at the coarse descriptive resolution (4 classes). Cases in row categories are classified into column predicted classes. Overall classification accuracy: 50%.

Useful acoustic variables - At the coarse classification level, the stepwise procedure selected three acoustic variables as best discriminating the four habitat types and discarded the rest as redundant for separating these classes (Table 6). The variables selected were the mean values of the class signatures of the 2 texture bands (Var_{100} and Var_{500}) and the class standard deviation of the 100 kHz frequency texture bands (SD_{100}). At the medium classification level, the stepwise procedure selected slightly different acoustic variables to best discriminate the 10 habitat classes (Table 6). The variables selected were the class mean values of the 500 kHz texture band (Var_{500}) and the class standard deviations of the 100 kHz and 500 kHz frequencies texture bands (SD_{100} and SD_{500}). At both classification levels, the mean signal intensity data bands (I_{100} and I_{500}) were discarded. Similarly, among the variables selected SD_{100} had the largest discriminant coefficients for both the first and second discriminant functions indicating that it was the most significant variable at both classification levels.

Coarse level	Medium level
Var_{100}	Var_{500}
Var_{500}	SD_{100}
SD_{100}	SD_{500}

Table 6. The acoustic parameters identified by the stepwise discriminant analysis to provide the best discrimination between habitat classes at the coarse and medium descriptive levels.

The standard deviation of the signature areas represents large-scale spatial variability in the images while the two texture (variance) layers indicate small-scale spatial variability in the backscatter signal. The kernel size of the moving window that produced the texture layers is significant in determining their texel values. These values would be directly related to the window homogeneity and the size of the objects in the original image would influence the choice of kernel size, e.g. in this case the size of coral mounds. Hence, it is perhaps inappropriate to measure all classes and objects with the same measure, and each data set and application should have different kernel sizes that maximize their discrimination. In the case of this study, the size of the moving window was optimized for the coral classes and this must have contributed to the improved classification accuracy, when compared to the other classes. Further research could investigate using more texture layers created by changing kernel sizes to optimize them for different habitat types. The applicability of variogram-derived texture measures using a moving window, the size of which is determined by the range of the variogram could also be investigated. Such variogram-derived texture measures have been extracted from microwave images of agricultural landscapes with partial success and warrants further investigation (Jakomulska & Stawiecka 2002)

The use of dual frequency sonograms improved texture and pattern recognition since products from both bands were selected by the DFA, even though the 500 kHz band seemed visually more noisy. The lower frequency (100 kHz) seemed more useful at the coarse classification level, while the higher frequency (500 kHz) was more useful for the discrimination of the detailed scale habitats. This may be partly due to the higher resolution provided by the 500 kHz band, resulting in more detailed sonograms, even after the re-sampling process. This would be translated in the textural bands of the higher frequency. Additional frequency bands could potentially further improve the discrimination between the classes and increase the classification rates.

The side scan sonar survey had a number of limitations which may have contributed to the misclassifications of some of the classes. Positional and locational errors are in general greater for acoustic data compared with satellite data (Malthus & Mumby, 2003). Positional errors may have been introduced from a variety of sources including inadequate positioning of the towfish in relation to the survey boat, and the approximate nature of the manual georectification of the sonograms. Similarly, system resolution, which dictates the minimum size of the feature identifiable at a particular distance from the survey instrument, is a function of both instrumental limitations (sonar instrument specifications) and practical/operational considerations which will be affected by navigational errors, location errors of the instrument and acoustic noise (Bates et al., 2002).

3.2 Optical and acoustic synergy results

With the inclusion of both optical and acoustic signatures in the DFA classification accuracy improved significantly compared to either method used in isolation, at both the coarse and medium level of descriptive resolution (Tables 7 and 8). The overall accuracy of the classifications improved to 52% at the medium level (10 classes) and to 61% at the coarse level (4 classes).

The textural information derived from the high resolution side scan sonar data made a significant improvement to the user's accuracies of each class at both discrimination levels compared to the original spectral discrimination performed used the IKONOS data alone (Table 10). All except one habitat class, showed an increase of at least 10% from their optical classification accuracy when the acoustic data were included in the DFA. The classes that benefited most from the inclusion of the textural acoustic data at the detailed level of resolution were the three coral classes (classes 1.2, 1.4 and 1.5), the green algae class and the sand class (3.4), where % increases were 27%, 19%, 38%, 23% and 28%, respectively. Even though the overall accuracy at this level is still low (52%) and probably inadequate for management purposes (Mumby and Edwards, 2002), at the individual class level, classes 1.4 and 3.4 were well discriminated. This separation is further illustrated in scatterplots of the first three discriminant functions in Figure 3.

Actual class	Predicted class										Row Total
	1.2	1.4	1.5	2.1	3.1	3.3	3.4	4.1	4.2	4.3	
1.2 Sheet corals (mainly *Agaricia*) > 1%	3						1				4
1.4 Massive and encrusting corals		9		2	3	3			3	1	21
1.5 Dead coral (Dead > live coral cover)	3	5	2					1		1	12
2.1 Green algae (≥ 50% algal cover, majority green algae)				2		2				2	6
3.1 Bedrock and rubble with dense gorgonians (> 50% bare)	1		1		2				1		5
3.3 Sand & rubble with some algae (> 50% bare)						6	3	3			12
3.4 Sand with some algae (> 50% bare)			1		1	4	21				27
4.1 Sparse seagrass and algae (<50%)							2	8		3	13
4.2 Medium density seagrass and algae (<50%)									5	3	8
4.3 Dense seagrass and algae (<50%)				1				3	4	6	14
Column Total	7	14	4	5	6	15	27	15	13	16	122
User classification accuracy (%)	**43**	**64**	**50**	**40**	**33**	**40**	**78**	**53**	**38**	**40**	

Table 7. Classification error matrix for combined acoustic and optical signatures extracted from the sonar and IKONOS imagery at the medium descriptive resolution (10 classes). Cases in row categories are classified into column predicted classes. Overall classification accuracy 52%.

Actual class	Predicted class				
	1. Coral classes	2. Algal dominated	3. Bare substratum	4. Seagrass dominated	Row total
1. Coral classes	25	3	5	6	39
2. Algal dominated		1	2	4	7
3. Bare substratum	7	7	28	2	44
4. Seagrass dominated		11	2	22	35
Column Total	32	22	37	34	125
User classification accuracy (%)	**78**	**5**	**76**	**65**	

Table 8. Classification error matrix for combined acoustic and optical signatures extracted from the sonar and IKONOS imagery at the coarse descriptive resolution (4 classes). Cases in row categories are classified into column predicted classes. Overall classification accuracy: 61%.

Coarse level	Medium level
Blue Ikonos	Blue Ikonos
Var_{100}	Var_{100}
Var_{500}	Var_{500}
SD_{100}	SD_{100}
	SD_{500}

Table 9. The acoustic and optical parameters identified by the stepwise discriminant analysis to provide the best discrimination between habitat classes at the coarse and medium descriptive levels.

The better discrimination achieved at the coarse level of descriptive resolution was demonstrated, with the exception of the algal class, by high class user's accuracy values close to or over 70% (Table 10, Figure 4). Similar to the medium level classification, the class that showed the greatest improvement when the textural parameters were added to the DFA was the coral class exhibiting an improvement in its accuracy from 0% to 78%. This was followed by the bare substratum class with an increase from 58% to 76%, and by the seagrass class where accuracy increased from 53% to 65%.

The only classes which did not benefit from the inclusion of the textural acoustic information in the DFA were dense seagrass class (medium classification level) and the algal class (coarse level). Dense seagrasses, along with the other seagrass classes, showed relatively poor discrimination based on their acoustic properties alone. At the coarse level, when all seagrass subclasses were amalgamated into one class, the sonar classification accuracy was higher (46%) which had the overall effect of improving the classification accuracy to 65% for the combined dataset. These results may indicate the inability of sonar data to differentiate between different seagrass densities, and demonstrates that if a class has very good discrimination on the basis of one dataset as in this case (100%) but low discrimination on the basis of the other, then it is best classified only on the basis of the single dataset that gives the best results.

The advantages to be gained from synergistic use of the two datasets is best illustrated by the fact that, for most classes, the discrimination achieved when both datasets were used in combination was equal to or greater than the best discrimination achieved on the basis of each dataset in isolation. This was the case for eight out of the total of ten classes at the medium level and for all classes at the coarse level. However, the results also indicate that some classes will not be successfully differentiated using either dataset (e.g. algal class 2 at the coarse level).

Mumby and Edwards (2002) used textural information to improve the spectral classification of IKONOS data for mapping coral reef habitats and found that their inclusion improved overall accuracy of the thematic map at the medium level by 9%, and at the coarse level by 7%. This study achieved a much greater improvement in accuracy: 23% at the medium level and 21% at the coarse level, attributable to the higher spatial resolution of the sonar data, its greater depth penetration, and the information contained in the sonograms regarding the structural organisation of the habitats. As with the present study, the high coral cover classes showed the greatest texture due to their greater structural heterogeneity, while sand had the least, but both benefited from improved classification accuracy with its inclusion.

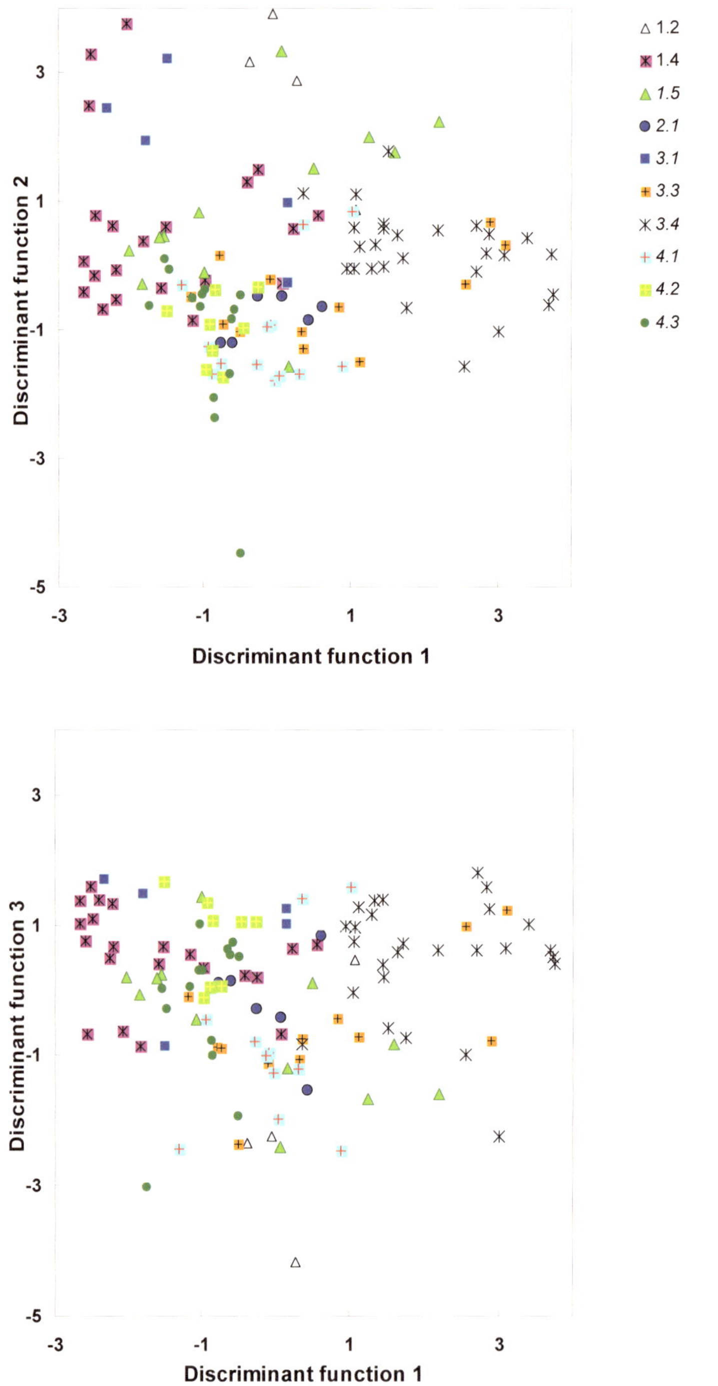

Fig. 3. DF scores from the analysis of the combined optical and acoustic signatures at the medium descriptive level projected in discriminant function space. First and second functions (top graph), first and third functions (bottom graph).

Classification	IKONOS data	Side Scan Sonar data	IKONOS & side scan sonar data
Medium resolution classification:			
1.2 Sheet corals (mainly Agaricia) > 1%	**16**	43	**43**
1.4 Massive and encrusting corals	**45**	40	**64**
1.5 Dead coral (Dead > live coral cover)	**12**	50	**50**
2.1 Green algae (≥ 50% algal cover)	**17**	25	**40**
3.1 Bedrock and rubble with dense gorgonians (> 50% bare)	**23**	40	**33**
3.3 Sand & rubble with some algae (> 50% bare)	**25**	40	**40**
3.4 Sand with some algae (> 50% bare)	**50**	40	**78**
4.1 Sparse seagrass and algae (<50%)	**43**	29	**53**
4.2 Medium density seagrass and algae (<50%)	**18**	26	**38**
4.3 Dense seagrass and algae (<50%)	**100**	22	**38**
Overall accuracy (%):	**29**	34	**52**
Coarse resolution classification:			
1. Coral classes	**0**	78	**78**
2. Algal dominated	**0**	5	**5**
3. Bare substratum	**58**	52	**76**
4. Seagrass dominated	**53**	46	**65**
Overall accuracy (%):	**40**	50	**61**

Table 10. Individual class user's accuracies and overall accuracies (%) from the discriminant function analysis of the optical, acoustic, and combined datasets at the medium and coarse descriptive levels.

The improvement in the discrimination of the dead coral class with the inclusion of the acoustic textural data is particularly significant for monitoring coral health. The results of the optical classification showed that diseased coral cannot be discriminated spectrally on the basis of IKONOS bands alone as, due to their rapid colonization by macroalgae, they are spectrally indistinguishable from macroalgal beds. This is evident in the misclassifications of the other classes (seagrass, sand with algae, and massive coral classes into this class) into the dead coral class (Table 2). Even after the inclusion of the sonar data the classification accuracy of this class is still not satisfactory (50%), but the combination of the two datasets shows potential for improving the discrimination of diseased or dead coral. This may be attributed to the acoustic signatures of algae overlying dead coral mounds; it still identifies the distinct texture of coral, even though spectrally the signature is similar to algal or seagrass classes.

Overall, the improvement in classification accuracies brought about by the inclusion of the acoustic data in the DFA was mainly due to the improved discrimination of spectrally similar classes but which had contrasting textural characteristics, or of classes whose distribution could not be resolved by the spatial resolution of the IKONOS imagery. A limitation of the combined dataset that may have resulted in misclassifications is the imperfection in the co-registration of the optical and sonar datasets. Scale differences

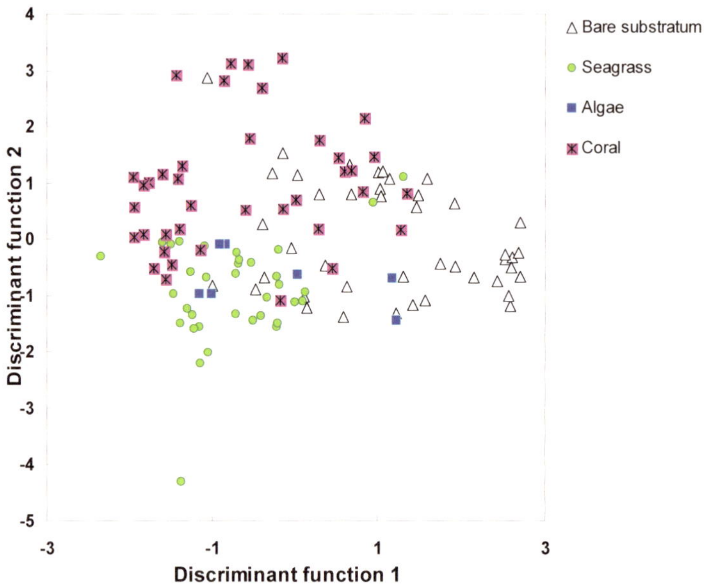

Fig. 4. DF scores from the analysis of the combined optical and acoustic signatures at the coarse resolution level projected in discriminant function space; first and second functions.

between the two datasets exacerbate the co-registration process. Further improvements in the classification accuracies reported here could be expected to be achieved by improved methods of co-registration, using a supervised classifier on the IKONOS image and sonograms themselves, which would allow contextual editing to be implemented, or by entering complementary data in the classification process, such as bathymetry.

4. Conclusion

IKONOS imagery and dual frequency side scan sonar data were acquired in the coastal waters of San Andres island encompassing diverse coral, seagrass, algal and sediment habitats. The characteristics of both data types were compared with the aim of determining if synergistic use of both methods improved the accuracy of classification of these habitats. The optical classification showed that only a few classes can be discriminated by their IKONOS spectral signatures alone, and the incorporation of spatial information, in the form of fine scale, acoustically-derived texture, greatly improved the accuracy of the classification at both the coarse (habitat) and medium (community) levels. The results indicate that the combined use of both techniques provides a means by which the rich diversity of tropical reef ecosystems can be mapped and monitored with significantly greater accuracy than with either technique alone.

In this study, greatest accuracies were achieved at both classification levels based on the Blue IKONOS water column corrected spectral band, and texture parameters derived from the dual frequency high spatial resolution sonograms, which best exploited the differences between classes, although fewer of these parameters were required at the coarse classification level of discrimination.

Overall, the improvement in classification accuracies brought about by the inclusion of the acoustic data in the DFA was due to the improvement of the individual class accuracies that were spectrally similar but had contrasting textural characteristics, or of classes whose distribution could not be resolved by the spatial resolution of the IKONOS imagery. Textural (spatial) information was of particular benefit for discriminating classes characterized by a complex spatial pattern, represented by heterogeneous acoustic response, and even though the overall classification accuracies were still not satisfactory (at 52% for the detailed level and 61% at the coarser level), the improvement from the optical classification of 23% at the fine level and 21% at the coarse level was very encouraging.

The advantages of the synergistic use of the two datasets was illustrated by the fact that, for many classes, when both datasets were used in combination, accuracies were greater than the discrimination achieved on the basis of each of the datasets in isolation. Significant increases in classification accuracies were noted with the inclusion of the acoustic textural data, for the highly textured coral classes in particular, where individual class accuracy levels at 78% (coarse level resolution) were very satisfactory. The improvement in the discrimination of the dead coral class, the differentiation of which is very problematic when based on spectral data alone, has significant implications for monitoring coral health.

The selection of a single IKONOS band for classification highlights the limited capacity of high and medium spatial resolution terrestrial satellite sensors to discriminate reef bottom types compared to higher spectral resolution systems (Maeder et al., 2002; Bouvet et al., 2003; Karpouzli, 2003). These results confirm that sensors with wavebands different to those used by conventional terrestrial satellites are required for detailed mapping of reef biotic systems. It can be expected that higher spectral resolution data would further improve the classification accuracies obtained when optical and acoustic data are combined. Thus, the need for increased spectral resolution is highlighted - a conclusion also reached by other investigators (Hochberg & Atkinson 2003).

The most obvious advantage of using acoustic and optical methods in combination is the different depth ranges to which each system operates. Knowledge of the upper and lower limits of habitats is important for management purposes (Malthus & Mumby, 2003), and the synergistic use of optical and acoustic data can be useful for such studies since optical systems perform best in shallow waters while sonar systems, are limited to depths generally over 2 m but can be used to depths of hundreds of metres, depending on the system employed. Similar conclusions were reached by Riegl and Purkis (2005) when investigating the synergy of IKONOS and single-beam sonar data.

A limitation of this analysis was that the overall and user's accuracies reported were not obtained from an independent dataset. Although these accuracies were useful for comparing relative accuracy levels between different classification levels and dataset, they do not necessarily reflect the accuracy with which another dataset would classify the same classes. This limits comparison with results from other studies where accuracies might be expected to be lower than those obtained here. However, as most studies report accuracies following supervised classification combined with contextual editing, it might be expected that the use of these techniques on combined optical and acoustic data may lead to greater accuracies than those achieved here using DFA.

Overall, the results of this study are particularly encouraging for the benefits to be gained from the synergistic use of optical and acoustic data. It is perhaps easy to understand why the combination of texture or coarseness, and morphological information (represented by acoustic data) and 'colour' characteristics would facilitate the discrimination of different habitats instead of one based on colour alone. Limitations, such as those related to the side scan sonar survey, can be reduced or removed, and hence accuracy levels of the combined dataset are likely to be higher. Discrimination of the habitats could be further improved with the use of contextual editing and the use of complementary data such as bathymetry. Few studies have used spectral and textural variables in conjunction to improve the classification of high spatial resolution images fewer still have derived textural parameters from high spatial resolution side scan sonar data. The lack of research in this area in general and the encouraging results presented here highlights the need for significant development in the synergistic use of optical remote sensing and acoustic data.

5. Acknowledgements

This research was supported by the Darwin Initiative, Onassis Foundation, Carnegie Trust, the University of Edinburgh, Reef U.K., and the Moray Development Fund. It was conducted in collaboration with CORALINA in San Andres. The assistance of several colleagues during fieldwork is gratefully acknowledged, and in particular of Anthony Mitchell Chui, Martha Ines Garcia, Natalia Restrepo, Phil Lovell, Callan Duck, Clare Cavers and Fran Taylor. Spectroradiometric equipment was obtained on loan from the NERC Field Spectroscopy Facility, UK and the sonar data collected by ESGEMAR, Spain.

6. References

Andréfouët, S.; Kramer, P.; Torres-Pulliza, D.; Joyce, K. E.; Hochberg, E. J.; Garza-Pérez, R., et al., (2003). Multi-site evaluation of IKONOS data for classification of tropical reef environments. *Remote Sensing of Environment*, 88, 128-143.

Barnhardt, W. A.; Kelley, J. T.; Dickson, S. M. & Belknap, D. F. (1998). Mapping the Gulf of Maine with side-scan sonar: A new bottom- type classification for complex seafloors. *Journal of Coastal Research*, 14, 646-659.

Bates, C. R.; Moore, C. G.; Malthus, T.; Harries, D. B.; Austin, W.; Mair, J. M. & Karpouzli, E. (2002). Broad scale mapping of sublittoral habitats in the Sound of Barra, Scotland. (pp. 100). Commisioned Report, Report Number F01AA401B, Scottish Natural Heritage, Edinburgh.

Bell, J. D. & Pollard, D. A. (1989). Ecology of fish assemblages and fisheries associated with seagrass beds. In A. W. D. Larkum, A. J. McComb & S. A. Shepherd (Eds.), *Biology of seagrasses: a treatise on the biology of seagrasses with special reference to the Australian region*. Canberra, Australia, Elsevier.

Blondel, P. & Murton, B. J. (1997). *Handbook of seafloor sonar imagery* (pp. 336), John Wiley & Son, Chichester.

Bouvet, G.; Ferraris, J. & Andréfouët, S. (2003). Evaluation of large-scale unsupervised clasification of New Caledonia reef ecosystems using Landsat 7 ETM+ imagery. *Oceanologica Acta*, 26, 281-290.

Buddemeier, R. W. (2001). Is it time to give up? *Bulletin of Marine Science*, 69, 317-326.

Capolsini, P.; Andréfouët, S.; Cedric, R. & Payri, C. (2003). A comparison of Landsat ETM+, SPOT HRV, Ikonos, ASTER, and airborne MASTER data for coral reef habitat mapping in South Pacific islands. *Canadian Journal of Remote Sensing*, 29, 187-200.

Clark, C. D.; Mumby, P. J.; Chisholm, J. R. M.; Jaubert, J. & Andréfouët, S. (2000). Spectral discrimination of coral mortality states following a severe bleaching event. *International Journal of Remote Sensing*, 21, 2321-2327.

Dekker, A.; Brando, V.; Anstee, J.; Fyfe, S.; Malthus, T.J. & Karpouzli, E. (2006). Remote sensing of seagrass ecosystems: Use of spaceborne and airborne sensors. In: Larkum, A.W.D.; Orth, R.J.; Duarte, C. M. (eds.). *Seagrass: biology, ecology and conservation*. Springer, Dordrecht, pp. 347-359.

Diaz, J. M.; Garzon-Ferreira, J. & Zea, S. (1995). Los arrecifes coralinos de la isla de San Andres, Colombia: Estado actual y respectivas para su conservacion, Academia Colombiana de Ciencias Exactas, Fisicas y Naturales, Santafe de Bogota.

Dustan, P.; Dobson, E. & Nelson, G. (2001). Landsat Thematic Mapper : Detection of shifts in community composition of coral reefs. *Conservation Biology*, 14, 892-902.

Geister, J. & Diaz, J. M. (1997). A field guide to the oceanic barrier reefs and atolls of the southwestern Caribbean (Archipelago of San Andres and Providencia, Colombia). In *8th International Coral Reef Symposium*, (pp. 235-236). Vol. 1 Panama City, Panama.

Green, E. P.; Mumby, P. J.; Edwards, A. J. & Clark, C. D. (1996). A review of remote sensing for the assessment and management of tropical coastal resources. *Coastal Management*, 24, 1-40.

Green, E. P.; Mumby, P. J.; Edwards, A. J. & Clark, C. D. (2000). *Remote Sensing Handbook for Tropical Coastal Management*. UNESCO, Paris.

Grigg, R. W. & Dollar, S. J. (1990). Natural and anthropogenic disturbance on coral reefs In *Coral Reefs*, Dubinsky Z. (ed), Elsevier Science Publishing Company, New York, pp. 439-452.

Hochberg, E. J. & Atkinson, M. J. (2000). Spectral discrimination of coral reef benthic communities. *Coral Reefs*, 19, 164-171.

Hochberg, E. J. & Atkinson, M. J. (2003). Capabilities of remote sensors to classify coral, algae and sand as pure and mixed spectra. *Remote Sensing of the Environment*, 85, 174-189.

Hochberg, E. J.; Atkinson, M. J. & Andréfouët, S. (2003). Spectral reflectance of coral reef bottom-types worldwide and impications for coral reef remote sensing. *Remote Sensing of the Environment*, 85, 159-173.

Holden, H. & LeDrew, E. (1998). The scientific issues surrounding remote detection of submerged coral ecosystems, *Progress in Physical Geography*, 22, 190-221.

Holden, H. & LeDrew, E. (1999). Hyperspectral identification of coral reef features, *International Journal of Remote Sensing*, 20, 2545-2563.

Holden, H. & LeDrew, E. (2001). Coral reef features: Multi- vs. hyper-spectral characteristics, *Sea Technology*, 42, 63-65.

Holden, H. & LeDrew, E. (2002). Measuring and modeling water column effects on hyperspectral reflectance in a coral reef environment. *Remote Sensing of Environment*, 81, 300-308.

Isoun, E.; Fletcher, C.; Frazer, N. & Gradie, J. (2003). Multi-spectral mapping of reef bathymetry and coral cover; Kailua Bay, Hawaii. *Coral Reefs*, 22, 68-82.

Jakomulska, A. M. & Stawiecka, M. N. (2002). Integrating spectral and textural information for land cover mapping, In: *Observing our environment from Space: New solutions for a new millenium*, Begni R. (ed), Swets & Zeitlinger, Lisse, pp. 347-354.

Jupp, D. L. B.; Kuchler, D. A.; Heggen, S. J. & Kendall, S. W. (1981). Remote sensing by Landsat as support for management of the Great Barrier Reef, In: *2nd Australasian*

Remote Sensing Conference Proceedings (pp. 9.5.1-9.5.6). Canberra, Australia, pp. 9.5.1-9.5.6.

Karpouzli, E.; Malthus, T. J., & Place, C. (2004). Hyperspectral discrimination of coral reef banthic communities in Western Caribbean. *Coral Reefs*, 23, 141-151.

Karpouzli, E. (2003). High resolution remote sensing of marine reef habitats: towards an integration of satellite and sonar imaging techniques (pp 294). PhD Thesis, Heriot-Watt University, School of Life Sciences, Edinburgh, Scotland, U.K.

Karpouzli, E. & Malthus, T.J. (2003). The empirical line method for the atmospheric correction of IKONOS imagery, *International Journal of Remote Sensing*, 24, 1143-1150.

Karpouzli, E.; Malthus, T.J. & Place, C.J. (2004). Hyperspectral discrimination of coral reef benthic communities in western Caribbean. *Coral Reefs*, 23:141-151.

Karpouzli, E.; Malthus, T. J.; Place, C.; Mitchell Chui, A. & Garcia, M. I. (2003). Underwater light characterisation for correction of remotely sensed images, *International Journal of Remote Sensing*, 24, 2683-2702.

Karpouzli, E.; Malthus, T. J.; Place, C. & Mair, J. M. (2000). A spatial model to predict water attenuation for the bathymetric correction of remotely sensed images, In: *12th Annual Colloquium of the Spatial Information Research Centre (SIRC 2000) Conference Proceedings*. (pp. 15-21). University of Otago, Dunedin, New Zealand.

Knowlton, N. (2001). The future of coral reefs. *Proceedings of the National Academy of Science*, 98, 5419-5425.

Lough, J. M. (2000). 1997-1998: Unprecedented thermal stress to coral reefs? *Geophysical Research Letters*, 27, 3901-3904.

Lucieer, V. (2007). Morphometric characterisation of rocky reef using multibeam acoustic bathymetric data. Proceedings *Proceedings of IGARSS 2007*, Barcelona, 23-27 July 2007.

Luczkovich, J. J.; Wagner, T. W.; Michalek, J. L. & Stoffle, R. W. (1993). Discrimination of coral reefs, seagrass meadows, and sand bottom types from space: a Dominican Rublic case study. *Photogrammetric Engineering and Remote Sensing*, 59, 385-389.

Maeder, J.; Narumalani, S.; Rundquist, D. C.; Perk, R. L.; Schalles, J.; Hutchins, K. & Keck, J. (2002). Classifying and mapping general coral-reef structure using Ikonos data. *Photogrammetric Engineering and Remote Sensing*, 68, 1297-1305.

Malthus, T. J., & Mumby, P. J. (2003). Remote sensing of the coastal zone: an overview and priorities for future research, *International Journal of Remote Sensing*, 24, 2805-2815.

Maritorena, S.; Morel, A. & Gentili, B. (1994). Diffuse-Reflectance of Oceanic Shallow Waters - Influence of Water Depth and Bottom Albedo. *Limnology and Oceanography*, 39, 1689-1703.

Maritorena, S. (1996). Remote sensing of the water attenuation in coral reefs: A case study in French Polynesia. *International Journal of Remote Sensing*, 17, 155-166.

McManus, J.; & Noordeloos, M. (1998). Toward a global inventory of coral reefs (GICOR): Remote Sensing, International Cooperation, and Reefbase, Symposium Conference *Proceedings: Fifth International Conference on Remote Sensing of the Marine Environment* (pp. 83-89). Vol. 1, San Diego, California.

Mitchell, N. C. (1993). A model for the attenuation of backscatter due to sediment accumulations and its application to determine sediment thickness with GLORIA sidescan sonar. *Journal of Geophysical Research*, 98, 22477-22493.

Mora, C., Andrefouet, S., Costello, M.J., Kranenberg, C., Rollo, A., Veron, J., Gaston, K.J. & Myers, R.A. (2006). Coral reefs and the global network of marine protected areas. Science, 312:1750-1752. 23 June 2006.

Mumby, P. J. (1997). Coral reef and seagrass assessment using satellite and airborne remote sensing: An ecological approach (pp 240). PhD Thesis, Department of Geography, University of Sheffield, Sheffield.

Mumby, P. J. & Edwards, A. J. (2002). Mapping marine environments with IKONOS imagery: enhanced spatial resoltion can deliver greater thematic accuracy. *Remote Sensing of Environment*, 82, 248-257.

Mumby, P. J.; Green, E. P.; Edwards, A. J. & Clark, C. D. (1997). Coral reef habitat-mapping: how much detail can remote sensing provide? *Marine Biology*, 130, 193-202.

Mumby, P. J. & Harborne, A. R. (1999). Development of a systematic classification scheme of marine habitats to facilitate regional management and mapping of Caribbean coral reefs. *Biological Conservation*, 88, 155-163.

Palandro, D.; Andréfouët, S.; Dustan, P. & Muller-Karger, F. E. (2003). Change detection in coral reef communities using Ikonos satellite sensor imagery and historic aerial photographs. *International Journal of Remote Sensing*, 24, 873-878.

Pasqualini, V.; Pergent-Martini, C. & Pergent, G. (1999). Environmental impact identification along the Corcican coast. *Aquatic Botany*, 65, 311-320.

Piazzi, L.; Acunto, S. & Cinelli, F. (2000). Mapping of Posidonia oceanica beds around Elba Island (western Mediterranean) with integration of direct and indirect methods. *Oceanologica Acta*, 23, 339-346.

Prada, M. (2002). Mapping benthic habitats in the southwest of Pueto Rico as determined by side scan sonar.(pp. 146) . PhD Thesis, Department of Marine Sciences, Univesity of Puerto Rico, Mayaguez.

Powis, B. J. & Robinson, K. I. M. (1980). Benthic macrofaunal communities in the Tuggerah Lakes, NSW. *Australian Journal of Marine and Freshwater Resources*, 31, 803-815.

Riegl, B. M. & Purkis, S. (2005). Detection of shallow subtidal corals from IKONOS satellite and QTC View (50, 200 kHz) single-beam sonar data (Arabian Gulf; Dubai, UAE). *Remote Sensing of the Environment*, 95, 96-114.

Sabol, B. M.; Melton, R. E.; Chamberlain, R.; Doering, P. & Haunert, K. (2002). Evaluation of a digital echo sounder system for detection of submersed aquatic vegetation, *Estuaries*, 25, 133-141.

Siljestrom, P.; Moreno, A.; Cara, J., Carbo, R. & Rey, J. (2002). Selectivity in the acoustic response of Cymodocea nodosa (Ucria) Ascherson. *International Journal of Remote Sensing*, 23, 2869-2876.

Siljestrom, P. A.; Rey, J. & Moreno, A. (1996). Characterization of phanerogam communities (Posidonia oceanica and Cymodocea nodosa) using side-scan-sonar images. *Journal of Photogrammetry and Remote Sensing*, 51, 308-315.

Sheppard, C. R. C. (2003). Predicted recurrences of mass mortality in the Indian Ocean. *Nature*, 425, 294-297.

Veron, J. E. N. (1995). *Corals in space and time*. Cornell University Press, New York.

Vollmer, L. (1997). The History of the settling process of the Archipelago of San Andres, Old Providence and St. Catherine. (pp 120). Ediciones Archipelago, San Andres, Isla.

Ward, D. H.; Markon, C. J. & Douglas, D. C. (1997). Distribution and stability of eelgrass beds at Izembek Lagoon, Alaska. *Aquatic Botany*, 58, 229-240.

White, W. H.; Hardborne, A. R.; Sotheran, I. S.; Walton, R. & Foster-Smith, R. L. (2003). Using an acoustic ground discrimination system to map coral reef benthic classes. *International Journal of Remote Sensing*, 24, 2641-2660.

Zea, S.; Geister J.; Garzon-Ferreira, J. and Diaz, J. M. (1998). Biotic changes in the reef complex of San Andres Island (Southwestern Caribbean sea, Colombia) occurring over nearly three decades. *Atoll Research Bulletin*, 456, 1-32.

Air-acoustics Sonar Systems

9

Outdoor Sonar Sensing

Fernando J. Álvarez Franco[1] and Jesús Ureña Ureña[2]
[1]*University of Extremadura*
[2]*University of Alcalá*
Spain

1. Introduction

High-precision applications of airborne ultrasonic sensory systems have traditionally been restricted to indoor environments. The number of studies that have proposed the use of ultrasonic sensors outdoors is very limited and most times in a similar way that it has been made indoors. In applications like the guidance of vehicles and mobile robots, ultrasonic sensors are usually integrated in a more complex multi-sensory system, where they have been assigned low-precision tasks. Among these tasks are, for example, the detection of very close obstacles with which there exists an imminent risk of collision, or the coarse ranging of large navigation landmarks (Langer & Thorpe, 1992), (Maeyama *et al.*, 1994) (Guo *et al.*, 2002).

Sonar's reputation as an unreliable sensing technology for outdoor applications is mainly due to the large influence that meteorological parameters have on the propagation of ultrasonic signals, which is a direct consequence of the mechanical nature of these waves. Changes in temperature and humidity, the presence of fog or rain in the atmosphere, and wind-induced refraction can cause strong variations in the attenuation of acoustic waves. As a result, a classical sonar system based on threshold detection of the signal envelope can provide very different results depending on the operating conditions. Furthermore, acoustic noise sources are more likely to be found outdoors. Aircrafts, pneumatic drills, bridge vibrations or even corona effects in high voltage cables are examples of ultrasonic sources which could render sonar in systems completely useless in certain environments. To overcome these problems, the research with ultrasonic sensors outdoors, has evolved to the use of special signal processing techniques such as the continuous transmission frequency modulated (CTFM) ultrasonic sensor used for path edge detection (Ratner & McKerrow, 2003), the use of crosscorrelation with transmitted patterns for an outdoor sonar (Tanzawa *et al.*, 1995) or the wind compensation method based on differential emitters for a positioning system (Jiménez & Seco, 2005).

The first part of this chapter describes in detail the phenomenology associated with the propagation of acoustic waves in the atmosphere, placing an emphasis on the effects of the different mechanisms over the entire range of ultrasonic frequencies used in air. In the second part of the chapter, an outdoor sonar prototype is presented as an example of how the most recent signal coding and pulse compression techniques can be used to improve the reliability of sonar systems when operating under varying meteorological conditions. It will be seen that the transmission of ultrasonic encoded signals in the atmosphere entails a new challenge because of the effects of turbulence on the amplitude and phase of these signals. This problem is examined and results worthy of future exploration are pointed out in this new, fascinating, and open field of research.

2. Propagation of ultrasonic waves in the atmosphere

Sound propagation in the atmosphere has been the subject of intensive research since the second half of the twentieth century. This research has been mainly motivated by the increasing necessity to control noise generated by an ever more industrialized society.

There are several phenomena which have a significant effect on the propagation of acoustic waves through the atmosphere and these phenomena can be divided into three groups:

- Attenuation mechanisms
- Mechanisms affecting propagation speed
- Turbulence

Regarding the first group, useful conclusions for ultrasonic signals can be drawn by merely extrapolating the results already well established in the case of audio frequencies. A more detailed analysis is necessary when dealing with the other two categories.

2.1 Attenuation mechanisms

Geometrical spreading, atmospheric absorption and the attenuation caused by the presence of fog or rain are included in this group of phenomena.

Geometrical spreading is defined as the amplitude decay of an elastic wave caused by the expansion of its wave-front away from the source. Therefore, it does not depend on the propagation medium but on the features of the transducer employed. The acoustic field generated by a transducer with a certain level of symmetry in a point of spherical coordinates (r, ϕ, θ) can be written as

$$P(r,\varphi,\theta) = P_{ax}(r)\cdot D(\varphi,\theta) \tag{1}$$

where $P_{ax}(r)$ is the acoustic pressure over the transducer ($P_{ax}(r) = P_0/r$) and $D(\varphi,\theta)$ is the directional factor whose value depends on the symmetry that characterizes the transducer. For a typical circular piston-shaped transducer, the directional factor is given by the first order Bessel function of the first kind $J_1(\cdot)$ as

$$D(\varphi,\theta) = D(\theta) = \frac{2\cdot J_1(k\cdot a\sin\theta)}{k\cdot a\sin\theta} \tag{2}$$

where a is the piston radius and k is the wave number ($k = 2\pi/\lambda$). In the region close to the transducer axis, $D(\theta) \approx 1$ and the attenuation is that of spherical waves $P = P_0/r$ (6 dB loss when distance to the source is doubled).

In addition to the loss associated with geometrical spreading, when an acoustic wave propagates through the atmosphere part of its energy is dissipated into thermal energy causing the exponential decay of pressure with travelled distance. Thus, the acoustic field can be written as

$$P(r,\varphi,\theta) = P_{ax}(r)\cdot D(\varphi,\theta)\cdot e^{-\alpha_{abs}\cdot r} \tag{3}$$

where α_{abs} is the absorption coefficient, whose value depends on the mechanisms involved in this phenomenon. Air is one of the polyatomic gases which has been the subject of more in-depth study, and currently it is known that there are basically two different mechanisms that cause absorption of acoustic waves: the viscothermal losses (or classical absorption) and the oxygen and nitrogen molecular relaxation processes. The theoretical analysis of both processes led to a set of equations that have been later experimentally adjusted to increase agreement with real data. Today, these equations are grouped into the ISO-9613 standard (ISO, 1993). This rule establishes that the absorption coefficient can be calculated as

$$\alpha_{abs}\ (\mathrm{Np/m}) = f^2 \left\{ 18.4 \cdot 10^{-12} \left(\frac{P}{P_r}\right)^{-1} \left(\frac{T}{T_{20}}\right)^{\frac{1}{2}} + \right. \qquad (4)$$

$$\left. + \left(\frac{T}{T_{20}}\right)^{\frac{-5}{2}} \left[0.01275 \frac{e^{-2239.1/T}}{f_{ro} + f^2 / f_{ro}} + 0.1068 \frac{e^{3352/T}}{f_{rn} + f^2 / f_{rn}} \right] \right\}$$

where f is the sound frequency in Hz, P is the atmospheric pressure in Pa (P_{ref} = 101 325 Pa), T is the absolute temperature in K (T_{ref} = 293.15 K), and f_{rO}, f_{rN} represent the relaxation frequencies of oxygen and nitrogen, respectively. As can be seen from Eq. (4), atmospheric absorption of acoustic waves depends on four parameters: wave frequency, temperature, humidity and pressure, although the dependence with the latter parameter is negligible if one takes into account the range of variation of this magnitude in practice. Fig. 1(a) shows the dependence of α with frequency for T = 20 °C, H = 70 % and P = 1 atm, together with the individual contribution of the different phenomena involved in this process. As can be seen, atmospheric absorption increases rapidly with frequency, being the vibrational relaxation of oxygen the dominant mechanism between 2 kHz and 100 kHz. Fig. 1(b) shows the dependence of α with temperature and humidity for a fixed frequency of 50 kHz and a pressure of 1 atm. The values shown in this figure range between a minimum of 0.37 dB/m for T = −20°C and H =0%, and a maximum of 2.55 dB/m for T= 50°C and H = 13%. In the design of an ultrasonic sensory system intended for outdoor operation, one must take into account that the signal absorption in a warm summer afternoon can be more than six times greater than that of a cold winter morning.

Geometrical spreading and atmospheric absorption are always present when an acoustic signal is transmitted through the air. However, when this transmission is performed outdoors, there are other phenomena such as the presence of rain, fog or turbulence that can cause an additional attenuation of these signals. The effect of turbulence deserves special attention, and will be analysed later in this section. As for the other phenomena, a theoretical expression that can be used to estimate the attenuation caused by fog is that obtained by Cole and Dobbins (Cole and Dobbins, 1970):

$$\alpha_{fog} = k \cdot C_m \left\{ \frac{(C_4 - C_8) + (C_1 - C_6)\frac{C_5}{C_2}\overline{\tau}_t^2}{C_2\overline{\tau}_t \left[1 + \left(\frac{C_5}{C_2}\overline{\tau}_t \right)^2 \right]} + \frac{\overline{\tau}_d}{1 + \overline{\tau}_t^2} \right\} \ (\mathrm{Np/m}) \qquad (5)$$

where coefficients C_i are thermodynamic parameters depending on temperature, k is the wave number, C_m is the ratio between the drops mass and the mass of the gaseous phase per unit volume, and τ_t, τ_d are adimensional relaxation times characterizing the processes of energy and momentum interchange responsible for the attenuation –see (Cole and Dobbins, 1970) for details–. These last three parameters depend on the droplet radius R and on the number of droplets per unit volume N:

$$C_m = \frac{4\pi R^3 N \rho_l}{3\rho_m} \qquad \overline{\tau}_t = \frac{\omega R^2 \rho_l}{3\eta_l \rho_m} \qquad \overline{\tau}_d = \frac{2\omega R^2 \rho_l}{9\nu \rho_m} \tag{6}$$

where ρ_l is the droplets density, ρ_m is the gaseous phase density, and η_l and ν are the thermal diffusivity and cinematic viscosity coefficients of air respectively. From these expressions it can be concluded that, at a given temperature, the attenuation undergone by an acoustic wave propagating in a foggy atmosphere depends basically on three parameters: the droplets concentration, their radius and the wave frequency. In a given fog (R and N constants), this attenuation increases with frequency until it reaches a saturation value that remains stable for higher frequencies. Since atmospheric absorption increases with the frequency to the power of two, a limit frequency exists above which atmospheric absorption is the dominant mechanism. Fig. 2 shows the attenuation values given by Eq. (5) in the range of frequencies 100 Hz - 100 kHz for T = 23 °C. Two types of fog are represented in this figure, a dense fog (N = 2000 drops/cm^3 with R = 6 μm) and a light fog (N = 400 drops/ cm^3 with R = 10 μm). The attenuation associated with atmospheric absorption (Eq. (1)) has also been included in the same figure. Clearly, the dominant mechanism above 10 kHz is atmospheric absorption, although a dense fog can cause an additional attenuation as large as ≈ 0.1 dB/m in this range of frequencies.

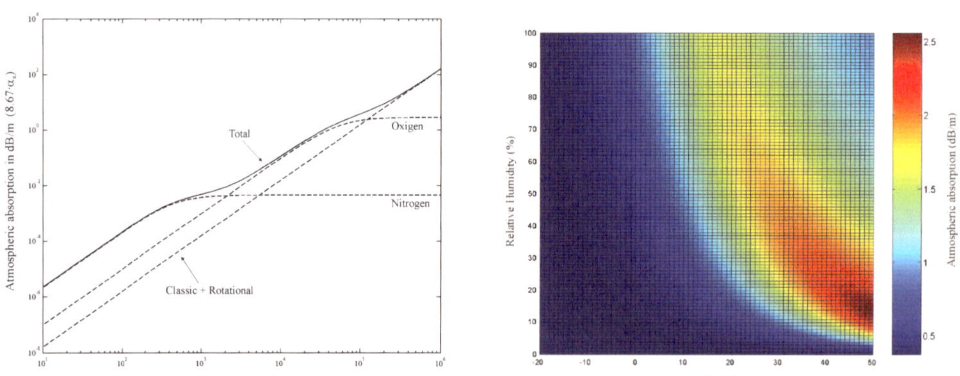

(a) (b)

Fig. 1. Dependence of the atmospheric absorption coefficient α_{abs} on different parameters. (a) Dependence on frequency for fixed temperature (T = 20ºC), humidity (H = 70%) and pressure (P = 1 atm). (b) Dependence on temperature and humidity for fixed frequency (f = 50 kHz) and pressure and (P = 1atm).

With respect to rain, the drop size is so large in this case that the acoustic wave propagates through it without significant perturbation. Only a very intense rain can cause an appreciable attenuation. For frequencies greater than 20 kHz Shamanaeva (Shamanaeva,

1988) provides a very simple expression for the attenuation as a function of frequency and rain intensity I (mm/h):

$$\alpha_{rain} = 1.63 \cdot 10^{-28} f^{4.9} I^{1.5} \quad \text{(Np/m)} \tag{7}$$

Figure 2 also shows the results provided by Eq. (7) for an intense rain of 80 mm/h and a light rain of 5 mm/h. As can be seen, an intense rain can cause in a 50 kHz ultrasonic wave an attenuation similar to that caused by a dense fog (≈ 0.1 dB/m), and this attenuation is even greater for higher frequencies, although it is always less than atmospheric absorption. For frequencies below 50 kHz or in less intense rains this attenuation is negligible in practice.

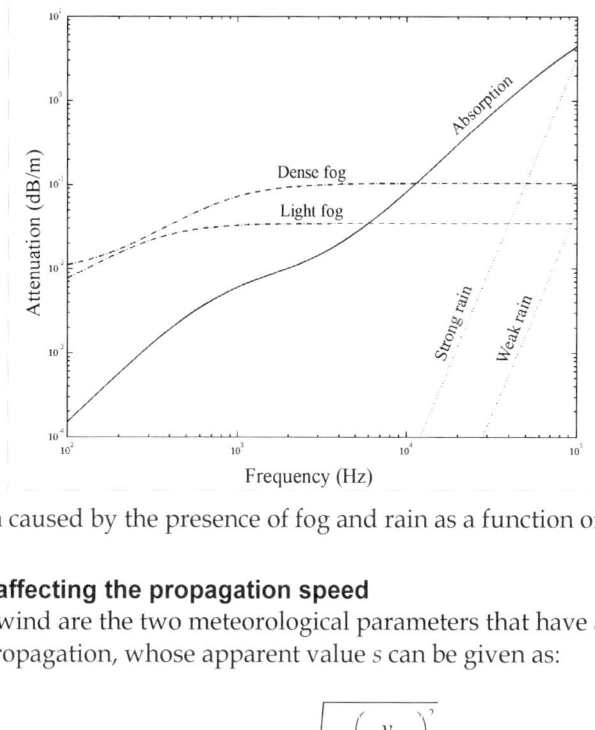

Fig. 2. Attenuation caused by the presence of fog and rain as a function of frequency.

2.2 Mechanisms affecting the propagation speed

Temperature and wind are the two meteorological parameters that have a stronger influence on sound speed propagation, whose apparent value s can be given as:

$$s = v_p + c(T) \cdot \sqrt{1 - \left(\frac{v_n}{c(T)} \right)^2} \tag{8}$$

where $c(T)$ is the propagation speed of the undulatory phenomenon (temperature dependent); and v_p and v_n are respectively, the parallel and normal to the direction of propagation components of wind which is responsible for convective transportation. Assuming that the normal component of the wind v_n is small compared to the wave speed c, a reasonable approximation for the apparent speed is

$$s \approx v_p + c(T) = v_p + 331.6 \cdot \sqrt{1 + \frac{T}{273.15}} \tag{9}$$

where T is the temperature in Celsius. In the atmosphere, both temperature and wind speed are functions that are strongly dependent on height, a dependency inherited by sound speed that causes the refraction of ultrasonic waves propagating outdoors. Figure 3 shows the effects of temperature-induced refraction. During the day the ground is heated by solar radiation, and this heat is transferred to the lower layers of air that progressively cool with increasing height z. This negative gradient of temperature causes the upward curvature of the acoustic rays, a situation depicted in Fig. 3a. The opposite situation occurs at night, when the ground rapidly cools by radiation and air temperatures are colder near the ground. In this case, the acoustic rays are bent downwards as shown in Fig. 3b.

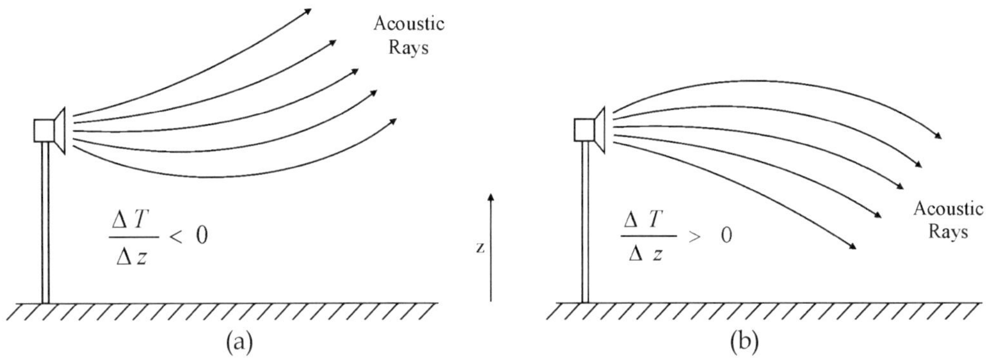

Fig. 3. Temperature-induced refraction.

A similar effect is caused by wind speed dependence on height. Due to the friction between the ground and the moving air, wind speed increases from zero at ground level to a practically constant value at an altitude of several hundreds of meters, a phenomenon known as wind shear. According to Eq. (9) an acoustic wave propagating downwind will be bent downwards, because the apparent sound speed is smaller, and the opposite effect will occur with a wave propagating upwind. Both situations are depicted in Fig. 4.

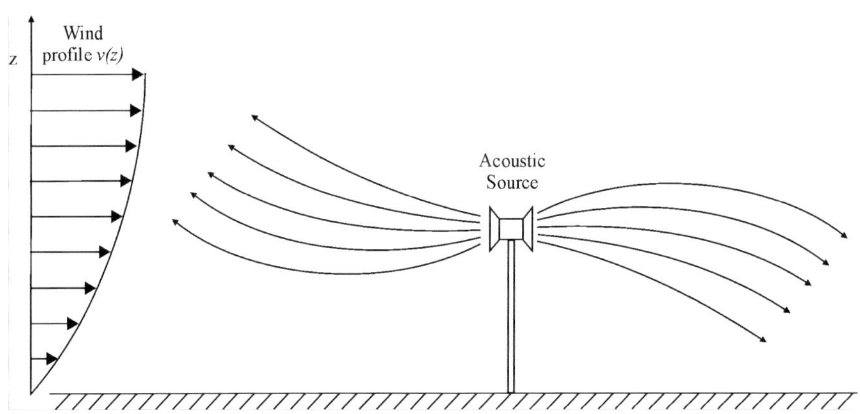

Fig. 4. Wind-induced refraction.

The quantitative analysis of this phenomenon requires an expression for the apparent sound speed as a function of height, $s = s(z)$. Then, it will be possible to calculate the trajectory followed by an acoustic ray using Snell's law

$$\cos\theta = \frac{s(z)}{s_0} \tag{10}$$

where s_0 represents the propagation speed at a reference height and θ is the angle between the acoustic ray and the horizontal plane. The expression for $s(z)$ can be obtained from Eq. (9) if the profiles for temperature $T(z)$ and wind velocity $v(z)$ are known. In the surface layer of the atmosphere, whose height may vary from 10 m on clear nights to 100 m on windy days, these profiles can be obtained from the Monin-Obukhov similarity theory (Monin & Obukhov, 1954):

$$T(z) = T(z_0) + \frac{T^*}{K_a}\left[\ln\left(\frac{z}{z_0}\right) - \psi_T\left(\frac{z}{L_{mo}}\right)\right] - \frac{G_0}{c_P}(z - z_0) \tag{11a}$$

$$v_l(z) = \frac{v^*}{K_a}\left[\ln\left(\frac{z}{z_0}\right) - \psi_v\left(\frac{z}{L_{mo}}\right)\right] \tag{11b}$$

In these expressions K_a is Von Karman's constant, whose value is approximately 0.4; c_p is the air specific heat at constant pressure; and G_0 is the gravitational acceleration module. Constants T^* and v^* represent two scale values for temperature and wind speed respectively and their values are usually experimentally adjusted. Constant z_0 is the *roughness length*, and it is a measure of the minimum height below which the profiles above are not valid. Its value depends on the roughness characteristics of the terrain and can be obtained with reasonable accuracy from tables by simple inspection of the terrain - see for example (Panofsky & Dutton, 1984) -

Functions ψ_T and ψ_v appearing in the equations above depend on the type of atmosphere. These functions have been empirically adjusted to obtain the following expressions:

$$\psi_T\left(\frac{z}{L_{mo}}\right) = \begin{cases} 2\ln\left[0.5\cdot\left(1+x^2\right)\right] & \text{Unstable} \\ 0 & \text{Neutral} \\ -5\dfrac{z}{L_{mo}} & \text{Stable} \end{cases} \tag{12a}$$

$$\psi_v\left(\frac{z}{L_{mo}}\right) = \begin{cases} \ln\left[\left(\dfrac{1+x^2}{2}\right)\left(\dfrac{1+x}{2}\right)^2\right] - 2\arctan x - \dfrac{\pi}{2} & \text{Unstable} \\ 0 & \text{Neutral} \\ -5\dfrac{z}{L_{mo}} & \text{Stable} \end{cases} \tag{12b}$$

being $x = (1 - 16z/L_{mo})^{1/4}$.

Finally, L_{mo} is the *Monin-Obukhov length* that basically depends on the vertical flux of heat at the surface. Since this flux cannot be easily measured experimentally, L_{mo} is not directly calculated in practice. Instead, its value is inferred from the relation of this parameter with the roughness length and Turner classes found by Golder (Golder, 1972). Turner classes constitute a classification method for different types of atmospheres, which is based on the relative importance that thermal convection and mechanical turbulence have on a particular atmosphere. Seven classes exist numbered from 1 to 7. The first three classes correspond to unstable atmospheres, where thermal convection prevails over mechanical turbulence (sunny days with light winds), class 4 represents neutral atmospheres where thermal convection does not exist and only mechanical turbulence is present (days and nights with strong winds), and the last three classes correspond to stable atmospheres characterized by an inversion of the temperature gradient (clear nights with light winds). The class a certain atmosphere belongs to, may be directly calculated from the solar altitude (which determines the intensity of the radiation received), the cloud cover, and the wind speed at a reference height. The relation experimentally obtained by Golder between z_0, L_{mo} and Turner classes is represented in Fig. 5

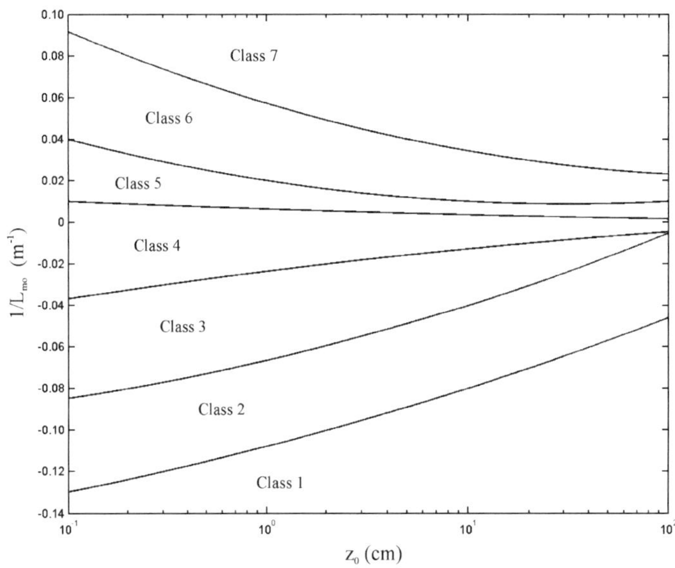

Fig. 5. Relation between z_0, L_{mo} and Turner classes (Golder, 1972).

Let us now focus on the influence that this phenomenon has on the signal emitted by a sonar system. Due to atmospheric absorption, a signal horizontally emitted can cover a maximum distance of several tens of meters before reaching an object, and the maximum vertical deviation caused by refraction is not expected to exceed a few meters even in the most unfavorable conditions. In such a situation, it can be generally assumed that the apparent sound speed depends linearly on height, and then, the departure angle of a ray that is detected by a receiver placed at a distance r from the emitter in the same horizontal plane is given by:

$$\theta_s = \arctan\left(\frac{r \cdot g}{2 \cdot s_0}\right) \tag{13}$$

where $g = ds/dz$ is the constant speed gradient and s_0 is the apparent sound speed at the height of the sensor. The constant gradient can be obtained as the average sound speed variation in the first Fresnel ellipsoid. This ellipsoid is represented in Fig. 6, and it is defined as the region of the space where the differences between the length of the direct path and the length of any diffracted path is less than a half wavelength:

$$r_1 + r_2 - r = \frac{\lambda}{2} \tag{14}$$

From this figure, the constant gradient can be expressed as:

$$g = \frac{s(h_{max}) - s(h_{min})}{h_F} \tag{15}$$

where h_F represents the ellipsoid height, which depends both on the wavelength λ and on the emitter-receiver distance r as shown

$$h_F = \sqrt{\frac{\lambda}{4} \cdot \left(r + \frac{\lambda}{4} \right)} \tag{16}$$

The evaluation of Eq. (13) is a tedious task which involves many previous calculations that are necessary to obtain the constant speed gradient g. An algorithm performing all these calculations has been used in (Alvarez *et al.*, 2008) to study a particular ultrasonic system installed at a height of 150 cm over a terrain a short grass where maximum wind speeds of 7.6 m/s were registered during a period of 6 months. This algorithm predicted a maximum departure angle of 1.6° under the worst conditions of wind, assuming a propagation distance of 14 m. This angle causes a negligible additional attenuation of 0.2 dB in the signal received at the same height if a Polaroid series 600 transducer (Polaroid, 1999) is employed in the emission.

This conclusion should not be generalized, though. If transducers with narrower emission patterns are used, longer propagation distances are considered, or if the system is installed in locations where greater speed gradients can occur then the effect of refraction might not be insignificant.

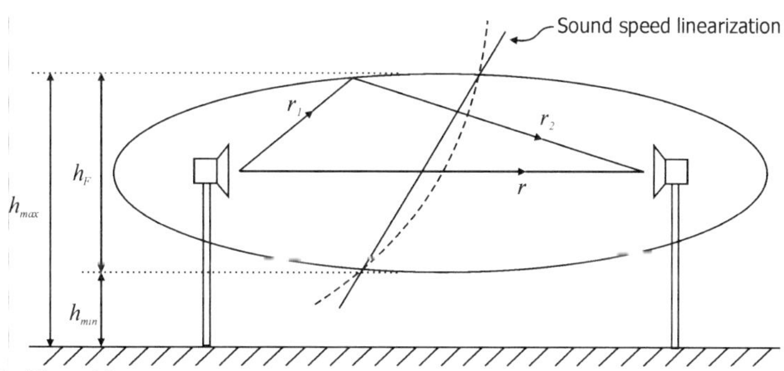

Fig. 6. Fresnel ellipsoid.

2.3 Turbulence

Wind is rarely stationary in the lower layers of the atmosphere and almost always random fluctuations of its behavior appear in the form of highly rotational fluxes. These *turbulent eddies* cause local random fluctuations in wind velocity and temperature, two parameters that have a strong influence on the propagation velocity of acoustic waves as was seen above. Each one of these eddies causes a sudden change on the refraction index and the subsequent scattering of part of the wave energy. An immediate consequence of this scattering is the additional attenuation undergone by a wave propagating through a turbulent medium. The average value of this attenuation was obtained by Brown and Clifford (Brown & Clifford, 1976) as

$$A_{turb} = 10\log\left(1 + 0.48k^{12/5}d_0^2 r^{6/5} C_n^{12/5}\right) \quad \text{(dB)} \tag{17}$$

where k is the wave number; d_0 is the diameter of the transducer; r is the travelled distance and C_n^2 is the refractive index structure parameter which characterizes the turbulence strength. Equation (17) provides attenuation values far below those associated with atmospheric absorption in the range of frequencies 100 Hz – 100 kHz, as shown in Fig. 7.

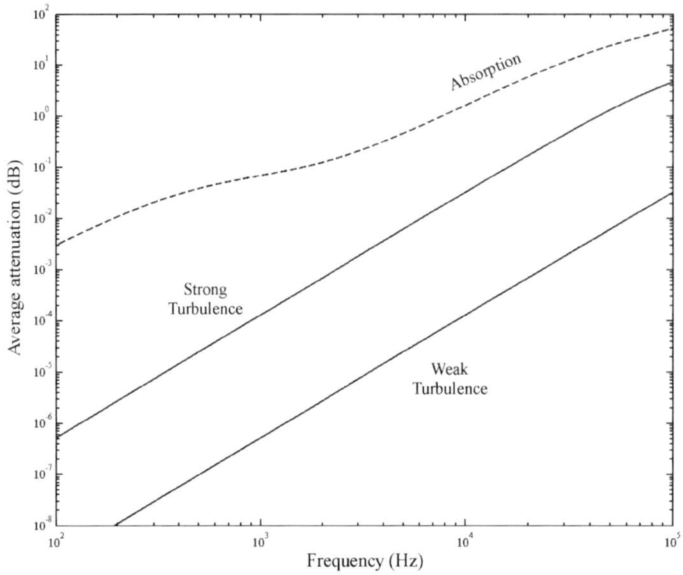

Fig. 7. Attenuation associated with the presence of turbulence as a function of frequency.

Nevertheless, it is important to note that this attenuation is an average value and that the instantaneous values can vary widely around it. When an acoustic wave propagates through a turbulent region, it encounters a variety of eddies with different sizes, velocities and temperatures. Their combined effect alters the initial coherence of the wavefronts, which will no longer be spherical nor have identical amplitude after crossing the turbulent region. This situation is depicted in Fig. 8. A receiver placed at a certain distance from the emitter will record random fluctuations in the amplitude and phase of the acquired signals. This effect can be characterized through a *coherence time* t_c, defined as the time during which the characteristics of an acoustic wave propagating through a turbulent region remain essentially invariant.

Turbulence theory provides a means to estimate the value of this time theoretically. Random fluctuations of wind velocity and temperature in a turbulent atmosphere cannot be described as stationary random fields through the spatial correlation function

$$C_u(r_1, r_2) = \left\langle \left[u(r_1) - \langle u(r_1)\rangle\right] \cdot \left[u(r_2) - \langle u(r_2)\rangle\right] \right\rangle \tag{18}$$

where u represents a generic magnitude; $< \bullet >$ is the spatial average operator; and r_1, r_2 two arbitrary positions. These meteorological fields do not have constant mean and Eq. (18) is not invariant under translations. In this case, the three dimensional behavior is best described through the so-called *structure functions*, first introduced by Kolmogorov (Kolmogorov, 1941)

$$D_u(r_1,r_2)=\left\langle \left\{\left[u(r_1)-\langle u(r_1)\rangle\right]-\left[u(r_2)-\langle u(r_2)\rangle\right]\right\}^2\right\rangle \tag{19}$$

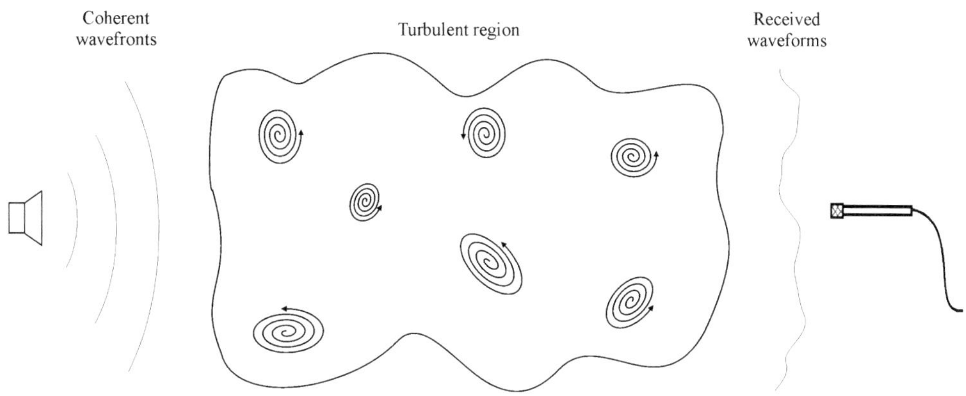

Fig. 8. Propagation of an acoustic wave through a turbulent medium.

Although the statistical properties of these random fields depend on location, it can be assumed that difference between values taken at any two points depends on the distance $|r_1 - r_2|$ whenever this distance is not excessively large, a property known as *local homogeneity*. Then, if isotropy is also assumed Eq. (19) takes the simpler form:

$$D_u(r)=\left\langle\left[u(r_0+r)-u(r_0)\right]^2\right\rangle \tag{20}$$

This structure function is related to the three dimensional *power spectrum* $\Phi_u(k)$ (Fourier transform of the correlation function) according to the following expression

$$\Phi_u(k)=\frac{1}{4\pi^2k^3}\int_0^\infty \frac{\sin(kr)}{kr}\frac{d}{dr}\left[r^2\frac{d}{dr}D_u(r)\right]dr \tag{21}$$

where $k = 2\pi/\lambda$ is the spatial wavenumber. The spectral analysis of random fields in a turbulent medium is very useful because such a regime can be characterized by several length scales. The longer scale is named *outer scale* or *integral scale* L_0 and represents the size of the larger eddies in the atmosphere. These eddies are very energetic and low dissipative, and their size is in the range of some tens of meters, although it may vary according to local conditions. The energy of these eddies is redistributed without loss to eddies of decreasing size until it is converted into heat by viscous dissipation (see Fig. 9). The size of the smaller eddies is named *inner scale* or *microscale* l_0 and it measures approximately a few milimeters. Taking into account that the amount of kinetic energy per unit mass is proportional to v^2, and assuming that the rate of transfer of energy from the largest eddies of size L is proportional to v/L, then the rate of energy supply to the small-scale eddies is in the order of v^3/L. The range of distances between the outer scale and the inner scale is named the *inertial subrange* and, as the energy is transported from large eddies to small eddies in this range without piling up at any scale, the rate of energy supply must be equal to the dissipation rate ε, and then $v^2 \propto L^{2/3}$. This dimensional analysis was carried out by Kolmogorov who concluded that in the inertial subrange the structure functions of some meteorological magnitudes such as wind velocity, temperature and refraction index must exhibit the same dependence on distance, i.e

$$D_u(r) = C_u^2 \cdot r^{2/3} \tag{22}$$

C_u^2 is a constant called *structure parameter* of magnitude u fluctuations, and it depends on the turbulence strength. Combining Eq. (22) and Eq. (21) the Kolmogorov power spectrum is obtained

$$\Phi_{u,K}(k) = 0.033 \cdot C_u^2 \cdot k^{-11/3} \tag{23}$$

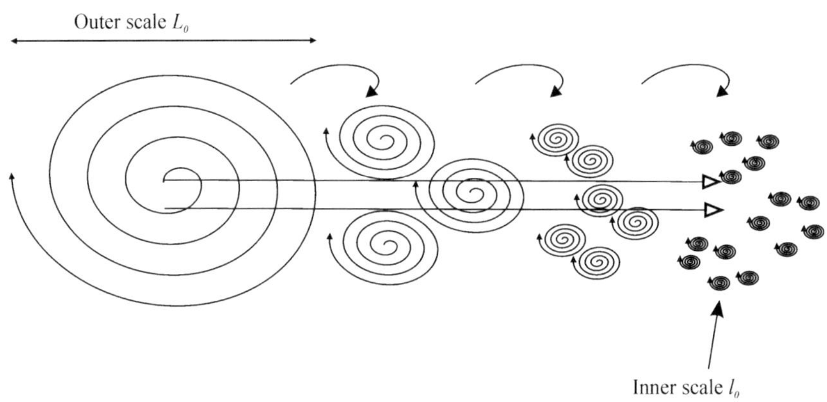

Fig. 9. Kolmogorov model of Turbulence.

The spatial coherence of an acoustic wave is usually described by means of the *mutual coherence function* (*MCF*), defined as the cross-correlation function of the complex pressure field in a plane perpendicular to the direction of propagation

$$MCF(\boldsymbol{\rho}, r) = \left\langle P(\mathbf{r} + \boldsymbol{\rho}, r) \cdot P^*(\mathbf{r}, r) \right\rangle \tag{24}$$

where ρ is the separation along the plane that is perpendicular to the direction of propagation and located at a distance r from the source. Tatarskii (Tatarskii, 1971) showed that a simple relation exists between the *MCF* and the phase and log-amplitude structure functions given by D_Φ and D_χ respectively

$$MCF(\boldsymbol{\rho}, r) = \exp\left\{ -\frac{1}{2}\left[D_\chi(\boldsymbol{\rho}, r) + D_\phi(\boldsymbol{\rho}, r) \right] \right\} \tag{25}$$

From this expression, and assuming that scattering angles are small, Tatarskii obtained the following expression valid for spherical waves propagating in an isotropic medium

$$MCF(\boldsymbol{\rho}, r) = \exp\left\{ -4\pi^2 k^2 r \times \int_0^\infty \left[1 - \int_0^1 J_0(K\rho u)du \right] \phi_n(K) K dK \right\} \tag{26}$$

where J_0 is the zero order Bessel function of the first kind. A further simplification of Eq. (26) requires selecting a model for atmospheric turbulence through the power spectrum of the refraction index $\Phi_n(k)$. The model to be selected is determined by the size of the eddies that have a larger scattering effect, which are the ones whose size is about one Fresnel Zone $(\lambda \cdot r)^{1/2}$. For an ultrasonic wave of about 50 kHz traveling over some tens of meters, the Fresnel Zone is clearly within the inertial region and $\Phi_n(k)$ must be selected as a Kolmogorov spectrum (Eq. (23)). Introducing Eq. (23) into Eq. (26) provides a method to measure the size of the coherent wavefront as the lateral separation value ρ_0 where the MCF is $1/e$ times lower than its on-axis value, i.e. $MCF(\rho_0,r)=1/e$. This length is called *lateral coherence length* and can be calculated as follows

$$\rho_0 = \left(0.545 \cdot k^2 \cdot C_n^2 \cdot r\right)^{-3/5} \quad r_i < r < r_c \tag{27}$$

where the lengths r_c and r_i defining the range of validity of the above expression are given by

$$r_c = \left[0.4k^2 C_n^2 \left(L_0 / 2\pi\right)^{5/3}\right]^{-1} \tag{28a}$$

$$r_i = \left[0.4k^2 C_n l_0^{5/3}\right]^{-1} \tag{28b}$$

These expressions were first obtained by Yura (Yura, 1971) for optical waves, but they are still valid for acoustic if the structure parameter of the refraction index is replaced with a new efficient structure parameter defined as (Ostashev, 1997)

$$C_n^2 = \frac{C_T^2}{T^2} + \frac{22}{3}\frac{C_v^2}{c^2} \tag{29}$$

where T is the temperature; c is the sound speed; and C_T^2, C_v^2 represent the structure parameters of temperature and wind velocity respectively.
From the lateral coherence length, it is possible to estimate the temporal coherence of a received wave if a "Frozen Model" for the turbulent atmosphere is assumed. This model establishes that in a turbulent atmosphere the collection of eddies remains frozen in relation to one another while the entire collection moves with mean wind velocity v. In this case a direct relation exists between the spatial and temporal behaviors of the statistical fields:

$$u(\mathbf{r}, t + t') = u(\mathbf{r} - \mathbf{v} \cdot t', t) \tag{30}$$

In a first approximation, the effects derived from the longitudinal displacement of the pattern are negligible when compared to those derived from the transversal one. Then, the time for which a signal received at a certain distance remains coherent is equal to the time that the pattern of the eddies takes to travel over the lateral coherence length. This coherence time is given by (Álvarez *et al.*, 2006)

$$t_c = \frac{1}{v_n \cdot \left(0.545 k^2 C_n^2 r\right)^{3/5}} \tag{31}$$

where v_n is the transversal component of the wind. For an ultrasonic signal of 50 kHz and a reference propagation distance of r = 14 m, this expression yields a minimum value of 8.2

ms for the coherence time when strong turbulence conditions ($C_n^2 = 10^{-5}$ $m^{-2/3}$) and $v_n = 10$ m/s are assumed. For greater values of wind velocity small scattering angles cannot be assumed and Eq. (31) is no longer valid.

3. Reliable outdoor operation: signal coding and pulse compression

It is clear from the results obtained in the previous section that a classical sonar system whose echoes are detected when they or their envelopes first exceed a certain threshold, cannot reliably operate outdoors. The amplitude of the echo generated by the same object at the same distance can vary largely depending on meteorological conditions of temperature, humidity, rain, fog and wind, giving rise to very different measurements of time-of-flight (*TOF*). This problem could be partially overcome by adapting the detection threshold to the average attenuation associated to current values of these conditions. However, the random and wide variations of amplitude and phase induced by turbulence are very difficult to deal with.

Classical systems are very sensitive to ultrasonic noise too, since such noise is added to the received echo modifying the instant in which it or its envelope exceeds the threshold. High intensity noise could even be confused with real echoes giving rise to phantom reflectors (artifacts). Obviously, robustness to noise can be improved by increasing the energy of the emissions, but there is always a physical limit for the maximum amplitude than can be transmitted with a given transducer. If envelope detection is used, another alternative would be to increase the duration of the emissions, but at the expense of degrading the system precision (two overlapped echoes cannot be distinguished by these systems).

Signal coding and pulse compression techniques emerge as an attractive alternative in the development of reliable outdoor sonars. These techniques have been already used in the design of high performance indoor sonars that are capable of simultaneously measuring the *TOF* of echoes coming from different emissions with a precision of microseconds. Instead of ultrasonic pulses, these systems emit modulated binary codes with good correlation properties that are detected through matched filtering. Thus, a correlation peak is obtained only when the code matched to the corresponding filter is received, the relative height of this peak being proportional to the length (and not the amplitude) of the code. With these techniques the system has a resolution similar to that obtained with the emission of short pulses, still maintaining the high robustness to noise achieved with the emission of long pulses (hence the name pulse compression). Strong variations in the amplitude of the received echoes modify the height, but not the position, of the correlation peak and the results provided by the system would be the same regardless of the attenuation of the signal if the detection threshold for this peak is sensitive enough. Moreover, an adequate selection of codes with low values of cross-correlation allows a system composed of several transducers to perform simultaneous measurements under exactly the same operating conditions.

However, random fluctuations of amplitude and phase induced by turbulence are still a problem. If the coherence time that characterizes this phenomenon (Eq. (31)) is much shorter than the length of the emitted code, this code could be completely distorted before arriving at the receiver and it could not be properly detected by matched filtering. The following sections deal with the analysis of this phenomenon.

3.1 Outdoor prototype

The outdoor propagation of encoded signals has been studied with the help of the prototype shown in Fig. (10). In this system, a Polaroid series 600 electrostatic transducer (Polaroid, 1999) placed 1.5 meters over the ground is used as the emitter. A high-frequency microphone placed at 14 meters from the emitter in the same horizontal plane has been used as the receiver. In order to minimize the filtering effect associated with the acoustic pattern peculiarities of both the emitter and the receiver, a laser pointer was used to align their axes. All the process has been controlled by a PC equipped with an acquisition board that simultaneously sends the emission pattern to the amplifier driving the transducer and acquires the signal coming from the microphone. Figure 10a shows a picture of the experimental site with the emitter in the background. This emitter, together with the anemometer, can be seen in more detail in Fig. 10b.

(a) (b)

Fig. 10. Experimental site (a) and detailed view of the transducer with the anemometer (b).

This system has been used to conduct the emission of complementary sets of four sequences with different lengths. A set of 4 binary sequences $\{x_i[n],\ 1 \le i \le 4\}$, whose elements are either +1 or -1, is a complementary set if the sum of their aperiodic autocorrelation functions ϕ_{xixi} equals zero for all non-zero time shifts

$$\phi_{x1x1}[n]+\phi_{x2x2}[n]+\phi_{x3x3}[n]+\phi_{x4x4}[n]=\begin{cases}4\cdot L & n=0\\ 0 & n\neq 0\end{cases} \tag{32}$$

where L is the length of the sequences. One main advantage that complementary sequences have against other codes used for pulse compression, such as Barker codes or m-sequences, is the existence of orthogonal families. Two sets with the same number of sequences $\{x_i[n],\ y_i[n];\ 1 \le i \le 4\}$ are said to be orthogonal when the sum of the corresponding cross-correlation functions equals zero

$$\phi_{x1y1}[n]+\phi_{x2y2}[n]+\phi_{x3y3}[n]+\phi_{x4y4}[n]=0 \quad \forall n \tag{33}$$

This property allows the simultaneous emission of different signals with ideal null interference. The other main advantage of using these sequences is the existence of an efficient correlation system called *Efficient Sets of Sequences Correlator* (*ESSC*), that notably reduces the total number of operations carried out from 4 x 2 x (L-1) to 4 x $\log_2 L$ in order to perform the correlations with the four sequences in a set (Álvarez *et al.*, 2004). This digital filter, shown in Fig. 11, is formed by $\log_2 L$ similar stages, each one with 3 delay elements and 8 adders/substracters. Coefficients $W_{i,j}$ appearing in this figure take values +1 or -1 and are not implemented as amplifiers in practice.

In our prototype, the four sequences composing the set have been simultaneously transmitted through the Polaroid transducer first by interleaving these sequences to generate a new 4L-bit sequence defined as:

$$s_i = \left[x_1[1] \;\; x_2[1] \;\; x_3[1] \;\; x_4[1] \;\; \cdots \;\; x_1[L] \;\; x_2[L] \;\; x_3[L] \;\; x_4[L] \right] \quad (34)$$

and then by implementing the BPSK modulation of the new sequence with a symbol formed of two cycles of a 50 kHz carrier. The modulated signal has a centralized bandwidth of about 12.4 kHz that allows its efficient transmission through the ≈20 kHz bandwidth of this transducer. The total duration of the emission is proportional to the length of the sequences and is given by:

$$t_e = 4L \text{ (bits) } \times 2 \text{ (cycles)} \times 20\mu s = 160L \; \mu s \quad (35)$$

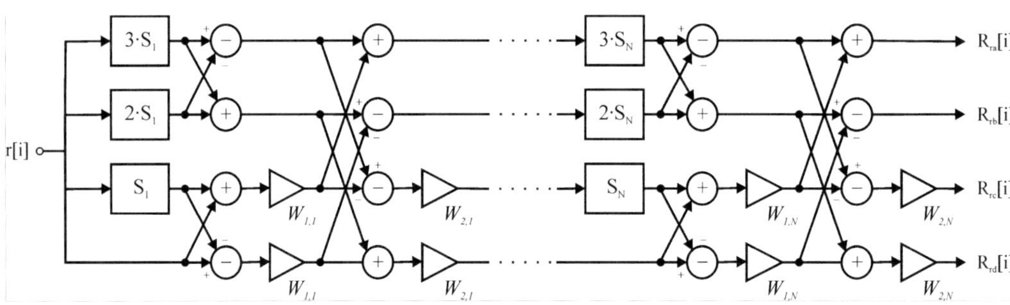

Fig. 11. Efficient Sets of Sequences Correlator (*ESSC*).

The signal received by the microphone placed at 14 m from the transducer is first sampled at a rate of 800 ksps, and then demodulated with a digital correlator matched to the modulation symbol (32 samples). Actually, this filter correlates the acquired signal with a binarized version of the symbol, simplifying the operations at the expense of a nearly negligible decrease in the output *SNR*. The signal from the demodulator is an interpolated version of the sequence obtained by interleaving the complementary sequences (Eq. (34)), with an interpolation factor of 32. Thus, in this signal, two samples belonging to the same sequence are 4 x 32 = 128 samples apart, and it is necessary to decimate the signal by the same factor prior to carrying out the correlations. This decimation can be easily achieved just by multiplying the values of all the delay stages in the *ESSC* by 128. Finally, taking into account that the bits of the interleaved sequences are delayed 32 samples, it is necessary to add three additional delay stages at the outputs of the *ESSC* in order to perform the in-phase sum of the autocorrelation functions.

In the precise moment in which the last sample of the set matched to the correlator is acquired, a maximum value, ideally proportional to $\phi_{x1x1}[0]+\phi_{x2x2}[0]+\phi_{x3x3}[0]+\phi_{x4x4}[0]$, is obtained at the input of the peak detector. However, as a consequence of the asynchronism characterizing the detection process, this maximum value always appears with self-induced noise depending on the shape of the modulation symbol. A parameter commonly used to measure the quality of this signal is the *Sidelobe-to-Mainlobe Ratio* (*SMR*), defined as the ratio between the higher value obtained outside the vecinity of the main peak and the value of this peak.

3.2 Experimental results

In order to investigate the effect of turbulence on the performance of the prototype described in the previous section, a continuous emission of the codes has been conducted. Fig. 12 shows the received and processed signals when complementary sequences of 64 bits (t_e = 10.24 ms) are emitted under very weak and very strong turbulent conditions. As can be seen in this figure, in both cases all the sets are properly detected, although the scattering of energy caused by turbulence is evident in Fig 12b, with the consequent deterioration of the *SMR*. Figure 13 shows the same signals when sequences of 256 bits (t_e = 40.96 ms) are emitted. As can be seen in Fig. 13a, under weak turbulent conditions the sets are still property detected, but when the coherence time is clearly shorter than the emission time spurious peaks appear that may confuse the system. In this case, when the received signal is compared to the emission pattern, slight compressions and expansions can be clearly visualized which are deteriorating the phase coherence required by the correlation process. This comparison is represented in Figs. 14 and 15 for very weak and very strong turbulence respectively.

The observed increase in *SMR* has been experimentally studied under different conditions of turbulence, and the results are presented in Table 1. The *SMR* remains below 0.3 even under very strong turbulence when 64-bit sequences are transmitted, showing the good performance of the system in all cases. The same result is true with 256-bit sequences except

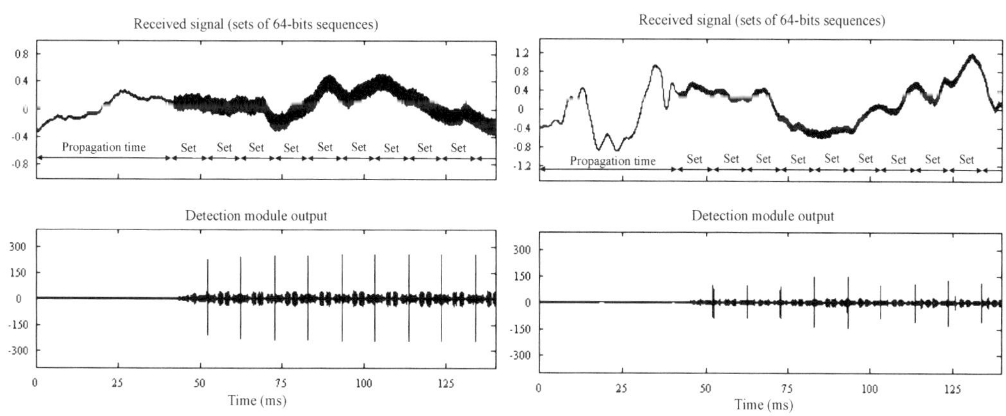

(a) (b)

Fig. 12. Detection process of 64-bit sequences (t_e = 10.24ms) under very weak (a) and very strong (b) turbulence conditions.

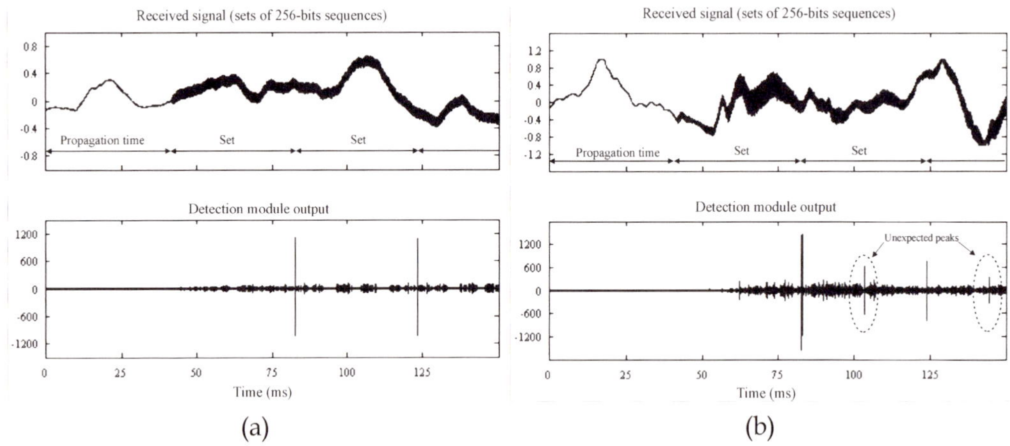

Fig. 13. Detection process of 256-bit sequences (t_e = 40.96 ms) under very weak (a) and very strong (b) turbulence conditions.

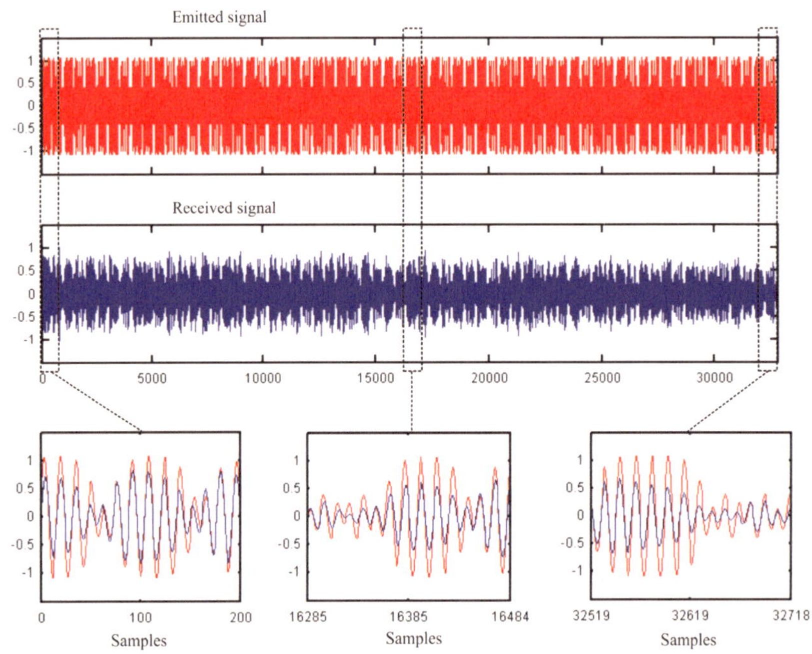

Fig. 14. Comparison between the emitted and the received signals under very weak turbulence conditions (L = 256).

under very strong turbulence, when the coherence time is clearly shorter than the emission time. In this case the average *SMR* raises to 0.45 and, even more remarkable, the standard deviation is nearly three times this value. In this case, represented in Fig. 13b, very large sidepeaks are obtained whose height is occasionally higher than that of the corresponding main peak, given rise to values of *SMR* above 1. The system cannot reliably operate under these circumstances since these sidepeks could be erroneously validated as sets arrivals.

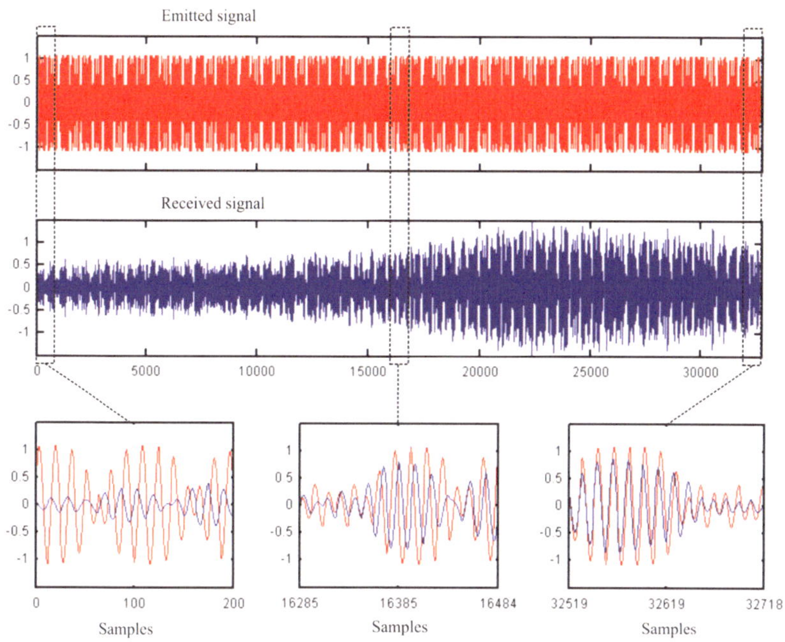

Fig. 15. Comparison between the emitted and the received signals under very strong turbulence conditions ($L = 256$).

TURBULENCE	64 bits		256 bits	
	Mean	Std	Mean	Std
Very weak (cloudy, wind < 2m/s)	0.1809	0.0225	0.1311	0.0268
Weak (sunny, wind < 2m/s)	0.1979	0.0441	0.1611	0.0660
Medium (cloudy, 2<wind < 4 m/s)	0.2032	0.0561	0.1630	0.0484
Strong (sunny, 2<wind < 4 m/s)	0.2396	0.1042	0.1924	0.0863
Very Strong Wind > 4 m/s	0.2622	0.1289	0.4507	1.3012

Table 1. Average Sidelobe-to-Mainlobe Ratio (*SMR*) under different turbulence conditions.

In addition to the appearance of spurious peaks, another phenomenon was observed during the experimentation. Although the arrivals of the sets are still correctly detected even when the coherence time is shorter than the emission time, the peaks associated to these arrivals appear more shifted from their expected positions with increasing turbulence intensity. The quantitative analysis of this phenomenon is summarized in Table 2. Note that when 64-bit sequences are transmitted, the average shift is always below 5 μs (4 samples!), although in all cases the dispersion is significant. The shift is even smaller with 256-bit sequences under similar conditions of turbulence, except again, when the coherence time is shorter than the emission time. In this case, the average shift jumps to nearly 60 μs (3 carrier cycles) and the deviation goes up to 313 μs, showing that the actual values of shift vary wildly between a few μs and several tenths of ms.

TURBULENCE	64 bits		256 bits	
	Mean (μs)	Std (μs)	Mean (μs)	Std (μs)
Very weak (cloudy, wind < 2m/s)	0.1762	0.4356	0.5842	0.6539
Weak (sunny, wind < 2m/s)	0.2513	0.5016	0.4755	0.6377
Medium (cloudy, 2<wind < 4 m/s)	1.2304	4.0418	0.7609	0.7852
Strong (sunny, 2<wind < 4 m/s)	2.1508	5.5043	1.8503	2.2120
Very Strong Wind > 4 m/s	4.8890	9.3993	55.927	313.42

Table 2. Average shift of the autocorrelation peaks.

4. Discussion and future research

The design of high performance outdoor sonars seems to require the use of encoding and signal processing techniques to ensure the reliable operation of the system under changing meteorological conditions. This task introduces new problems mainly due to turbulence phenomenon and its random effect on the amplitude and phase of the emitted signals. We have seen that this effect can be characterized through a coherence time that is a measure of the time during which the features of the received signal remain essentially invariant. Some experimental results have been conducted showing that emissions below this time can be properly detected through matched filtering. Also, the deterioration of the correlated signal with decreasing coherence times has been verified by measuring the increase of its *SMR* and the autocorrelation peak shift.

Although we know that turbulence causes random variations on the amplitude and phase of acoustic signals, little is known about the statistical properties of these variations. An accurate model for this phenomenon would allow a precise prediction of the effects that a turbulent atmosphere has on the performance of advanced sonar systems where encoded signals are transmitted. It would also allow to clearly define the limits of operation for these systems. Moreover, this model could be used to determine the type of encoding and modulation schemes that would be more appropriate to operate under adverse conditions. Future research in this field should be focused on obtaining this model, as well as experimentally determining the accuracy of Eq. (31) that gives the coherence time dependence on frequency, turbulence intensity, normal component of wind and travelled distance.

5. References

Álvarez, F. J., Ureña, J., Mazo, M., Hernández, A., García, J. J. & Jiménez, J. A. (2004). Efficient generator and pulse compressor for complementary sets of four sequences. *IEE Electronics Letters,* Vol. 40, No. 11 (May 2004), pp. 703-704.

Álvarez, F. J., Ureña, J., Mazo, M., Hernández, A., García, J. J. & De Marziani, C. (2006). High reliability outdoor sonar prototype. *IEEE Transactions on Ultrasonics, Ferroelectrics and Frequency Control*, Vol. 53, No. 10 (October 2006), pp. 1862-1871.

Álvarez, F. J., Ureña, J., Hernández, A., Mazo, M., García, J. J. & Jiménez, A. (2008). Influence of atmospheric refraction on the performance of an outdoor pulse compression system. *Applied Acoustics*, Vol. 69, No. 11 (November 2008), pp. 994-1002.

Brown E. H. & Clifford, S. F. (1976). On the attenuation of sound by turbulence. *Journal of the Acoustical Society of America*, Vol. 60, No. 4 (October 1976), pp. 788–794.

Clifford, S. F & Brown, E. H. (1970). Propagation of sound in a turbulent atmosphere. *Journal of the Acoustical Society of America*, Vol. 48, No. 5 (November 1970), pp.1123–1127.

Cole, J. E & Dobbins, R. A. (1970). Propagation of sound through atmospheric fog. *Journal of the Atmospheric Science*, Vol. 27, No. 3 (May 1970), pp. 426–434.

Golder, D. (1972). Relations among stability parameters in the surface layer. *Boundary-Layer Meteorology*, Vol. 3, No. 1 (September 1972), pp. 47–58.

Guo L., Zhang Q. & Han S. (2002). Agricultural Machinery Safety Alert System Using Ultrasonic Sensors. *Journal of Agricultural Safety and Health*, Vol. 8, No. 4, (November 2002), pp. 385–396.

ISO/TC 43 Technical Comittee, Acoustics, Sub-Comittee SC1, Noise (1993). Attenuation of sound during propagation outdoors. Part 1: Calculation of the absorption of sound by the atmosphere, Tech. Rep. ISO 9613-1:1993 (E), *International Organization for standarization* (1993).

Jiménez A. R. & Seco, F. (2005). Precise localisation of archaeological findings with a new ultrasonic 3D positioning sensor. *Sensors and Actuators A, Physical.* Vol. 123-124 (September 2005), pp. 224–233.

Kolmogorov, A. N. (1941). The local structure of turbulence in incompressible viscous fluid for very large Reynolds numbers. *Proceedings of the USSR Academy of Sciences*, Vol. 30, pp. 299-303.

Langer, D. & Thorpe, C. (1992). Sonar based outdoor vehicle navigation and collision avoidance. *Proceedings of IEEE/RSJ International Conference on Intelligent Robots and Systems*, Raleigh, NC, (July 1992), pp. 1445–1450.

Maeyama S., Ohya A. & Yuta S. (1994). Positioning by tree detection sensor and dead reckoning for outdoor navigation of a mobile robot. *Proceedings of MFI'94, IEEE*, Las Vegas, NV, (October 1994), pp. 653–660.

Monin A. S. & Obukhov, A. M. (1954). Basic laws of turbulent mixing in the ground layer of the atmosphere. *Transactions Geophys. Inst. Akad., Nauk USSR*, Vol. 24, No. 151 (1954), pp. (163–187).

Ostashev, V. E. (1997). Acoustics in Moving Inhomogeneous Media. E & FN SPON (An Imprint of Thompson Professional), London.

Panofsky, H. A. & Dutton, J. A. (1984). Atmospheric Turbulence. Models and Methods for Engineering Applications. John Wiley & Sons, Inc, New York, 1984

Polaroid Corporation (1999). 600 Series. Instrument Grade Electrostatic Transducers. Technical Specification, 1999.

Ratner D. & McKerrow P. (2003). Navigating an outdoor robot along continuous landmarks with ultrasonic sensing. *Robotics and Autonomous Systems*, Vol. 45, No. 2 (November 2003), pp. 73–82.

Shamanaeva, L. G. (1988). Acoustic sounding of rain intensity. *Journal of the Acoustical Society of America*, Vol. 84, No. 2 (August 1988), pp.713–718.

Tanzawa, T., Kiyohiro N., Kotani S. & Mori K. (1995) The Ultrasonic Range Finder for Outdoor Mobile Robots. *Intelligent Robots and Systems 95. 'Human Robot Interaction and Cooperative Robots', Proceedings*. 1995. pp. 368-373. Pittsburgh, PA, USA.

Tatarskii, V. I. (1971). The Effects of the Turbulent Atmosphere on Wave Propagation. *Israel Program for Scientific Translations, Ltd*, Jerusalem, 1971.

Yura, H. T. (1971). Atmospheric turbulence induced laser beam spread. *Applied Optics*, Vol. 10, No. 12 (December 1971), pp. 2771–2773.

10

Mobile Robot Localization using Particle Filters and Sonar Sensors

Antoni Burguera, Yolanda González and Gabriel Oliver
Universitat de les Illes Balears
Spain

1. Introduction

1.1 The localization problem

Nowadays, nearly all mobile robotic tasks require some knowledge of the robot location in the environment. For example, those tasks involving the robot to reach a specific target require knowledge about the current robot pose in order to plan a path to the goal. Also, exploration tasks require some estimate of the robot pose in order to decide whether a specific region has been already visited by the robot or not. The problem of computing the robot pose is known as the *mobile robot localization* problem.

The mobile robot localization problem appears in many flavours. In some cases, only a *qualitative* pose estimate is needed. For example, for high level spatial reasoning, the robot may only need to know if a certain area, such as a room, has been previously visited or not. This kind of localization is commonly named *weak localization*. In some other cases, *quantitative* pose estimates with respect to a fixed reference frame are required. For example, to build metric maps, such as occupancy grids, the robot needs accurate numerical estimates of its pose in the space. This approach to localization is usually referred to as *strong localization*.

Both weak and strong localization problems can be defined in a global or in a local context, constituting the so called *global localization* and *local localization* problems respectively. The former refers to the obtention of the robot pose without an *a priori* estimate of its location. It is called global localization by analogy with global function minimization, whereby an optimum must be found without a reliable initial guess. On the contrary, local localization, sometimes named *pose maintenance*, refers to a continuous refinement of the robot pose, starting with an initial guess.

This chapter focuses on the strong localization problem in the local context. From now on, in the context of this document, the terms localization and mobile robot localization will refer to the strong localization problem in the local context. A common approach to confront this localization problem is the use of *exteroceptive* sensors, such as range finders or cameras, measuring the external environment. Exteroceptive sensor data is correlated at subsequent robot poses to compute displacement estimates, usually based on initial guesses provided by *proprioceptive* sensors, such as odometers or inertial units. As a consequence of this, the quality of the pose estimates is strongly related to the quality of the measurements provided by the exteroceptive sensors.

1.2 The sonar sensors

Many recent localization strategies rely on accurate sensors, such as laser range scanners (Hähnel et al. 2003; Biber & Straβer 2003; Montesano et al. 2005; Minguez et al. 2006). Today, off the shelf laser sensors provide thousands of readings per second with a sub degree angular resolution. Other sensors, such as standard Polaroid ultrasonic range finders, are only able to provide tenths of readings per second, with angular resolutions one or two orders of magnitude worse than laser. Moreover, effects such as multiple reflections or cross-talking are very frequent in sonar sensing, producing large amounts of readings not corresponding to real objects in the environment. Figure 1 compares the sets of readings gathered simultaneously by a laser scanner and an ultrasonic range finder along the same trajectory. The laser set contains 150 times more readings than the sonar set. This provides a clear idea of the sparse sets of readings provided by ultrasonic range finders when compared to laser scanners. Also, it can be observed how the ultrasonic set contains large amounts of wrong readings, due to the previously mentioned effects.

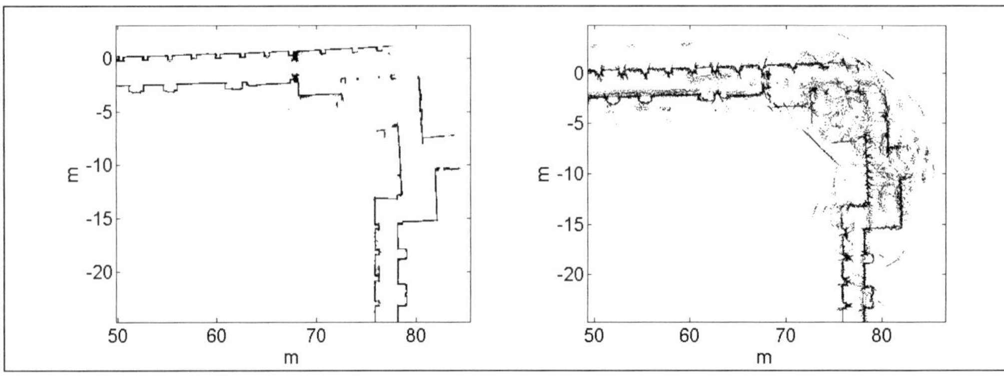

Fig. 1. Example of readings gathered by a laser scanner (left) and a ring of 16 ultrasonic range finders (right).

Nevertheless, ultrasonic range finders have interesting properties that make them appealing in the mobile robotics community (Lee 1996). On the one hand, their size, power consumption and price are better than those of laser scanners. Consequently, they are well suited for low cost and domestic robots, such as automatic vacuum cleaners. On the other hand, their basic behaviour is shared with underwater sonar sensors, which are vastly used in underwater and marine robotics. Thus, typical underwater sonar, although being far more complex than standard Polaroid ultrasonic range finders, can take profit of those localization techniques accounting for sonar limitations. In the context of this work, the terms *sonar*, *ultrasonic range finder* and *ultrasonic sensor* will be used interchangeably, and will refer to standard time-of-flight Polaroid ultrasonic range finders.

The validity of ultrasonic range finders to perform localization has been demonstrated by different researchers. For instance, Tardós et al. (Tardós et al. 2002) use a perceptual grouping technique to identify and localize environmental features, such as lines and corners. These features are correlated using robust data association to perform SLAM with sonar sensors. Also, Groβmann et al. (Groβmann & Poli 2001) confront the sonar localization problem by means of the Hough transform and probability grids to detect walls and corners. Burguera et al. (Burguera et al. 2008a) adopt a different approach, named spIC (*sonar probabilistic Iterative Correspondence*), not requiring environmental features to be detected. They have shown that scan matching localization, even in its most basic expression (Burguera et al. 2005), can be applied to sonar sensors if their limitations and uncertainties are appropriately taken into account. Also, a recent laser scan matching technique, the NDT (*Normal Distributions Transform*) (Biber & Straβer 2003) has been adapted to work with sonar sensors (Burguera et al. 2008b).

1.3 Probabilistic methods

It is broadly accepted that probabilistic methods are the most promising ones to deal with sensor and pose uncertainties in real-time (Thrun et al. 2005). A key concept in probabilistic robotics is that of *belief*. The beliefs, which reflect the robot's internal knowledge about its state, are represented by probability distributions. In the localization context, the beliefs reflect the robot's internal knowledge about its pose in the space. The most general approach to compute beliefs is the *Bayes filter*, which determines the belief distribution from control data and measurements. However, the general Bayes filter algorithm is not a tractable implementation for continuous state spaces, such as the one of robot poses.

Gaussian filters are one of the earlier tractable implementations of Bayes filters. One particular Gaussian filter, the *Kalman filter* (Kalman 1960), has been vastly used to perform mobile robot localization and SLAM. A Kalman filter represents the belief by means of a normal distribution. Mainly due to the normal distribution assumption, Kalman filters fail to represent ambiguities and to recover from localization (Neira & Tardós 2001; Castellanos et al. 2004). These problems are especially relevant when dealing with ultrasonic range finders. On the one hand, the low sonar angular resolution may lead to ambiguous robot pose estimates. On the other hand, wrong readings due to multiple reflections and cross-talking combined with the low measurement rate of ultrasonic range finders may lead the filter to unrecoverable localization failures.

An alternative tractable implementation of the Bayes filter is the *Particle filter* (Metropolis and Ulam 1949, Doucet et al. 2001). In particle filters, the belief distribution is represented by a set of samples, called *particles*, randomly drawn from the belief itself. The particle filter is in charge of recursively updating the particle set. Dellaert et al. (Dellaert et al. 1999) and Fox et al. (Fox et al. 1999) introduced particle filters in the localization context, defining the so-called MCL (*Monte Carlo Localization*). Since then, particle filters have been successfully applied to SLAM (Montemerlo et al. 2002; Hähnel et al. 2003), multi-robot localization (Fox et al. 2000) and localization given an *a priori* map both using laser (Yaqub & Katupitiya 2007) and sonar sensors (Thrun et al. 2001), among many other applications.

Particle filters are nonparametric implementations of Bayes filters. They are said to have two important advantages. First, they can approximate a wide range of probability distributions, even multimodal. When compared to Kalman filters, which can only deal with normal distributions, this feature constitutes an important benefit. These filters are much better suited than Kalman filters to represent ambiguities and to cope with localization failures. The second advantage of particle filters is that, even its most straightforward implementation exhibits very good results when applied to localization. Thus, a particle filter constitutes an excellent tool to perform localization using ultrasonic range finders as exteroceptive sensors.

A key point in a particle filter is the so called *measurement model*. Broadly speaking, the measurement model is in charge of determining how likely the current sensor readings can be explained by each particle. This is usually accomplished by means of an *a priori* map. The current sensor readings are matched against the map and the degree of matching defines the measurement model.

Silver et al. (Silver et al. 2004) proposed a method not requiring any *a priori* map and dealing with underwater sonar sensors. Their proposal was to store a small local history of range readings for each particle, so that the current readings could be matched against each of the local histories. In order to compute the degree of matching, they borrowed a concept from the scan matching community. They compute the degree of matching as a function of the ICP (*Iterative Closest Point*) scan matching error (Lu & Milios 1997). Although this approach was only tested in simulation, it does not require any *a priori* map and exhibits very good results.

1.4 Chapter overview

In this document we describe the use of the ICP scan matching as a measurement model in a particle filter to perform mobile robot localization using standard, terrestrial, Polaroid ultrasonic range finders. The particles are augmented with local environment information. This local information is recursively updated at each time step, allowing the localization process to be performed without any *a priori* map. Also, the aim of this local information is to deal with the sparseness of the sets of sonar readings.

In order to validate and measure the quality of this approach, sonar and laser data has been simultaneously gathered in different environments. Using the laser readings, a ground truth has been constructed. Then, the sonar-based particle localization is evaluated by comparing its results to the ground truth. The presented evaluation method takes into account the whole robot trajectory, instead of only its end points. The experiments evaluate different algorithm's parameters. The experimental results show how the proposed approach to sonar-based localization is able to provide robust and accurate robot pose estimates.

This chapter is structured as follows. The general particle filter operation, as well as the specific details of the sonar-based particle localization, is provided in Section 2. The notation used along the paper is also presented in this section. Section 3 focuses on the measurement model and the introduction of local environment information on the particles. A quantitative evaluation method is presented in Section 4. Also in this section, the experimental results evaluating the algorithm are shown and discussed. Section 5 concludes the chapter, and some proposals of future work are given in Section 6.

2. The particle filter

2.1 Overview and notation

The key idea in mobile robot localization is to estimate the robot pose from sensor data. Thus, in the localization context, the robot pose is the state to be estimated. However, the robot pose is usually not directly observable by sensors. In consequence, the robot pose has to be inferred from sensor data. In order to accomplish this state estimation process, at least two models are necessary. On the one hand, a model describing the evolution of the robot poses with time. On the other hand, a model that relates the sensor measurements to the robot poses. The former is commonly named the *motion model*, although it is also referred to as the *system model* or the *plant model*. The latter is the so called *measurement model*.

As stated previously, Bayes filters constitute a general, widely used, approach to state estimation. They are recursive estimators consisting on two main steps: the *control update* and the *measurement update*. The control update predicts the robot pose one time step forward, by means of the motion model. The measurement update uses the latest sensor readings and the measurement model to modify the prediction. This update is based on the Bayes theorem, and that is why these recursive estimators are named Bayes filters.

The general Bayes filter algorithm involves integration over the whole state space. In localization, the state space is the continuous space of robot poses. In consequence, the general Bayes filter algorithm is not computationally tractable in most of the localization problems. Nevertheless, there exist several tractable implementations of the Bayes filter. Among them the particle filter is capturing the attention of the localization community. The key idea is to represent the belief by a set of weighted random samples called particles. Thanks to this, the filter is able to deal with arbitrary probability distributions, not necessarily unimodal. Thus, they are particularly well suited to perform sonar-based localization, and they will focus the attention of this chapter. Additionally, the presented approach does not require an *a priori* map of the environment. The particular implementation of the particle filter to perform such task is described in detail in Section 2.2. Now, some notation is provided.

Let z_t and u_t denote the sonar measurements and the control vector, respectively, at time step *t*. In the context of this chapter, u_t corresponds to odometry data. Let the set of particles at time step *t* be defined as follows:

$$X_t := \{(x_t^{[m]}, w_t^{[m]}, s_t^{[m]}), 1 \leq m \leq M\} \tag{1}$$

where *M* is the number of particles. Each $x_t^{[m]}$ is a concrete instantiation of the robot pose $[x, y, \theta]^T$ at time *t*. Each $w_t^{[m]}$ is the particle *importance factor*, also referred to as *weight*, so that $w_t^{[m]} \propto p(z_t|x_t^{[m]})$. A key issue in the presented approach is $s_t^{[m]}$, a short history of the most recent *k* sets of sonar readings. This history constitutes the particle *local map*, and it is intended to cope with the low amount of readings provided by ultrasonic range finders. Also, these local maps are recursively updated during the filter operation, letting the localization process work without the use of *a priori* maps. The use and on-line building of $s_t^{[m]}$ is one of the novelties of the presented approach, and will be described in detail in Section 3. In the context of this chapter, the terms *local map*, *history* and *readings history* will be used interchangeably and will refer to $s_t^{[m]}$.

Let $\bar{x}_t^{[m]}$ denote the relative robot motion from time step *t-1* to time step *t* according to the particle *m*. Finally, let the operators $\ominus$ and $\oplus$ denote the inversion and the compounding transformations, similarly to those defined by Smith et al. (Smith et al. 1990). These operators will be now described, together with two additional compounding operators for transforming the references of a point (Tardós et al. 2002) and of a set of points.

Only to perform such description, the following notation will be used. Let $x_B^A = [x_1, y_1, \theta_1]^T$denote the location of a coordinate frame *B* relative to a coordinate frame *A*. Let $x_C^B = [x_2, y_2, \theta_2]^T$be defined similarly. Finally, let $x_p^B = [x_3, y_3]^T$ denote the location of the point *p* relative to the coordinate frame *B*. The compounding $x_C^A = x_B^A \oplus x_C^B$ denotes the location of the coordinate frame C relative to *A*, and is computed as follows:

$$x_C^A = x_B^A \oplus x_C^B = \begin{bmatrix} x_1 + x_2 \cos\theta_1 - y_2 \sin\theta_1 \\ y_1 + x_2 \sin\theta_1 + y_2 \cos\theta_1 \\ \theta_1 + \theta_2 \end{bmatrix} \tag{2}$$

The inversion $x_A^B = \ominus x_B^A$ denotes the location of coordinate frame A relative to B as follows:

$$x_A^B = \ominus x_B^A = \begin{bmatrix} -x_1 \cos\theta_1 - y_1 \sin\theta_1 \\ x_1 \sin\theta_1 - y_1 \cos\theta_1 \\ -\theta_1 \end{bmatrix} \tag{3}$$

The compounding $x_p^A = x_B^A \oplus x_p^B$ denotes the location of the point p relative to the coordinate frame A as follows:

$$x_p^A = x_B^A \oplus x_p^B = \begin{bmatrix} x_1 + x_3 \cos\theta_1 - y_3 \sin\theta_1 \\ y_1 + x_3 \sin\theta_1 + y_2 \cos\theta_1 \end{bmatrix} \tag{4}$$

Finally, if the right-hand operand of the compounding transformation is a set of points, the transformation is applied individually to each point and, thus, the compounding returns the resulting set of points. Figure 2 summarizes the notation used along this chapter.

2.2 Sonar-based particle localization

A particle filter builds the particle set X_t recursively from the particle set X_{t-1} one time step earlier. Thus, it is necessary to start the recursion by defining the initial particle set X_0. If an *a priori* map is available, this initialization is accomplished by uniformly distributing $x_0^{[1:M]}$ over the free space in the map. However, the presented approach uses local maps to avoid the need for previous information. In consequence, the particle set initialization has to be in charge of building $s_0^{[1:M]}$.

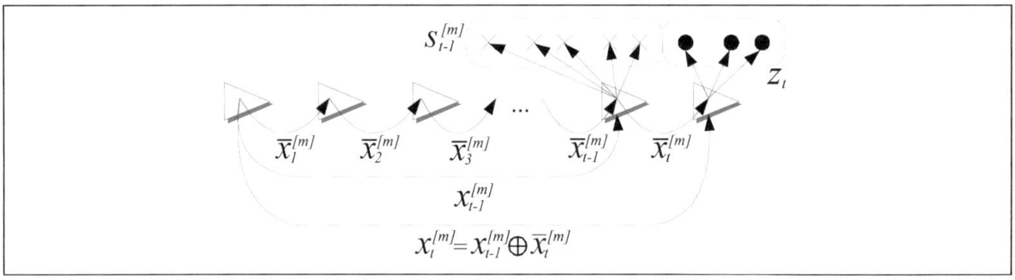

Fig. 2. Notation used for sonar-based particle filtering

To initialize the particle set, the robot has to move during k time steps computing its pose using odometry. Then, the robot pose $x_0^{[1:M]}$ for all particles is set to the odometric pose estimate after the mentioned k time steps. During this initialization, k sets of sonar readings are gathered. Their $[x, y]^T$ coordinates are represented with respect to a coordinate frame located in $x_0^{[1:M]}$ using the odometry estimates. The initial local map $s_0^{[1:M]}$ for all particles is then set to the mentioned k sets of sonar readings.

Although this initial dependence on odometry may seem problematic, it is not if the value of k is appropriately chosen. Different values for this parameter will be tested and experimentally evaluated along this chapter. It will be shown that good values for this parameter are around k=*100*. So, let us assume for the moment this value. Let also assume a mobile robotic platform providing odometric and sonar readings at steps of 100ms. This time step is quite common. In this case, the robot has to rely solely on odometry during the first 10s of operation. In many robot applications, the odometric error accumulated during 10s is negligible if compared to the whole mission execution.

After the initialization process, the sonar-based particle filter localization algorithm is executed at each time step. Figure 3-a shows the algorithm, where two loops involving the whole particle set can be observed. The first loop, from line 1 to line 3, predicts the robot pose by means of the motion model. Thus, it constitutes the control update. The second loop, from line 4 to line 8, updates the particle set according to the sensor readings and constitutes the measurement update. It is important to remark that, although the measurement model is executed in line 3, the weights it computes are not used until line 5. Thus, the measurement update is performed from line 4 to 8 although the measurement model appears in line 3.

Line 2 is in charge of sampling the motion model. As stated previously, in order to perform a state estimation process, a motion model is necessary. In general terms, the motion model describes the evolution of the robot pose with time. Thus, in general, the motion model provides an estimate of where the robot is at time *t*, given its previous pose at time *t-1* and the current control vector u_t. However, our proposal for the motion model is not to compute the absolute robot pose at time *t*, but relative motions from one time step to the next. In consequence, line 2 generates hypothetical robot motions $\bar{x}_t^{[m]}$ from time step *t-1* to time step *t* using a stochastic motion model that does not depend on $x_{t-1}^{[m]}$. This step involves sampling from the distribution $p(\bar{x}_t|u_t)$, where $\bar{x}_t$ represents a robot motion from time step *t-1* to time step *t*. This distribution depends on the specific robot configuration, and it is out of the scope of this chapter to discuss it. Relevant information on this subject can be found on (Thrun et al. 2005).

Algorithm: Sonar-Based Particle Localization

Input:
M: Number of particles
X_{t-1}: Particle set $\{(x_{t-1}^{[1:M]}, w_{t-1}^{[1:M]}, s_{t-1}^{[1:M]})\}$
u_t: Odometry
z_t: Sonar readings

Output:
X_t: Particle set $\{(x_t^{[1:M]}, w_t^{[1:M]}, s_t^{[1:M]})\}$

begin
1 **for** $m \leftarrow 1$ **to** M **do**
2 $\quad\bar{x}_t^{[m]} \leftarrow$ sample motion model(u_t)
3 $\quad w_t^{[m]} \leftarrow$ measurement model$(z_t, \bar{x}_t^{[m]}, s_{t-1}^{[m]})$
4 **for** $m \leftarrow 1$ **to** M **do**
5 $\quad$draw i with probability $\propto w_t^{[i]}$
6 $\quad x_t^{[i]} \leftarrow x_{t-1}^{[i]} \oplus \bar{x}_t^{[i]}$
7 $\quad s_t^{[i]} \leftarrow$ update history$(s_{t-1}^{[i]}, z_t, \bar{x}_t^{[i]})$
8 $\quad X_t \leftarrow X_t \cup \{(x_t^{[i]}, w_t^{[i]}, s_t^{[i]})\}$

end

(a)

Algorithm: Low-Variance Sampling

Input:
M: Number of particles
X_t: Particle set $\{(x_t^{[1:M]}, w_t^{[1:M]}, s_t^{[1:M]})\}$

Output:
X_t': Resampled particle set

begin
1 $c \leftarrow \emptyset$
2 $X_t' \leftarrow \emptyset$
3 $c(1) - w_t^{[1]}$
4 **for** $m \leftarrow 2$ **to** M **do**
5 $\quad c(m) \leftarrow c(m-1) + w_t^{[m]}$
6 $u \leftarrow rand(0; M^{-1})$
7 $k \leftarrow 1$
8 **for** $m \leftarrow 1$ **to** M **do**
9 $\quad$**while** $u > c(k)$ **do**
10 $\quad\quad k \leftarrow k + 1$
11 $\quad X_t' = X_t' \cup \{(x_t^{[k]}, w_t^{[k]}, s_t^{[k]})\}$
12 $\quad u \leftarrow u + M^{-1}$

end

(b)

Fig. 3. Sonar-Based Particle Localization algorithm (a) and Low Variance Sampling algorithm (b).

The second model necessary to perform the state estimation is the measurement model, which relates the sonar readings to the robot pose. Line 3 uses the measurement model to incorporate the current readings z_t into the particle set by computing the importance factor $w_t^{[m]}$. Those particles with the relative motions that better explain the current readings will have better weights. This is accomplished by correlating z_t and $s_{t-1}^{[m]}$ by means of an ICP-like approach and will be described in Section 3.

Line 5 executes the so called *resampling*, also referred to as *importance sampling*. The importance sampling is the core of the measurement update step. At this point, the algorithm draws with replacement M particles. The probability of drawing each particle is proportional to its importance factor. Differently speaking, during the importance sampling, those particles with better weights have higher probability to remain in the particle set.

There is a problem in particle filters directly related to the importance sampling. The statistics extracted from the particles may differ from the statistic of the original density, because the particle set only holds a finite number of random samples. This problem may lead to a *degeneracy phenomenon* through repetitive resampling (Sanjeev Arulampalam et al. 2002). Degeneracy appears when, after a number of resampling steps, all but one particle have negligible weights. Among the existing resampling strategies, the *low variance sampling* has proved to be very efficient, in computational terms, while reducing the degeneracy phenomenon. The underlying idea is to select the samples in a sequential stochastic process instead of independently. A comprehensible description of the algorithm is available in (Thrun et al. 2005). Because of the mentioned advantages, the low variance sampling has been adopted in the present work. The algorithm is presented in Figure 3-b.

Going back to Figure 3-a, the line 6 is in charge of updating the global robot pose for each particle selected during the resampling step. This is accomplished by compounding the global robot pose at time *t-1* with the relative motion from time step *t-1* to time step *t*. This idea is illustrated in Figure 2.

Line 7 is in charge of building the new local map of each particle, by adding the current set of sonar readings and discarding to oldest readings so that the map size remains constant along the whole mission execution. This process will be described in Section 3.

Finally, line 8 constructs the new particle set X_t. After this step, depending on the specific robot application, the particle set may be treated in different ways. For instance, some applications need a single vector $[x, y, \theta]^T$ informing the most likely robot pose. In that cases, the mean of $x_t^{[1:M]}$ may be used. Some other applications require a continuous probability density function to be extracted from the samples. In those cases, techniques such as *Gaussian approximation*, *K-Means* or *Kernel Density Estimation* can be used. The reader is directed to (Thrun et al. 2005) to learn more about these density extraction techniques.

3. Matching sets of readings

3.1 Overview and notation

In particle filters, the measurement model is in charge of computing the weights of the particles. In particle filter localization, the weights represent the likelihood of having the current set of readings z_t at the robot pose $x_t^{[m]}$. Thus, $w_t^{[m]} \propto p(z_t|x_t^{[m]})$. This dependence on the absolute robot pose is useful if an *a priori* map is available, because the range readings can be matched against the global map using the absolute robot pose.

However, one advantage of the presented approach is that it does not require a previously constructed map. Instead, local maps are recursively built during the mission execution. For a given particle, the local map $s_{t-1}^{[m]}$ is represented with respect to the coordinate frame in $x_{t-1}^{[m]}$ (see Figure 2). Also, the presented motion model generates $\bar{x}_t^{[m]}$, the relative motions from time step *t-1* to time step *t*. Taking into account that the current set of readings z_t has been gathered at time *t*, the particle weight can be computed by evaluating the degree of matching between $\bar{x}_t^{[m]} \oplus z_t$ and $x_{t-1}^{[m]}$. Figure 2 clarifies this point. Thus, in our approach, $w_t^{[m]} \propto p(z_t|\bar{x}_t^{[m]}, s_{t-1}^{[m]})$. Broadly speaking, the idea is to weight the particles according to the existing overlap between the current set of readings and the stored maps. Computing the overlap between two sets of range readings is a common practice in the scan matching community. Thus, some scan matching concepts will be used in this section. Next, some notation is introduced.

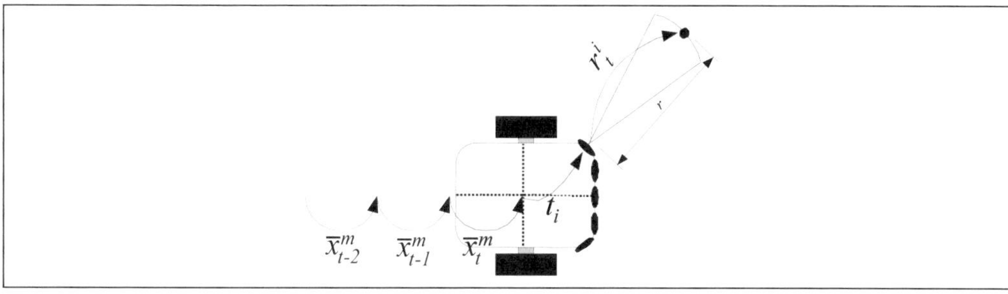

Fig. 4. Relations between the coordinate frames used by the measurement model. The circular sector represents the sonar beam. The dashed cross is the robot coordinate frame.

Let r_t^i represent the range reading provided by the *i-th* sonar sensor at time step *t*. Let this reading be represented with respect to a coordinate frame located on the sonar sensor and aligned with the ultrasonic acoustic axis. Thus, r_t^i has the form $[r, 0]^T$, where *r* is the raw range provided by the sensor.

Let t_i denote the relative position of the sonar sensor *i* with respect to the robot reference frame. Ultrasonic range finders are assumed to be at fixed positions on the robot body. Consequently, t_i does not change over time. That is why the subindex *t* has been dropped. Figure 4 illustrates the notation.

3.2 Building the local maps

At time *t*, the array of ultrasonic range sensors provides a set of raw range readings. The set z_t is built from the raw range readings as follows:

$$z_t = \{t_i \oplus r_t^i, \forall i \in VS_t\} \tag{5}$$

where VS_t is the set of sonar sensors that have generated a reading during the time step *t*. Each item in z_t will be denoted z_t^i, meaning that it was gathered at time *t* and produced by the *i-th* sonar sensor.

Let S_{new} be defined as the set of readings in z_t represented with respect to the coordinate frame of $x_{t-1}^{[m]}$ using the relative motion $\bar{x}_t^{[m]}$ proposed by the particle:

$$S_{new} = \bar{x}_t^{[m]} \oplus z_t = \{\bar{x}_t^{[m]} \oplus z_t^i, \forall i \in VS_t\} \tag{6}$$

Each item in S_{new} will be denoted by p_t^i, meaning that it has been generated from z_t^i.
To ease notation, let S_{old} denote the local map $s_{t-1}^{[m]}$. It was stated previously that all the readings in S_{old} are represented with respect to the coordinate frame of $x_{t-1}^{[m]}$. This is accomplished by building S_{old} as follows:

$$\begin{aligned} S_{old} = \ & \{z_{t-1}, \\ & \ominus\bar{x}_{t-1}^{[m]} \oplus z_{t-2}, \\ & \ominus\bar{x}_{t-1}^{[m]} \oplus (\ominus\bar{x}_{t-2}^{[m]}) \oplus z_{t-3}, \ldots, \\ & \ominus\bar{x}_{t-1}^{[m]} \oplus \ldots \oplus (\ominus\bar{x}_{t-k+1}^{[m]}) \oplus z_{t-k}\} \end{aligned} \quad (7)$$

where *k* is the local map size. By observing the previous equation it is easy to see that $s_t^{[m]}$ can be obtained recursively from $s_{t-1}^{[m]}$, z_t and $\bar{x}_t^{[m]}$. This recursive update, which is performed by the update history function in line 7 of Figure 3-a, can be expressed as follows:

$$s_t^{[m]} = ((\ominus\bar{x}_t^{[m]}) \oplus s_{t-1}^{[m]}) \cup z_t \quad (8)$$

First, the readings in $s_{t-1}^{[m]}$ are represented with respect to the coordinate frame of $x_t^{[m]}$ by compounding them with $\ominus\bar{x}_t^{[m]}$. Then, the new set of readings z_t is added. Finally, although not been represented in Equation (8), the oldest readings in the resulting set have to be deleted so that the size of the local maps remains constant along the whole mission execution.

3.3 The measurement model

There exist many algorithms to match sets of range readings in the scan matching literature] (Lu & Milios 1997; Rusinkiewicz & Levoy 2001; Pfister et al. 2004; Burguera et al. 2008a). Most of them follow the structure proposed by the ICP algorithm. The key step in the ICP algorithm is the establishment of point to point correspondences between readings in two consecutive range scans. These correspondences are established by means of the Euclidian distance, and they give information about the degree of matching between two sets of readings. Our proposal is to measure the degree of matching between S_{new} and S_{old} in that way. This will constitute our measurement model.
Let p_i and q_j be points in S_{new} and S_{old} respectively. To decide whether a correspondence between p_i and q_j can be established or not, the Euclidian distance is used:

$$d(p_i, q_j) = \sqrt{(p_i - q_j)^T (p_i - q_j)} \quad (9)$$

For each $p_i \in S_{new}$, the closest point $q_j \in S_{old}$ according to the distance in Equation (9) is selected to be the corresponding point. Thus, the set *C* of correspondences is defined as follows:

$$C = \{(p_i, q_j), \forall p_i \in S_{new} | q_j \in S_{old}, q_j = \arg\min(d(p_i, q_j))\} \quad (10)$$

Broadly speaking, the idea is to establish correspondences between the points in S_{new} and S_{old} that are closer in the Euclidian sense. This is commonly referred to as the *closest point rule*.

The sum of Euclidian distances between pairs of corresponding points is a good indicator of the degree of matching between S_{new} and S_{old}: the worse the matching, the bigger the sum of distances. However, the importance factor represents the opposite idea: particles that produce better matching should have higher weights. In consequence, the importance factor for a particle m is computed as follows:

$$w_t^{[m]} = (\sum_{(p_i,q_j)\in C} d(p_i, q_j))^{-1} \tag{11}$$

In order to avoid numerical problems, those situations where the sum of distances is close to zero should be especially taken into account. However, experimental results suggest that, due to the noisy nature of sonar sensors, these situations are extremely unusual.

4. Experimental results

4.1 Experimental setup

In order to evaluate the presented approach, a Pioneer 3-DX robot, endowed with 16 Polaroid ultrasonic range finders and a Hokuyo URG-04LX laser scanner, has been used. The robot has moved in four different environments in our university, gathering various data sets. Each data set contains the odometry information, the sonar range readings and the laser range readings. The laser readings have only been used to obtain ground truth pose estimates. In order to obtain such ground truth, the ICP scan matching algorithm has been applied to the laser readings. Then, the wheel encoder readings have been corrupted with Gaussian noise ($\mu = 0m/s$ and $\sigma^2 = 0.0025$) to simulate worse floor conditions. Thus, the quality of our algorithm operating with noisy and sparse sets of sonar readings in bad floor conditions is compared to a well known localization algorithm operating with dense and high quality laser readings and good floor conditions.

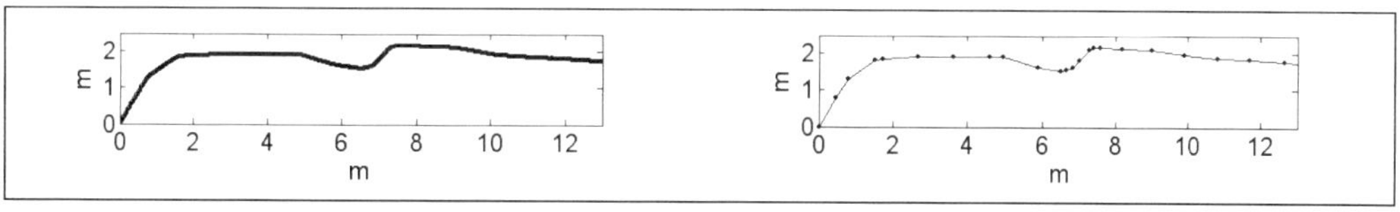

Fig. 5. Fragment of a real trajectory (left) and the polyline that approximates it (right). The dots represent the vertexes.

In order to quantitatively compare odometry and the different particle filter configurations, the following procedure has been used. First, the trajectories obtained by odometry, particle filter and ground truth are approximated by polylines. The vertex density of each polyline increases in those regions with significant amount of robot rotation. Also, the maximum robot motion between two vertexes has been set to 1m. This kind of approximation is useful to overcome the local perturbations in the individual motion estimates, both for odometry, particle filter and ground truth. Figure 5 exemplifies the polyline approximation. Then, the individual edges of the trajectory being evaluated are locally compared to those of the ground truth. The Euclidian distance between their end points is used as a measure of the edge error. Finally, the edge errors for the trajectory being evaluated are summed. This sum is normalized, using the path lengths between vertexes and the number of edges, and constitutes the *trajectory error*. Due to the mentioned normalization, the errors of different

trajectories can be compared. It is important to remark that, as a result of the mentioned procedure, the evaluation takes into account the whole trajectory, not only its end points.
Two different experiments have been performed. The first experiment evaluates our approach with respect to the number of particles, *M*. The second experiment evaluates our approach with respect to the local map size, *k*.

4.2 Evaluating the influence of the number of particles

The first experiment evaluates the quality and the execution time of our approach with respect to the number of particles. The values of *M* that have been tested are 10 and 50, to observe how the algorithm behaves with a low number of particles, and then 100, 200 and 400 particles. The local map sizes has been set to *k=100.* The trajectory error has been computed for odometry and particle filter using the mentioned number of particles.

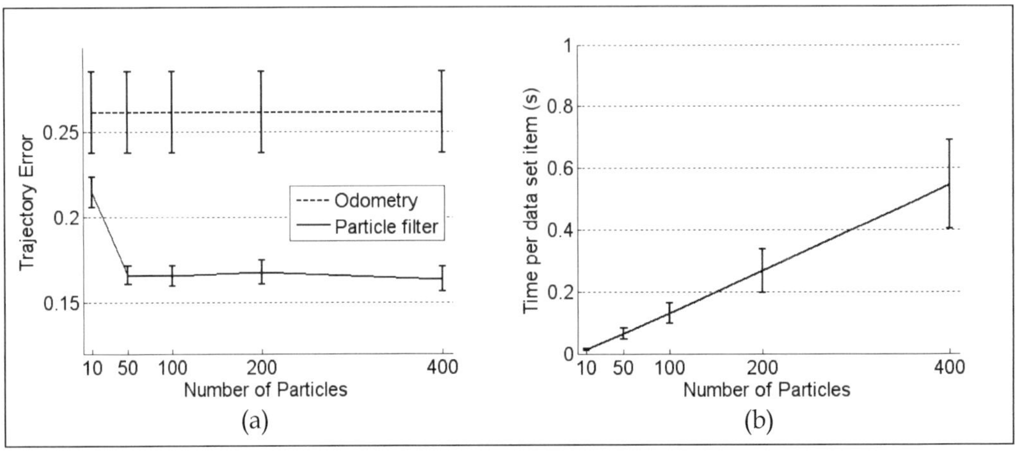

Fig. 6. Experimental results obtained using different numbers of particles and setting the history size to *k=100.* (a) Means and standard deviations of the trajectory errors. (b) Means and standard deviations of the execution time per data set item on a Matlab implementation.

Figure 6-a depicts the mean and the standard deviation of the obtained trajectory errors for all data sets. The graphical representation of the standard deviation has been reduced to a 20% to provide a clear representation, both for odometry and particle filter. Also, although the odometric error does not depend on the number of particles, it has been included on the figure for comparison purposes.
The first thing to be noticed is that the presented approach is able to reduce the odometric error in all cases. Even if only 10 particles are used, the resulting trajectory is, in mean, a 21.9% better than odometry. In the case of 400 particles, the resulting trajectory achieves, in mean, a 60% of improvement with respect to odometry. Also, the standard deviations of the particle filter errors are significantly lower than those of odometry. This suggests that the quality of the particle filter estimates is barely influenced by the initial odometric error.
The second thing to be noticed is that a large error reduction appears from 10 to 50 particles. From this number of particles onward, the error reduction is very small. This suggests that the behaviour of our algorithm does not strongly depend on the number of particles. It also suggests that using a number of particles between 50 and 100 would be a good choice, more if the execution times are taken into account.

Figure 6-b shows the mean and the standard deviation of the execution times per data set item, with respect to the number of particles. It is important to remark that these execution times correspond to a non optimized Matlab implementation. Thus, the absolute values are meaningless as a C++ implementation will greatly increase the execution speed. The interest of these results is that the execution time is strongly linear with the number of particles. This linear relation reinforces the idea that using between 50 and 100 particles is the better choice: the small improvement of using more particles does not compensate the increase in computation time.

4.3 Evaluating the influence of the local maps size

The second experiment evaluates the quality and the execution time of our approach with respect to the local maps size. Now, the number of particles is set to 100, as it has shown to be a good choice, and the history sizes k=25, k=50, k=100, k=200, k=400 and k=800 are tested.

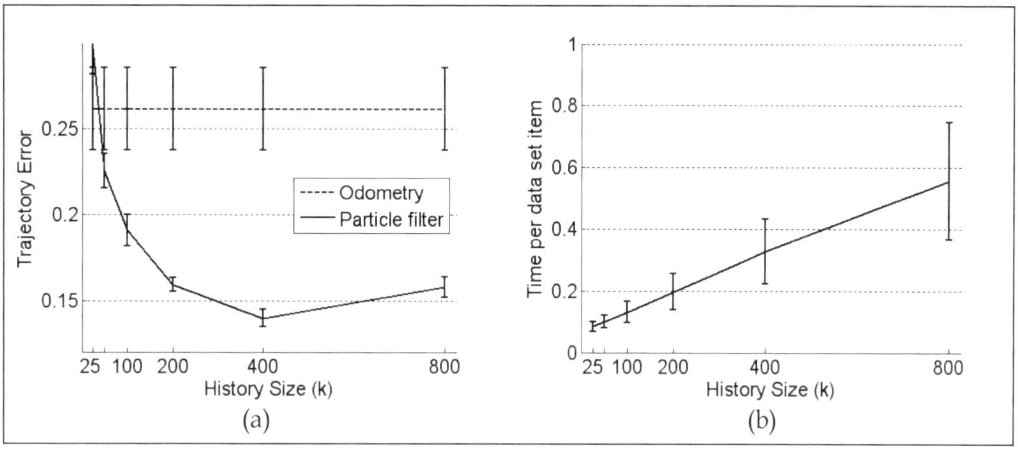

Fig. 7. Experimental results obtained using different local map sizes and setting the number of particles to *M=100*. (a) Means and standard deviations of the trajectory errors. (b) Means and standard deviations of the execution time per data set item on a Matlab implementation.

Figure 7-a shows the mean and the standard deviation of the trajectory errors, both for odometry and particle filter. The standard deviation has been graphically reduced to a 20% to provide a clear representation.

It can be observed how the effects of the history size are more noticeable than those of the number of particles. For example, if the very short history *k=25* is used, the resulting trajectory is worse than the one provided by odometry. The reason of this problem is that, using a very short history, the influence of spurious and wrong readings in the measurement model is not negligible. Also, it is clear that increasing the history size may lead to better results than increasing the number of particles. For instance, the trajectory obtained using *M=100* and *k=400* is an 87% better than the odometric one, while the trajectory obtained using *M=400* and *k=100* is only a 60% better.

It is important to remark that the quality of the particle filter slightly decreases for *k=800*. This quality reduction is mainly due to the initialization process. As stated previously, the time spent to build the initial particle set X_0 depends on the value of *k*. In our implementation, setting *k=800* means that the robot has to solely rely on odometry during 1

minute and 20 seconds at the beginning of its operation. This dependence on odometry is responsible of the mentioned quality reduction.
Figure 7-b shows the mean and the standard deviation of the execution times per data set item. As in the previous experiment, these times correspond to a non optimized Matlab implementation. Thus, the interest of the execution times does not reside on their absolute values but on their evolution with respect to the history size.
Similarly to the previous experiment, the execution time is strongly linear with the history size. Looking at the Figures 6-b and 7-b , it is clear that taking into account the time consumption, the better choice is to increase the history size rather than the number of particles. For instance, the errors for *M=400* and *k=100* are similar to those of *M=100* and *k=200*, but the mean execution time for the former is more than twice the execution time of the latter.

4.4 Qualitative evaluation

In order to provide a clear understanding of the results, some images are provided for visual inspection. Different trajectories have been plotted, as well as the sonar readings according to each trajectory.
Figure 8 visually depicts some of the results of the first experiment. The quality of the algorithm with respect to the number of particles can be observed. The first row shows the initial odometry estimates in four different environments. The second, third and fourth rows depict the results using an increasing number of particles (10, 100 and 400). All of them correspond to a history size of *k=100*. Finally, the fifth row shows the results of applying ICP to the laser readings. It is important to remark that, although ground truth trajectory has been obtained by matching laser range readings, the visual map shown in the last row has been plotted with the sonar readings to make the visual comparison easier.
It can be observed how, as the number of particles increases, the resulting trajectory becomes more similar to the ground truth. Even in the large environment of the fourth column, where the robot has moved more than 150m, the final pose estimate is very close to the ground truth. The environment in the third column deserves special attention. By observing the initial odometric estimate, it is easy to see that a significant error appears at the beginning of the trajectory. Because the initial particle set X_0 construction requires for the robot to be confident on odometry at the beginning of its operation, this initial error can not be fully corrected. That is why the particle filter provides a visual map rotated with respect to the ground truth. However, the shape of the trajectory is almost identical to the one of the ground truth.
The Figure 9 visually depicts some of the results of the second experiment. The quality of the algorithm with respect to the history size can be observed. The first and fifth rows, which correspond to the initial odometric estimates and the ground truth respectively, are the same that in Figure 8, and are plotted here again to provide a clear idea of the evolution of the pose estimates. The second, third and fourth row correspond to history sizes of *k=25*, *k=50* and *k=200*. In all of them, the number of particles used is *M=100*. Thus, the results for *k=100* can be observed in the third row of Figure 8.
It can be observed how the changes in the history size are clearly reflected in the quality of the resulting trajectory. Very accurate trajectories appear when a history size of 200 is used. As stated previously, the last row corresponds to the localization results of the well known

ICP algorithm applied to accurate and dense sets of laser range readings. On the contrary, our algorithm operates with the sparse and noisy sets of readings provided by standard Polaroid ultrasonic range finders. Moreover, our algorithm operated on a corrupted odometry, simulating bad floor conditions. Thus, it is remarkable that the presented approach is able to provide localization results close to the ones provided by a standard laser scan matching algorithm.

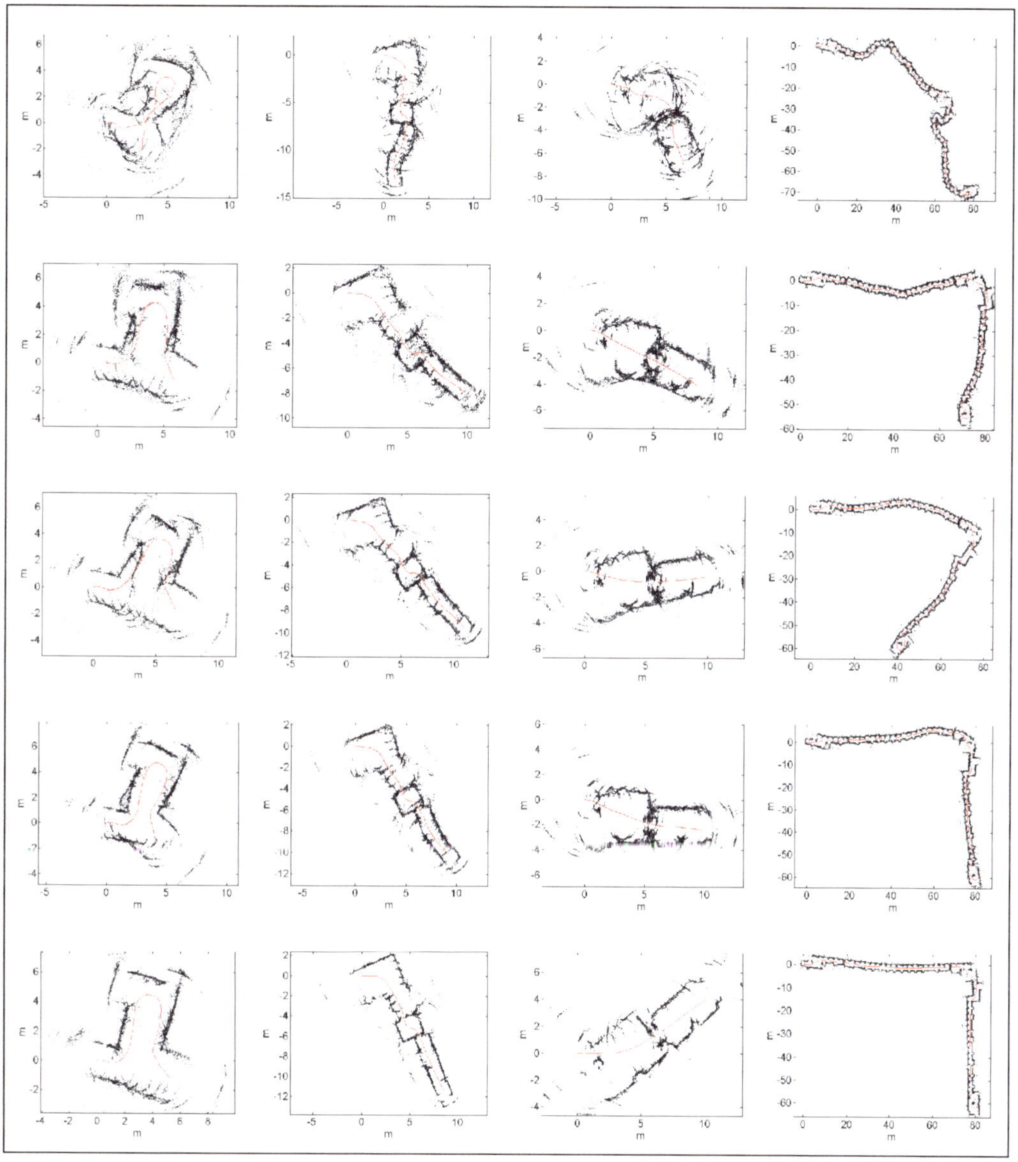

Fig. 8. Trajectories and sonar readings according to odometry (first row), particle filter using 10, 100 and 400 particles respectively (second to fourth row) and ICP laser scan matching (fifth row). The local map size used is *k=100*.

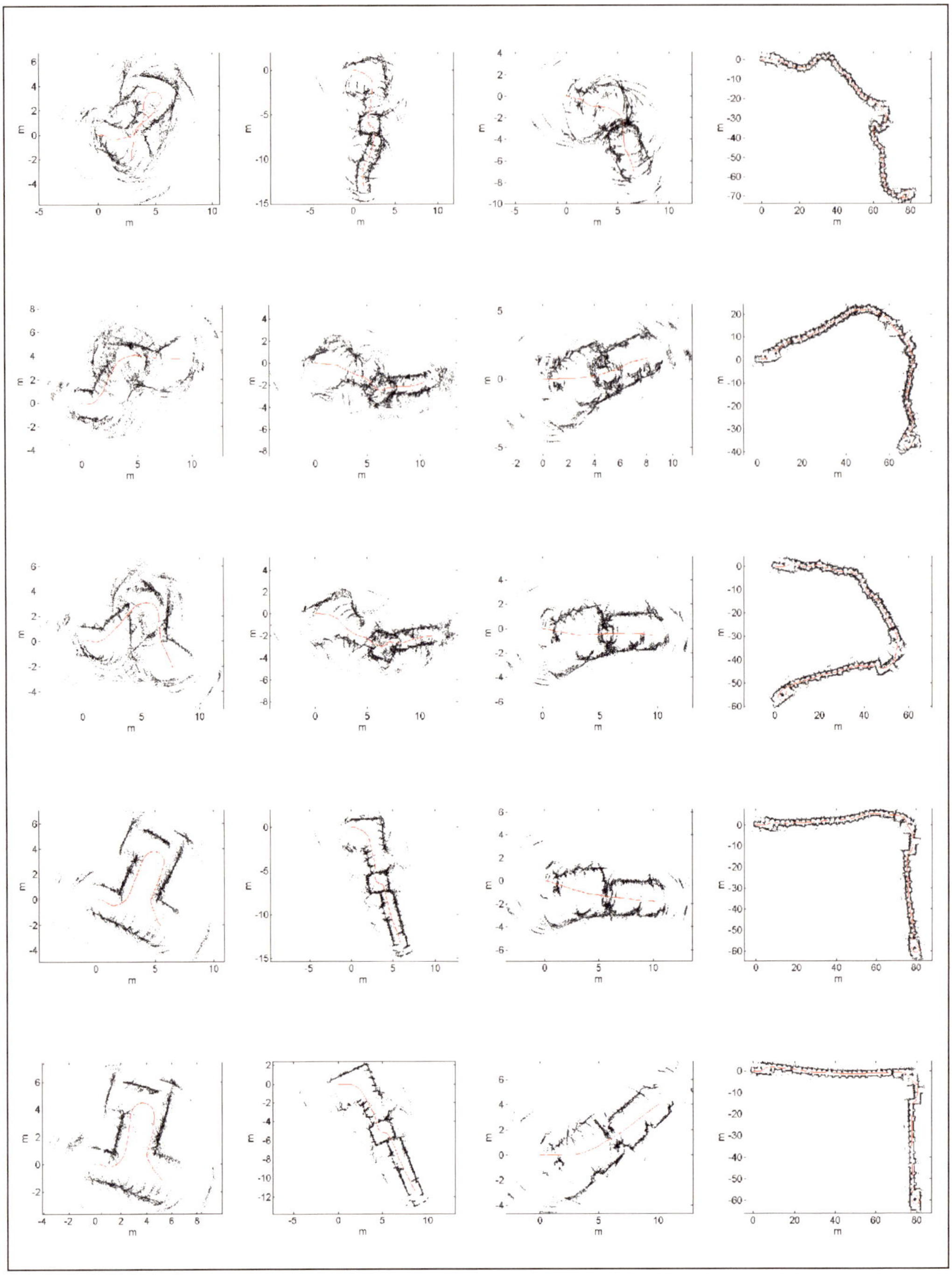

Fig. 9. Trajectories and sonar readings according to odometry (first row), particle filter using history sizes of *k=25*, *k=50* and *k=200* respectively (second to fourth row) and ICP laser scan matching (fifth row). The number of particles used is *M=100*.

5. Conclusion

Localization is a key issue in mobile robotics nowadays. Nearly all robotic tasks require some knowledge of the robot location in the environment. A common way to perform localization is to correlate exteroceptive sensor data at subsequent robot poses. This approach is strongly dependant on the exteroceptive sensor quality. Because of this, many localization algorithms rely on accurate laser range finders, providing dense sets of readings.

Standard ultrasonic range finders are not able to provide such dense and accurate information. That is why they are not frequently used in terrestrial mobile robot localization. However, they are appealing in terms of size, prize and power consumption. Moreover, their basic behaviour is shared with underwater sonar, which is extensively used in underwater and marine robotics. Consequently, a localization technique involving ultrasonic range finders is of great interest in the mobile robotics community.

In this chapter, particle filters have been proposed as a tool to perform localization using ultrasonic range finders. One of the advantages of the presented approach is that it does not require the use of previously constructed maps. Thus, it is suitable even for environments where no *a priori* knowledge is available. This is accomplished by recursively building local maps, which represent the local view that each particle in the filter has about the surrounding environment. Being the local map size constant, the time consumption required to deal with them is also constant.

The measurement model, which is in charge of computing the weights for the particles, has been defined similarly to the closest point rule of the ICP scan matching algorithm. The idea for the measurement model is to use the closest point rule to decide the amount of existing overlap between the current set of sonar readings and each of the local maps.

An experimental setup, involving the construction of a ground truth using accurate and dense laser readings, has been presented. Also, a technique to quantitatively compare different trajectories is discussed. By comparing different particle filter configurations with the ground truth, numerical error measures are obtained.

Two experiments have been defined. The first evaluates the effects of different sizes for the particle set. The second measures the effects of different sizes for the local maps. In both experiments, both the quality of the estimates and the time consumption has been observed. The results suggest that, thanks to the use of particle filters high quality localization results can be obtained using standard Polaroid ultrasonic range finders. These results are comparable to those obtained by standard scan matching algorithms applied to laser readings.

6. Future work

The presented measurement model is based on the ICP scan matching algorithm. This algorithm, which has been vastly used by the localization community, has also proved to be effective when applied to sonar readings (Burguera et al. 2005). However, recent works show that other matching approaches are able to provide more accurate and robust estimates (Burguera et al. 2008a; Burguera et al. 2008b). In consequence, it is reasonable to assume that the presented particle filter approach could benefit of these recent matching techniques in the measurement model.

7. Acknowledgment

This work is partially supported by DPI 2005-09001-C03-02 and FEDER funding.

8. References

Biber, P. & Straβer W. (2003). The Normal Distribution Transform: a new approach to laser scan matching, *Proceedings of the IEEE/RSJ International Conference on Intelligent Robots and Systems (IROS)*, pp. 2743-2748, October 2003.

Burguera, A.; González Y., & Oliver G. (2008a). A probabilistic framework for sonar scan matching localization. *Advanced Robotics*, Vol. 22, No. 11, August 2008, pp. 1223-1241, ISSN: 0169-1864

Burguera, A.; González Y., & Oliver G. (2008b). Sonar scan matching by filtering scans using grids of normal distributions. In *Intelligent Autonomous Systems 10*, Burgard, W; Dillman, R.; Plagemann, C. & Vahrenkamp, N. (Ed.), pp. 64-73, IOS Press, ISBN 978-1-58603-887-8, Netherlands

Burguera, A.; Oliver, G. & Tardós, J.D. (2005). Robust scan matching localization using ultrasonic range finders, *Proceedings of the Conference on Intelligent Robots and Systems (IROS)*, pp. 1451-1456, August 2005, Edmonton (Canada)

Castellanos, J.; Neira, J. & Tardós J.D. (2004). Limits to the consistency of EKF-based SLAM, *Proceedings of the 5th IFAC/EURON Symposium on Intelligent Autonomous Vehicles*, July 2004

Dellaert, F.; Burgard W. & Thrun, S. (1999). Monte Carlo Localization for mobile robots, *Proceedings of the IEEE International Conference on Robotics and Automation (ICRA)*, 1999

Doucet, A.; de Freitas, J. & Gordon, N.E. (2001). *Sequential Monte Carlo Methods in Practice*, Springer-Verlag, 2001

Fox, D.; Burgard, W.; Dellaert, F. & Thrun, S. (1999). Monte carlo localization: Efficient position estimation for mobile robots, *Proceedings of the National Conference on Artificial Intelligence (AAAI)*, 1999, Orlando (USA)

Fox, D.; Burgard W.; Kruppa, H. & Thrun, S. (2000). A probabilistic approach to collaborative multi-robot localization. *Autonomous Robots*, Vol. 8, No. 3, 2000.

Groβmann, A. & Poli, R. (2001). Robust mobile robot localisation from sparse and noisy proximity readings using Hough transform and probability grids. *Robotics and Autonomous Systems*, No. 37, 2001, pp. 1-18.

Hähnel, D.; Burgard, W., Fox, D. & Thrun, S. (2003). An efficient FastSLAM algorithm for generating maps of large-scale cyclic environments from raw laser range measurements, *Proceedings of the IEEE/RSJ International Conference on Intelligent Robots and Systems (IROS)*, pp. 206-211, October 2003.

Kalman, R. E. (1960). A new approach to linear filtering and prediction problems. *Transactions of the ASME. Journal of Basic Engineering*, Vol. 82, March 1960, pp. 35-45.

Lee, D. (1996). The Map-Building and Exploration Strategies of a Simple Sonar-Equipped Mobile Robot, Cambridge University Press, 1996.

Lu, F. & Milios, E. (1997). Robot pose estimation in unknown environments by matching 2D range scans. *Intelligent and Robotic Systems*. Vol. 18, No. 3, March 1997, pp. 249-275.

Metropolis, N. & Ulam, S. (1949). The Monte Carlo method. *Journal of the American Statistical Association*, No. 44, 1949, pp. 335-341.

Minguez, J.; Montesano, L. & Lamiraux, F. (2006). Metric-based iterative closest point scan matching for sensor displacement estimation, *IEEE Transactions on Robotics*, Vol. 22, No. 5, October 2006, pp. 1047-1054.

Montemerlo, M.; Thrun, S.; Koller, D. & Wegbreit, B. (2002). FastSLAM: A factored solution to the simultaneous localization and mapping problem, *Proceedings of the AAAI National Conference on Artificial Intelligence*, 2002.

Montesano, L.; Minguez, J. & Montano, L. (2005). Probabilistic scan matching for motion estimation in unstructured environments, *Proceedings of the International Conference on Intelligent Robots and Systems (IROS)*, pp. 1445-1450, August 2005, Edmonton (Canada)

Neira, J. & Tardós, J.D. (2001). Data association in stochastic mapping using the joint compatibility test, *IEEE Transactions on Robotics and Automation*, Vol. 17, No. 6, 2001, pp. 890-897.

Pfister, S. T.; Kriechbaum, K.L.; Roumeliotis, S. I. & Burdick, J. W. (2004). Weighted range sensor matching algorithms for mobile robot displacement estimation, *Proceedings of IEEE International Conference on Robotics and Automation (ICRA)*, pp. 1667-1674, May 2004.

Rusinkiewicz, S. & Levoy, M. (2001). Efficient variants of the ICP algorithm, *Proceedings of the International Conference on 3D Digital Imaging and Modeling (3DIM)*, 2001.

Sanjeev Arulampalam, M.; Maskell, S.; Gordon, N. & Clapp, T. (2002). A tutorial on particle filters for online nonlinear/non-gaussian bayesian tracking. *IEEE Transactions on Signal Processing*, Vol. 50, No. 2, February 2002, pp. 174-188.

Silver, D.; Bradley, D. & Thayer, S. (2004). Scan matching in flooded subterranean voids, *Proceedings of the IEEE Conference on Robotics, Automation and Mechatronics*, Singapore, December 2004.

Smith, R.; Self M., & Cheeseman P. (1990). Estimating uncertain spatial relationships in robotics. *Autonomous Robot Vehicles*, 1990, pp. 167-193, ISBN: 0-387-97240-4

Tardós, J. D.; Neira, J.; Newman, P.M. & Leonard, J.J. (2002). Robust mapping and localization in indoor environments using sonar data. *International Journal of Robotics Research*, Vol. 21, No. 4, April 2002, pp. 311-330.

Thrun, S.; Burgard, W. & Fox, D. (2005). *Probabilistic Robotics*. The MIT Press. ISBN 0-262-20162-3.

Thrun, S.; Fox, D. Burgard, W. & Dellaert, F. (2001). Robust monte carlo localization for mobile robots. *Artificial Intelligence*, Vol. 128, No. 1-2, 2001, pp. 99-141.

Yaqub, T. & Katupitiya, J. (2007). Laser scan matching for measurement update in a particle filter, *Proceedings of the IEEE/ASME International Conference on Advanced Intelligent Mechatronics*, pp. 1-6, September 2007.